SOCIAL ECOLOGY AFTER BOOKCHIN

DEMOCRACY AND ECOLOGY
A Guilford Series

Published in conjunction
with the Center for Political Ecology

JAMES O'CONNOR
Series Editor

Social Ecology after Bookchin

EDITED BY

Andrew Light

THE GUILFORD PRESS
New York   London

© 1998 The Guilford Press
A Division of Guilford Publications, Inc.
72 Spring Street, New York, NY 10012
http://www.guilford.com

Printed in the United States of America

This book is printed on acid-free paper.

Last digit is print number: 9   8   7   6   5   4   3   2   1

Library of Congress Cataloging-in-Publication Data

Social ecology after Bookchin / edited by Andrew Light.
        p.   cm.—(Democracy and ecology)
     Includes bibliographical references and index.
     ISBN 1-57230-379-4
     1. Bookchin, Murray, 1921–   2. Social ecology.
  I. Light, Andrew, 1966–   . II. Series.
  HM206.S55   1998
  304.2—dc21   98-35447
                                                          CIP

For my friends and colleagues
at The University of Montana.

Introduction to the Democracy and Ecology Series

This book series titled "Democracy and Ecology" is a contribution to the debates on the future of the global environment and "free market economy" and the prospects of radical green and democratic movements in the world today. While some call the post-Cold War period the "end of history," others sense that we may be living at its beginning. These scholars and activists believe that the seemingly all-powerful and reified world of global capital is creating more economic, social, political, and ecological problems than the world's ruling and political classes are able to resolve. There is a feeling that we are living through a general crisis, a turning point or divide that will create great dangers, and also opportunities for a nonexploitative, socially just, democratic ecological society. Many think that our species is learning how to regulate the relationship that we have with ourselves and the rest of nature in ways that defend ecological values and sensibilities, as well as right the exploitation and injustice that disfigure the present world order. All are asking hard questions about what went wrong with the worlds that global capitalism and state socialism made, and about the kind of life that might be rebuilt from the wreckage of ecologically and socially bankrupt ways of working and living. The "Democracy and Ecology" series rehearses these and related questions, poses new ones, and tries to respond to them, if only tentatively and provisionally, because the stakes are so high, and since "time-honored slogans and time-worn formulae" have become part of the problem.

JAMES O'CONNOR
Series Editor

Acknowledgments

This volume was a long time in coming, certainly too long. It originally began as an idea for a jointly authored book on Murray Bookchin's social ecology five years ago, to be penned by myself, Jay Moore, Jim O'Connor, and Alan Rudy. Many problems in orchestrating that volume convinced us to abandon the project, and so it evolved into the present anthology. Over the years since then, through many false starts and long delays, several prospective authors have come into and later left the volume; all of them have made some sort of mark by helping me to think through what sort of book this should be. Accordingly, I wish to thank Jay Moore, Stanley Aronowitz, Ynestra King, Steve Best, Elizabeth Carlassare, and especially, Enrique Leff, for their assistance along the way. To the rest of the authors remaining in this volume, I thank them for their perseverance, dedication, and most of all, patience. Other friends in environmental philosophy were always there to encourage the continuation of the process, especially Deane Curtin, Andrew Feenberg, Greta Gaard, and my dear friend and sometime collaborator, Eric Katz.

I also owe special thanks to Jim O'Connor, editor of the "Democracy and Ecology" series, Barbara Laurence, Managing Editor of *Capitalism, Nature, Socialism*, and Peter Wissoker, my editor and friend at The Guilford Press. Jim and Barbara expressed remarkable confidence in my being able to put this volume together, and Peter never failed in his ongoing support for the idea of the book. Jim's intellectual rigor, dedication, generosity, and character are always an inspiration. I may be mistaken in assuming that he, Barbara, and Peter never gave up hope that this volume would finally get done. In any event, their encouragement was always there, and I thank them for it.

I have dedicated this collection to my friends and colleagues at The University of Montana. It was there, during my all too short stay as a member of the philosophy department, that I finally found the time to finish this volume. For the first time in my nomadic academic career I felt at home, and surely this sense of being in place, if only for a brief time, made it possible for me to accomplish much. Of those in Missoula, I thank all of my colleagues in the philosophy department and elsewhere, especially Barbara Andrew, Irene Appelbaum, Lauren Bartlett, Albert Borgmann, Bill Chaloupka, Dan Flores, Carl Heine, Bruce Milem, Ron Perrin, Val Plumwood, and Deborah Slicer. A very singular thanks goes to my assistant at Montana, David Roberts, who never failed to make my working days more enjoyable and productive.

Finally, my biggest debts of gratitude go to two very different people, both of whom have been by my side in various ways since the beginning of this process. First, to Dorit Naaman, my partner for the last four years, and constant interlocutor on all projects, both great and small. Dorit's dedication to her own work continues to be an inspiration for me with mine. Second, to Murray Bookchin, who may indeed not welcome this collection but whom I hope realizes the honor we do him when we continue to press on and be inspired by his work and presence. Bookchin has already commented in an apparent reference to this volume prior to its publication that it is a "joint endeavor of Marxists, neo-Marxists, and deep ecologists" designed to denounce his work, and whose purpose "is to diminish the anarchist tendency in the ecology movement."[1] I hope that the charitable reader of this collection will see that it is not designed to "denounce" Bookchin or anarchism in general. (Nor should the papers in the volume be taken as a "joint endeavor" by the particular authors of the various chapters. Indeed, each author is responsible for the content of his or her own contribution and that only.) The purpose of this volume is to provide more and less supportive critiques and extensions of and challenges to social ecology, toward the goal of enriching the overall development of the broad array of views represented as political ecology. Certainly this task requires analysis of Bookchin's work, as that work is practically coextensive with the development of social ecology as we know it today. Even though we are all in some ways critical of his work, I hope that no

---

1. Murray Bookchin, "Whither Marxism? A Reply to Recent Anarchist Critics," available at http://www.pitzer.edu/dward/Anarchist_Archives/bookchin/whither.html, footnote 72, paragraph 4. This article is forthcoming in Bookchin's *Anarchism, Marxism, and the Future of the Left* (San Francisco and Edinburgh: AK Press, 1998).

reader will come away from this volume thinking that a social ecology after Bookchin means a social ecology without Bookchin. Bookchin, more than he may realize, has pressed us all to move forward and continue the conversation over his ideas into the next century whether we align ourselves in his camp or not. Surely, no author could ask for a greater tribute.

ANDREW LIGHT
Missoula, Montana
May 1998

Contents

Introduction

Bookchin as/and Social Ecology

ANDREW LIGHT

In 1993 I unwittingly stepped onto a battlefield. Just three years out of college, I was in graduate school studying philosophy in southern California, working on environmental ethics and political theory, largely in isolation. While most of the professional philosophers I interacted with thought that this topic (let alone something like political ecology) "wasn't really philosophy," I persisted, and at last, with the encouragement of friends, published a paper comparing the environmental political theories of Murray Bookchin and Herbert Marcuse. The piece came out in the relatively new journal of socialist ecology, *Capitalism, Nature, Socialism,* edited by Jim O'Connor.[1] That paper, completely rewritten and included in this volume as Chapter 11, was my second professional publication. I had no illusions that it would make a substantial contribution to the field.

To say that I was shocked when O'Connor called me a month later to tell me that Bookchin wanted to respond to my piece would be an understatement. I'm not sure that I would have been less surprised if Marcuse himself had risen from the dead to reply to my paper. This is not to say that I regarded Bookchin's career as comparable to Marcuse's at the time, but only to try to convey how amazed I was that not only had someone read the piece, but that the man whom I had written about had deigned to actually reply in print! I had no idea whether to feel honored or terrified. When I eventually saw Bookchin's response, my initial reaction was puzzlement. On the one hand, Bookchin had granted me respect incommensurate with my relative standing to his by replying forcefully to my paper (which, given the central role of antihi-

1

erarchical thinking in his work makes complete sense). On the other hand, it seemed to me that Bookchin had (1) almost willfully misread some of my essay (which I could explain away, given the relative unclarity of the piece), and (2) obliquely decided to augment his substantive claims with unnecessary aspersions. At various points in his reply, Bookchin had suggested that I was a deep ecologist, and therefore connected with the racism and xenophobia that he identified with that school of thought, and at another point he insinuated through a comparison with himself that my understanding of labor issues was learned only through the publication of the Frankfurt School and not through real experience.[2] Now, I knew that Bookchin couldn't have made these assessments with any certainty. I also knew that I wasn't a deep ecologist, and that, as the grandson of two West Virginia coal miners, I had learned a lot about the working class from sources other than the Frankfurt School. So, in short, I simply couldn't understand, first, why Bookchin had bothered to reply to what he thought was clearly such a bad paper, and second, why he felt it necessary to lampoon my character and background, about which he surely knew nothing.

Since then things have become much clearer to me. But to understand Bookchin's response I had to read all the accounts I could find about the beginnings of the deep ecology–social ecology debate, which was touted as pitting Bookchin's "confederalist municipalist" Left-Green Network against the fuzzy-headed followers of Norwegian philosopher Arne Naess. I also had to learn more about how Bookchin had arrived at his current political position. I was greatly aided in this task by looking at the opening volleys of the social ecology–deep ecology debate at the 1987 first open national greens gathering in Amherst, Massachusetts, and by reading the first draft of a chapter of a biography of Bookchin written by philosopher Steven Best. From Best's work, I learned that Bookchin had been a red-diaper baby before the phrase was turned. Bookchin's political evolution evokes a twentieth-century history of radical politics in the United States: from communist, to Trotskyist, to anarchist.

Knowing what I do now about Bookchin's roots in New York City political streetfighting in the 1930s and 1940s, and the career that followed, I see that in my paper I had inadvertently stumbled onto perhaps the most heated intellectual–political exchange of the late 1980s. Murray Bookchin had been slowly developing over a lifetime of leftist thought and activism his own unique political and social theory. But he was not an ivory-tower sideline theorist. Having participated actively in politics in that context in the 1930s and 1940s, and having moved on to participate in the rise of the new social movements in the 1960s and 1970s, Bookchin had developed a revolutionary political theory in the terms of one of the

new movements: ecology. But while ecological activism was the vehicle
for this theory, its parameters, like the grand theories of the Old Left,
went well beyond narrow environmental concerns.

Then, in the mid-1980s, Bookchin found himself confronted with
what he took to be the worst possible alternative to his own work. On his
view, the hearts and minds of greens, especially in North America, were
beginning to be dominated by, in general, a mysticism, primitivism, spiritualism, and antirationalism, and in particular by the burgeoning deep
ecology movement inspired by Naess. Bookchin paid particularly close
attention to the self-styled deep ecology vanguard organization Earth
First!, especially the writings of Dave Foreman, Edward Abbey, and the
pseudonymous Miss Ann Thropy. Bookchin consistently bolstered his critique of deep ecology (lumping it together with all forms of "ecospiritual-
ism") by citing various claims by Earth First!ers that starvation in Ethio-
pia was a natural and acceptable process, that "Northern European"
American culture was in danger of "Latinization" from Central and
South American immigrants, and that AIDS was a self-regulating process
to stem world population growth.[3]

Bookchin described deep ecology in the early days of this confrontation charitably as "vague" and "formless," but later he threw down
the gauntlet by labeling it the "*same kind* of ecobrutalism [that] led
Hitler to fashion theories of blood and soil that led to the transport of
millions of people to murder camps like Auschwitz."[4] As there is no
doubt that Bookchin fervently believes such claims, and given the fact
that he took part in the communist antifascist political struggles of the
1930s and 1940s, it makes sense that he would be compelled to vigor-
ously attack any seeming proponent of deep ecology, especially those
who appeared to be targeting him. If we agree with Bookchin's interpretation of the implications of deep ecology, such defenses of his own
views and criticisms of others were not only justified, but were the only
morally responsible course of action. Had I been fully aware of this
background, I would have understood both the tone and the force of
Bookchin's response to my paper. For in my piece I made the mistake of
suggesting (in this climate!) that Marcuse could serve as a possible
point of rapprochement between the approach to ecological issues represented by deep ecology and the approach represented by social ecology. To Bookchin, I was suggesting nothing less than the political and
moral equivalent of collaboration with the Nazis. In his reply to my
piece, Bookchin was up front about the reasons behind his prevalent
argumentative style: "if I have argued, cajoled, and denounced in my
writing, it is because I was long engaged with harsh opponents in an
era when clashes of ideas had concrete and earthy meaning in movements that I and other people regard as potentially revolutionary."[5]

But, of course, this short history of a small part of the social ecology–deep ecology debate begs the question of whether we should accept Bookchin's interpretation of the implications of deep ecology. I have no interest in exploring that issue here. Along with the other authors represented in this collection, I think there is more to Bookchin than his disagreements with deep ecology, and much more to political ecology than the social ecology–deep ecology debate. Bookchin's charge of ecofascism against deep ecology has been defused by many theorists and activists, most notably by Michael E. Zimmerman, who perhaps more than anyone else has taken the charge of ecofascism against deep ecology seriously enough to warrant careful scholarly attention.[6] More interesting to me, while introducing this collection of constructively critical essays on Bookchin's work, is trying to ascertain if Bookchin's worldview—which sees the rhetoric of some greens as the same kind of brutalism (eco-, or not) that led to the Holocaust—is productive for the future of political ecology in general, or even the further evolution of social ecology in particular.

Bookchin's personal political history, as he is the first to admit, is the source of his unique approach to debates in political ecology. No one can fault him for that. Surely Bookchin has always gotten part of it right in his critical interjections. Many people, including myself, have not understood important parts of his argument, and have made mistakes in our commentary about his work. And surely, some deep ecologists have said some ridiculously awful things. But the question is whether the approach to political ecology that Bookchin champions, including a tendency to make judgments about interlocutors based on a few extreme examples, is what we need today. I am not entirely skeptical about the possibility of a rise of neofascism in America or anywhere else, but I do think that the people to worry about are not deep ecologists. The reason to try to reach beyond the extremes of any green position is that on the environmental frontlines, we need to be able to draw on as many people as possible in bolstering a set of claims that is not well received in the rest of society. Those of us, like myself, and like Bookchin, who come to political ecology through questions about human social inequality, and who espouse a tempered form of anthropocentrism, need to work toward the possibility of forming workable green alliances, rather than adopting a stance that too easily dismisses potential allies. I do not think that the history of the Old or New Left so far gives us reason to think that the old intransigent approach to each other's views, evidenced in the collapse of both of those periods of left activism, is what we want to see happen today. We need to be more compatible, within limitations. Political ecology must be democratic. Other political ecologists, such as Arne Naess, agree that too hard divisions, while good for developing sharper theories, have a dulling effect on developing ecological movements.[7] Has Bookchin had such a dulling effect? Has his Old Left style infected the development of political ecology as a body of theoretical works and as a movement? In some ways he and it has. But ironically, this effect is no clearer than as found in Bookchin's relationship with his own circle of supporters and fellow travelers. Some day, the complete history of political ecology will more accurately iterate the slow dissolution of social ecology down to the views of one person and one person only, Murray Bookchin. Social ecology is not a movement, and as a theory it has come to be represented almost exclusively by Bookchin's work. When social ecologists go too far afield from this theory, they are pushed out of the camp, or else leave voluntarily out of frustration. This tendency within social ecology should not be transferred to the wider green movement. If that were to happen, then the movement would become an archipelago of isolated cells not cooperating with each other out of spite. But this is not to say that social ecology should be abandoned because of its own social history. All of the authors in this volume are united in the conviction that social ecology is one of the most important bodies of theory available to political ecologists. To isolate it from other conversations would be, and is today, a real tragedy. But if social ecology is to be reserved only to the work of Bookchin the individual, then this may well be the fate of this contribution to environmental thought. After briefly describing Bookchin's body of work, I will return to this issue.

Bookchin's Social Ecology

Well before the development of a wide public environmental consciousness, Bookchin wrote about the social, psychological, and health consequences of urbanization; the use of industrial chemicals in food production; and a variety of other antiecological consequences of modern industrial society.[8] The key texts here are his *Our Synthetic Environment* (1962)[9] and *Crisis in Our Cities* (1965),[10] both published under the pseudonym Lewis Herber.[11] These books represent an initial development of Bookchin's ecological and anarchist perspectives during his deep involvement in the struggles for civil rights and a wide variety of other social movements during the 1950s and early 1960s. Since the mid-1960s Bookchin has written widely on ecological and social issues and the ways each relate to the other. His writings and debates with social activists during the 1960s and 1970s are published in *Post-Scarcity Anarchism*[12] and *Toward an Ecological Society*.[13] A more recent collection of articles entitled *The Modern Crisis* was published in 1986.[14]

In 1982 Bookchin published perhaps his best-known book, *The Ecology of Freedom*,[15] which had taken more than a decade to write and research. This book was widely read by both theorists and practitioners in the ecological movement. It is here that Bookchin most extensively develops his view that the social domination of human by human leads first to the misleading idea of the domination of nature, and then to the destruction of nature. This novel perspective, linking modern social relations of domination with the destruction of global ecology, made Bookchin one of the most widely read ecological thinkers in the last thirty years. These positions from *The Ecology of Freedom* and his other publications have been further developed in *The Rise of Urbanization and the Decline of Citizenship*,[16] summarized in *Remaking Society*,[17] and extended theoretically in *The Philosophy of Social Ecology*.[18]

While I cannot hope to adequately summarize Bookchin's views here, in general his social ecology posits a spontaneous and teleological evolution of matter toward increasing complexity and consciousness.

> The universe bears witness to an ever-striving, *developing*—not merely "moving"—substance, whose most dynamic and creative attribute is its ceaseless capacity for self-organization into increasingly complex forms. Natural fecundity originates primarily from growth, not from spatial "changes" in location.[19]

This fecund self-organized growth of planetary life, the atmosphere, and the land and sea is thought to be based on symbiosis,[20] or what might be called evolutionary cooperation.

Bookchin sees nature's evolution generating "its own natural philosophy and ethics."[21] Healthy ecological and social differentiation is possible only under conditions of natural and social "spontaneous development" which "unfold and actualize [nature's] wealth of possibilities."[22] The relationship between society and nature is, for social ecologists, not one that looks "upon [nature] as a necessitarian, withholding, or 'stingy' redoubt of blind 'cruelty' and harsh determinism, but rather as the wellspring for social and natural differentiation."[23]

On Bookchin's views, the link between the evolution of external nature and social nature is profound. He argues that the "very natural processes that operate in animal and plant evolution along the symbiotic lines of participation and differentiation reappear as social processes in human evolution, albeit with their own distinctive traits, qualities and gradations or phases of development."[24] As humans and human societies emerge, Bookchin believes that "it is the logic of differentiation that makes it possible to relate the mediations of nature and society into a continuum."[25] For Bookchin, what "makes unity and diversity in nature more than a suggestive ecological metaphor for unity in diversity in society, is the underlying fact of wholeness."[26]

Human societies, though a continuation of natural evolution, are nonetheless quite different than the animal communities from which they evolved. For Bookchin, animal "communities are not societies . . . they do not form those uniquely human contrivances we call institutions . . . [they have] genetic rigidity . . . not contrived rigidity."[27] For Bookchin, the initial evolution of human institutions generated "organic" preliterate societies in which internal social relations and relations with the external world were organized around mutualistic practices that supported social and ecological differentiation. Historically, the conditions within which these societies existed are seen to have been disturbed by nascent, increasingly institutionalized social relations of domination and hierarchy.

Relations of hierarchy and domination, for Bookchin, are inherently social and

> must be viewed as *institutionalized* relationships, relationships that living things literally institute or create but which are neither ruthlessly fixed by instinct on the one hand nor idiosyncratic on the other. By this, I mean that they must comprise a clearly *social* structure of coercive and privileged ranks that exist apart from the idiosyncratic individuals who seem to be dominant within a given community, a hierarchy that is guided by a social logic that goes beyond individual interactions or inborn patterns of behavior.[28]

Hierarchy and domination, as institutions, can only be found in human societies, and, for Bookchin, cannot be said to exist in animal communities. Domination, as a coercive social relation, is seen to work against spontaneity and, unlike other social relations in organic societies, counters the processes of evolution. Given Bookchin's distinction between animal and plant communities and human societies, domination becomes the primary destructive relationship within society in that it destroys social "participation and differentiation."[29]

The ecologically destructive character of hierarchy and domination emerges from ideologies of the domination of nature that themselves spring from the real domination of human by human. On Bookchin's view, the history of social and natural evolution has become the history of two competing logics: the logic of spontaneous mutualistic ecological differentiation and the logic of domination, which works against everything represented by the other. Bookchin's historical work explores how these two logics work themselves out as spontaneous organic societies are transformed into cities, city-states, nations, nation-states, and capitalist political economies increasingly organized through domination and hierarchy. History, on Bookchin's view, largely becomes the story of the battle between communities committed to "freedom" and elites committed to domination.

Out of this view of history, and this analysis of the relationship between humans and nature, Bookchin builds a thoroughgoing positive theory of the reconstruction of society. Foremost among the steps necessary for the reconstitution of a more organic society is a rebuilding of local human association and the reconstruction of human beings within a libertarian political ecocommunity. Reempowering communities and people demands the abolition of the state, and the nation-state in particular. Politically, social ecology emphasizes decentralization and the return to the local level of the "resources and . . . potential for development" within society.[30] Importantly, Bookchin does believe that these communities should be linked in confederations.

For Bookchin, there "can be no politics without community."[31] The preferred social relationship for decision making is democratic politics, not logistical administration. Bookchin envisions human-scale communities encompassing a political and productive division of labor that is participatory and skilled. The community makes decisions about how it will administer its own decisions. He advocates for this process a strong form of direct democracy.

Programatically, Bookchin arrives at four basic principals for the regeneration of society and ecology. These coordinating principles focus on (1) "the revival of the *citizens assembly*,"[32] (2) the confederation of assemblies,[33] (3) the construction of communal and confederal politics as a "school for genuine citizenship,"[34] and (4) the economic empowerment of communities via the "municipalization of property," and the formulation of communal productivity policy by public democracy.[35] For Bookchin, it is only through social institutions such as these that we can overcome the social and ecological consequences of the centralized state, modern market economies, and the technologies they spawn. Overcoming these institutions are the necessary conditions for reestablishing social and ecological differentiation.

Bookchin's Social Economy

Bookchin is regarded by many activists as important not only for his emphasis on the interaction between deep-seated social relations and their effects on the external, "natural," world, but also for his utopian visions for the future. His concerns encompass far more than those of either liberal environmentalists or traditional leftists. His libertarian vision of confederalized, independent, human-scale, and self-reliant communities has proven to be attractive to many. More importantly, Bookchin's critique of domination, which includes theories of technology, politics, and social conflict, strikes a deep chord among progressive ecologists in many countries, including the German Greens and the Left-Green Network in the United States.

But it seems that over the years Bookchin's views have grown more and more isolated. While deep ecologists seem to proliferate into new schools of thought with each passing year, the number of Bookchin's political and theoretical associates shrinks.[36] In the process of the isolation of Bookchin, social ecology itself is less and less discussed as a viable form of critical ecological thought. Perhaps it is the very tight connection between Bookchin the person and social ecology as a body of thought that has cemented this process. Of course, it is clearly the case that Bookchin's work is undeniably at the center of social ecology, and justifiably so. No doubt, it is the sweeping and ambitious scope of his project—including a reading of prehistory to the present—that in part warrants this connection between Bookchin and social ecology. While other schools of thought—such as the collaborative projects of the Frankfurt School—seemingly required multiple personalities in order to meet the aspirations of the theoretical enterprise, Bookchin has risen to the task of filling in a complete body of theory largely by himself. But though one could claim that social ecology began with Bookchin, the question is whether it will end with him as well. How will social ecology continue, and in what form?

John Clark, a long-time associate of Bookchin, devotee of social ecology, and editor of a festschrift for Bookchin, has attempted to extend social ecology by reinterpreting, or perhaps rediscovering, its past. Clark begins a recent outline of this history in an article cleverly titled "*A Social Ecology*" (emphasis added) with a general description of social ecology rather than an immediate identification of the view with Bookchin. "As a philosophical approach, a social ecology investigates the ontological, epistemological, ethical and political dimensions of the relationship between the social and the ecological, and seeks the practical wisdom that results from such reflections."[37] Clark goes on to discuss the history of social ecology as beginning in the nineteenth century with Peter Kropotkin and more importantly for Clark, Elisée Reclus, and continuing into the twentieth century with Patrick Geddes, Lewis Mumford, and Bookchin.

But if one disagrees with Clark's expansion of social ecology—for example, for possibly illegitimately appropriating other thinkers to its camp—the problem remains of the theoretical isolation that Bookchin has tended to bring upon social ecology. On this issue, Clark lays the blame for the diminished appeal of social ecology solely at Bookchin's feet:

> Although Bookchin develops and expands the tradition of social ecology in important ways, he has at the same time also narrowed it through dogmatic and non-dialectical attempts at philosophical systems-building, through an increasingly sectarian politics, and through intemperate and divisive attacks on "competing" ecophilosophies and on diverse expressions of his own tradition. To the extent that social ecology has been identified with Bookchinist sectarianism, its potential as an ecophilosophy has not been widely appreciated.[38]

But is this a fair claim to make? Clark is going beyond a fairly standard criticism of Bookchin—that his attacks on competing theories lack charity—to the more worrisome suggestion that his criticisms of other social ecologists will contribute to a restriction of the theory.

In a dramatic example of this possible problem, a recent issue of the social ecology-friendly journal *Democracy and Nature* featured both an extensive (43-page) criticism by Bookchin of John Clark's revisionist social ecology and a resignation letter by Bookchin from the advisory board of the same journal.[39] Tellingly, in Bookchin's resignation letter, he criticizes Takis Fotopoulos, founding editor of *Democracy and Nature,* for, among other things, the latter's interpretation of the extent and limits of confederal or libertarian municipalism. Concluding this section of his critique, Bookchin suggests: "I did not propound this theory of politics to see it mutate into Bernsteinian evolutionary social democracy."[40] But one must wonder at the implications of such a statement. It is one thing to disagree with the *content* of the interpretation of one's views, but quite another to disagree with the *act* of interpretation of one's views, or the act of disagreement, as Bookchin seems to be doing here. In a concrete sense, Bookchin seems to own the ideas of social ecology in such a way that this school of thought sometimes appears to be solely coextensive with his own thought and no one else's. If Bookchin disagrees with the assertion of a challenge to social ecology, then it can apparently be deemed off-base by authority. Clearly, it is this narrowing identification of social ecology that Clark is trying to mitigate in his reconstruction of a larger history of social ecology.

It would be wrong to weigh in on one "side" or another on the issue of whether Bookchin should or should not have resigned from the editorial board of *Democracy and Nature.* Certainly, according to the rest of his resignation letter, Bookchin had other issues concerning scholarly courtesy that helped to prompt his action. But even without getting into these concerns, one can see a warning in this episode about the possible stagnation of social ecology if it becomes ideologically isolated as the work of one man alone, or if it fails to develop through a critical and constructive dialogue with other views. The editorial board of *Democracy and Nature* certainly express something like this problem in their reply to Bookchin's resignation.

While they maintain that the journal's scope goes well beyond social ecology the editors nonetheless admit that the journal has "been justifiably criticized as biased toward social ecology."[41] As such, they are still dismayed that Bookchin has "frequently criticized the journal in the past for the fact that it hosted other theoretical trends."[42] Continuing in this vein, the editors also assert that Bookchin has "frequently" criticized "the very presence on the International Advisory Board [of the journal] of people expressing alternative theoretical views."[43] But so far, the problem does not seem to be so much that Bookchin has objected to the discussion of his own work, or to critical engagement with it, but that he has had problems with the discussion of other kinds of theories in a journal that the editors admit has evolved into a journal largely associated with social ecology, or rather, with Bookchin's own views.

More to the point of the concern raised previously, the editors speak directly to the theoretical differences raised in Bookchin's resignation letter to those expressed by Fotopoulos. The thrust of their argument is that appreciation of disagreement about social ecology itself should follow from the established premises of social ecology.

> As for our philosophical differences with Murray are concerned, we think that no one can—and no one should expect to "end" philosophical questioning, including Bookchin. . . . We regard that, for one to avoid dialogue and resign using as his/her argument the idea that in this journal the views of social ecology are not promoted enough or, worse, that some of the published articles are indirectly critical of social ecology's aspects, is as much patronizing to our readers as it is incomprehensible, particularly so when it comes from an intellectual whose thought crucially centres around the needs for the creation of a society without enlightened leaders, gurus, gods, or bosses.[44]

Fotopoulos, in an addendum to the editors' reply to Bookchin, drives this point home by suggesting that the presentation of social ecology as a "closed system" is itself a reason to reject social ecology. This is not to say that Fotopoulos is rejecting social ecology per se, but rather an interpretation of it by Bookchin as the exclusive ground for "a new radical Left."[45]

Now, certainly, it would be a mistake to claim that a plurality of interpretations of social ecology, or challenges and amendments to it, is valuable in exactly the same way as, say, a diversity of species is valuable. We would not want to endorse a view that any interpretation of someone's work is valuable just because it offers another interpretation of the work. Certainly, there is room to claim, for example, that today there are too many deep ecologies, too many spinoffs from the original philosophical turn championed by Naess, so that it is unclear what the term "deep ecology" refers to, or whether there is anything like a "school" of thought under which all deep ecologists can be grouped. Such a proliferation of interpretations and elaborations of Naess's work may in fact have contributed to just the excesses upon which Bookchin hung his stern critique implicating all deep ecologists.

But if we confine ourselves for the moment solely to the question of the continuation of a dialogue about the core ideas of deep ecology, a clear case emerges for a proliferation of views. And if not for a proliferation of interpretations of a body of work in general, then at least for some body of supportive interpretation rather than none. Clearly, no individual can carry on the expression of a body of work indefinitely. As has been demonstrated time and time again in the history of philosophy and political theory, it is that work around which a dialogue forms that continues to live as a viable set of ideas. If the purpose of a political theory is not only to contribute a theoretical set of claims to an intellectual discussion, but also to offer a body of work useful for the reformation of society, then it must not become stagnant.

One thing is sure: it is only through constructive and critical exchange on Bookchin's work that a dialogue on social ecology can move forward into the next century. And certainly the operative word here is "critical." The editors of *Democracy and Nature* agree: "If dialogue does not involve a critical assessment of alternative theoretical views, then it is obviously meaningless."[46] It is into this breach, between the social ecology that could continue through dialogue and the social ecology that could stagnate, that the papers in this collection step. Some are more critical, some are more constructive; all, however, are united in a determination to continue, or perhaps begin to form, a new dialogue about social ecology.

Contents of the Volume

The essays in Part I of this volume examine the ethical and moral implications of Bookchin's work ranging from a general discussion of the relation between Bookchin's style and his philosophy to specific considerations of the content of his ethical theory. In "Negating Bookchin," a comprehensive critique of Bookchin's thought and rhetorical style, Joel Kovel distinguishes between two narratives that comingle throughout Bookchin's work, narratives referred to in the chapter as B1 and B2.[47] This heuristic device allows Kovel to find the source of the "bombast, vagueness, confusion, violent polemicism, and distortion" that he believes mar Bookchin's work. B1 is the positive doctrine of social ecology, the story of hierarchy and its overcoming. It includes the historical narrative of hierarchy/domination and the ways in which it has thwarted Enlightenment rationality, universality, and the peaceful coexistence of humanity and nature. B1 also includes the utopian aspirations for localized municipalism and the dissolution of all hierarchical political or social relations. B2 is a more fragmented narrative, rising to the surface frequently in overheated rhetoric. In B2, the messianic character of Bookchin's project plays itself out, leading to the one-sided hegemony of one form of social resistance: social ecology. Thus, the "running battles" Bookchin wages with his perceived adversaries, principally found elsewhere in the ecological Left, are, according to Kovel, more than hyperbolic diversions: they are indicators of a fundamental contradiction between the professed openness to natural differentiation and fecundity, on the one hand, and a devotion to the one true path, social ecology, on the other. Kovel explores the consequences of this contradiction through an analysis of several of Bookchin's themes, among them the rejection of Marxism, the notion of hierarchy "as such," the priority of freedom over justice, and the denial of dialectical negativity implicit in the denunciation of myth. The essay ends with a picture of a "negative ecology" that allows for imaginative dialectical progression.

Chapter 2 is a pair of essays. In the first (the only reprint in the collection), "Divining Evolution: The Ecological Ethics of Murray Bookchin," Robyn Eckersley critiques Bookchin's ethics from the point of view of a biocentric, nonanthropocentric ethic. Bookchin claims that his ethics are objectively grounded in the teleological unfolding of nature, a dialectic that human beings can understand and, importantly, should encourage. To adopt this ecological ethic, Bookchin holds, will produce the widest realm of freedom. However, Eckersley believes that Bookchin cannot deliver on this promise. She centers her arguments on Bookchin's separation of "first nature" and "second nature." Second nature, or humanity, has the unique capacity to affect its environment rationally, and thus bears an ethical responsibility to accelerate evolutionary process by fostering diversity and mutualism. However, says Eckersley, this lays Bookchin open to the same charge he makes against previous "naturalists" such as Aristotle and Herbert Spencer, namely, that they project human ends and ideas onto nature in order to justify a political stance. Bookchin's arguments with regard to the "objective" presence of his favored principles in nature are sketchy and, Eckersley asserts, inadequate. Also troubling is Bookchin's lack of detail on the subject of how much and in what way humans are to involve themselves in encouraging evolution. He dismisses the "quietism" of biocentric views—unfairly, according to Eckersley—but fails to note the danger of unintended consequences, a danger that has been illustrated in several human attempts to manage ecosystems. Finally, Eckersley believes that the ethic of social ecology is predicated on the presumptuous, arrogant belief that human beings are enlightened enough to divine the course of evolution, a belief that is demonstrably unfounded. Eckersley concludes her essay by asserting that, ironically, a biocentric ethic is more able to fulfill Bookchin's promise than his own.

In the second essay of Chapter 2, "Respecting Evolution: A Rejoinder to Bookchin," Eckersley offers a new, belated response to Bookchin's criticisms of "Divining Evolution." Bookchin had claimed that Eckersley misrepresents his position, had reiterated his critique of deep ecology, and then had employed his version of dialectics to criticize what he calls Eckersley's "Humean skepticism." It is this third set of arguments that are addressed in the essay. Eckersley argues that her ecocentric position is not an atomistic empiricism like Hume's, but that it does indeed imply a sense of humility and skepticism with regard to humans' ability to project the future course of nature. In her argument, she once again defends the fact/value distinction, claiming not that the two are unrelated, but simply that the former does not of itself determine the latter. She also defends ecocentrism against Bookchin's repeated charges that it implies a "hands-off" management style—"skeptical quietude"—acknowledging that any human activity obviously has ecological repercussions. However, she maintains her claim that Bookchin's notion of active and ethically required intervention is dangerous and presumptuous, arguing again that ecocentrism's humility and nonanthropocentrism are preferable guidelines.

Next, in "Ethics and Directionality in Nature," Glenn A. Albrecht focuses more closely and more sympathetically on Bookchin's views on evolution. He begins with an explication of Bookchin's thesis of directionality, exploring its divergence from orthodox Darwinian theory and some of its ethical implications. The second section of the essay presents two of the principle criticisms of the thesis. The first is the fact/value distinction, or the "is/ought" problem, which was employed by Robyn Eckersley to critique Bookchin's ethical naturalism. Albrecht considers whether Bookchin might be said to avoid this problem entirely due to his holistic view of nature, an argument supported by the work of John Rodman, and suggests that Bookchin's ethics are advantageous in that they provide an objective basis for normative judgments that enables us to avoid relativism and nihilism. The second criticism of Bookchin's view comes from mainstream evolutionary theory, which questions the notion that evolution proceeds in any one direction, the definition Bookchin gives of "variety and complexity," and the link drawn between physical evolution and cultural development. In the third section, Albrecht defends the idea of directionality in nature, turning for support to two fledgling scientific inquiries, complexity theory and nonequilibrium dynamics. The remainder of the essay is an exploration of these inquiries and the way in which they offer support for Bookchin's notion of self-generated movement toward complexity and diversity, as well as some directions they might suggest for the future development of social ecology.

In "Social Ecology and Reproductive Freedom: A Feminist Perspective," Regina Cochrane moves to a more specific critique of the ethics of social ecology by analyzing its relationship to the issue of abortion and, more generally, women's reproductive freedom. In doing so, she hopes to contribute to the development of a dialectical feminism. As the ostensible focus of her piece, Cochrane attempts to reconstruct social ecology's position on abortion. Bookchin says nothing specific on the subject, but several of his central ideas, combined with the few explicit statements on the subject by fellow social ecologist Janet Biehl, allow Cochrane to characterize a distinctive stance. In the third section of her chapter Cochrane explores the notion of reproductive freedom and its centrality in the feminist project. Here Cochrane draws on diverse sources to establish that reproductive freedom presupposes not only freedom of choice in regard to abortion, but sexual and social freedom generally. This includes, importantly, freedom from the unequal onus borne by women in the private domestic sphere, a goal which, for feminists, makes the private sphere a political one as well. Next, Cochrane provides a feminist critique of the position of social ecology on abortion and the wider notion of reproductive freedom it draws on. Cochrane notes some specific issues of concern, then turns to a broader discussion of what she perceives as Bookchin's historical depiction of women as passive onlookers to the male public sphere and a critique of Biehl's subsequent separation of the public and private spheres, specifically its consequences for women's social autonomy. The last section of the paper gestures toward a more appropriate dialectical feminism, drawing on Adorno's critique of dialectics founded on an "identity theory."

In Part II of this volume, we move to a specific focus on the political issues involved in Bookchin's theories of communalism and technology. In "Municipal Dreams: A Social Ecological Critique of Bookchin's

Politics," John Clark attempts to distinguish social ecology as an "evolving, dialectical, holistic philosophy" from Bookchin's increasingly narrow version of it. Specifically, Clark offers a thorough critique of libertarian municipalism. The essay is divided into twelve sections, each of which focuses on a particular aspect of Bookchin's municipalism, ranging from self-identity to economics to civic virtue. Though his criticisms are diverse, Clark returns repeatedly to several central themes. Bookchin's notion of "education," Clark claims, is a narrow and conservative notion of dialectics. Bookchin fails to recognize that every concept draws from and is dependent on its negation, which leads him to rigid positions. Like Kovel, Clark takes Bookchin to task for his tendency to criticize any social reform movement (e.g., the Hawkins wing of the Left Greens) that does not adhere to his program. Clark encourages a genuinely dialectical, open, experimental approach to social restructuring. Also, Clark points out that Bookchin's utopian municipalism lacks a sense of historical and cultural context. Bookchin criticizes any type of reform that proposes less than the complete abolition of capitalism and the nation-state, but he does not provide a strategy for getting from the present, which is decidedly immersed in both, to his utopian vision of the future. According to Clark his is a form of "abstract idealism, and tends to divert the energies of its adherents into an ideological sectarianism and away from an active and intelligent engagement with the complex, irreducible dimensions of history, culture and psyche." Thus, for Clark, Bookchin fails to provide a reliable pathway from here to there, from today's statist capitalism to his utopian libertarian municipalism.

Adolf G. Gundersen focuses on Bookchin's ideal community in "Bookchin's Ecocommunity as Ecotopia: A Constructive Critique," analyzing the utopian vision of social ecology with an eye toward strengthening it against several common criticisms. After briefly examining the basic elements of Bookchin's notion of decentralization and the role it plays in his thought, Gundersen lays out three theses upon which he believes Bookchin's argument rests. The first thesis is that small communities lessen human impact on the environment and make that impact easier to monitor; the second is that small communities are ecologically educative in that they intensify interaction with nature and participation in common life; and the third is that utopianism is necessary for any ecological vision due to the precipitous nature of the current ecological crisis. Gundersen then critiques each thesis in turn, using both "internal" critiques, which point out internal inconsistencies, and "external" critiques, which bring in outside arguments against decentralization. These critiques compose the substantial portion of the first part of the essay, though they ultimately fail to add up to a refutation. In the remainder of the essay, Gundersen analyzes ways in which to-

day's social ecologists can strengthen these theses. Specifically, he focuses on Bookchin's ontology of "emergence," claiming that more careful attention to this idea can salvage each of the three theses. Again,
Gundersen considers the theses in turn, applying the concept of emergence to each in an attempt to make them more internally consistent
and resistant to external critique. In the essay's conclusion, Gundersen
presents a caveat: he remains unconvinced that the improved version of
social ecology will be compelling, though there is undoubtedly much in
it from which we can and should learn.

David Watson analyzes Bookchin's thoughts on modern industrial
technology in "Social Ecology and the Problem of Technology." According to Watson, Bookchin is "trapped within the transition from a
red to a green radicalism," torn between a Marxian instrumentalist
view of technology and a more thorough rejection of the entire modern
technological network. Despite this tension, says Watson, Bookchin ultimately favors the red, or Marxian, approach, which holds that our
problems with technology arise from the fact that the means of production are owned by bourgeois capitalists, not from any cultural or psychological effects inherent in the technics themselves. According to
Watson, Bookchin seems to think that size and scale are irrelevant, that
any technology may be beneficial in the right hands. In fact, Bookchin
has what Watson believes are "fevered dreams of dialectical gadget fetishism," an unbounded faith in the liberatory power of postscarcity
"ecotechnologies." Watson asserts that, in adhering to the idea that
technology is the answer to the destruction wrought by the modern industrial machine, Bookchin is both naïve with regard to the complexly
interwoven networks of technology's effects and insensitive to his own
belief in the healing power of close contact with nature. He is also, as
several other writers in this volume maintain, arrogant with regard to
humanity's ability to understand and positively effect the course of evolution. Watson draws heavily on the work of Ellul and Mumford to
criticize Bookchin for failing to recognize that "mass technics have dramatically furthered a kind of psychic numbing and the fragmentation of
knowledge, undermining and complicating (though not entirely suppressing) human agency and responsibility." This type of technological
determinism is, Watson concludes, a more profitable foundation for social ecology than Bookchin's instrumental optimism.

Eric Stowe Higgs also takes up Bookchin's relationship to technology in " ' "Small" Is Neither Beautiful nor Ugly; It Is Merely Small':
Technology and the Future of Social Ecology." Higgs begins by analyzing the pluralism of Bookchin's accounts of technology, proposing three
possible readings: pluralism as confusion, as evolution (from the earlier
to the later work), and as pragmatism (of the Light/Katz variety). Concluding, like Watson, that confusion is the most likely source of

Bookchin's pluralism, Higgs then turns to the question of what more coherent theory of technology might best serve social ecology. After offering a general survey of the field and a critique of Bookchin's seeming instrumentalism, Higgs gives the work of Albert Borgmann a closer look. He elucidates Borgmann's central ideas and discusses briefly how the apparent political differences between Borgmann and social ecology might be mediated, concluding that Borgmann's communitarian theory of technology is the most promising in the field for future use by social ecology.

Part III turns to the more historical and comparative contexts of Bookchin's work, ranging from a specific inquiry into Bookchin's reading of history itself, and comparisons between Bookchin's thought and other theorists. Alan P. Rudy's "Ecology and Anthropology in the Work of Murray Bookchin: Problems of Theory and Evidence" focuses on Bookchin's account of preliterate, organic societies and the broader ecological theory upon which it is based. Rudy begins with a detailed presentation of Bookchin's anthropological account of the origin and institutionalization of domination and hierarchy. Rudy then turns to Bookchin's broader ecological theory of biosocial evolution, again arguing that Bookchin has simply placed himself in dualistic opposition to traditional theory, stressing mutualism and diversity while giving only lip service to competition and predation. Rudy points out several flaws, both theoretical and scientific, in this account. Throughout his critique, Rudy returns to several themes. One is that Bookchin's account is one of dualistic opposition rather than dialectical synthesis. Another is that the account is drastically overstated given the almost complete lack of empirical evidence to support it. Yet another is that the account is obviously driven by ideological ends, warping the evidence to reach particular sociological and ethical conclusions. In concluding, Rudy acknowledges the nobility of Bookchin's goals, but says that valid anthropological and evolutionary accounts must transcend the dualisms—"deterministic versus spontaneous and structuralist versus participatory"—that have ideologically obscured the real complexity of biosocial evolution. "The project, today," he says, "must be to analytically and practically understand the particular forms and general structures associated with contemporary [environmental] enablements and constraints so as to produce ecologically and socially appropriate responses to social and ecological crises."

David Macauley weighs in on the project of extending the history of social ecology by arguing that Bookchin writes within an eco-anarchist tradition that "antedates, influences and presently remains coextensive with Bookchin's own work." In "Evolution and Revolution: The Ecological Anarchism of Kropotkin and Bookchin," Macauley explores the connections between the two ecoanarchists, emphasizing the affinities in their work and claiming that Kropotkin has not received the attention from social ecologists that he warrants. In the essay's seven sections, Macauley analyzes Bookchin's and Kropotkin's respective (and often shared) positions on ecological mutualism, nature and human nature, ethical naturalism, anarcho-communism, evolution and revolution, decentralism and regionalism, and radical agriculture. In all these sections Macauley emphasizes the extensive parallels between the two, but he also points out where they diverge and discusses some of the stresses and strains in the work of each. Finally, the work of both thinkers is situated within the broader ecoanarchist tradition. In concluding, Macauley suggests the importance of the inspirations and insights in Kropotkin's wide-ranging body of work for the future of social ecology.

The collection ends with my own "Reconsidering Bookchin and Marcuse as Environmental Materialists: Toward an Evolving Social Ecology," as discussed at the beginning of this Introduction. I take the occasion of this volume to completely revise that previous argument directed at Bookchin and to further respond to criticism of this position by Bookchin. My response has taken the form of a thorough and extensive rewrite of the entire original argument. I admit several flaws in my initial interpretation of Bookchin's work, but nonetheless I argue that my original piece on Bookchin, involving a comparison of Bookchin's work with the critical theory of Herbert Marcuse, still holds true. The terms of that comparison involved two claims: (1) that both Bookchin and Marcuse could be classified in a similar way as political ecologists and (2) that knowing this could help to reform social ecology by helping it to recognize the importance of some of the issues of concern raised by deep ecologists. Originally, the stronger of the two was the first claim, specifically that Bookchin and Marcuse could be read as "environmental materialists," in a very specialized sense of the term "materialists." I define environmental materialists as simply those political ecologists who begin their description and prescription of environmental problems in social, institutional, or material terms. In contrast, "environmental ontologists," such as deep ecologists, argue that environmental problems have their basis in human ontological descriptions of themselves in relation to nature. A resolution of environmental problems on that approach must involve first and foremost a reform of that self-description. The second claim, however, concerning the reform of social ecology available through this distinction, has here evolved into a much stronger account of how Marcuse's work may serve as a bridge to a better, more complete, more evolved form of social ecology. But such an argument was not previously well received by Bookchin. I take great pains here to listen to each one of Bookchin's objections and to critically assess their merit in this new execution of my argument. The result is a better demonstration of how problems with Bookchin's work open up the possibility of improving social ecology through an interaction with critical theory.

Taken together, all of these essays speak to a variety of important questions involving Bookchin's work in itself and the next steps that social ecology can now take beyond the narrowing of the field as evidenced in Bookchin's work. It will be up to the eventual inheritors of the social ecology tradition to decide whether to emulate Bookchin's style and stick to the substance of his arguments alone, or to take social ecology into serious dialogue with other ideas in a way that risks changing the structure of social ecology itself. All of us who have contributed to this volume hope that social ecology will take the next bold step and become an integral part of a hoped-for coalition of effective environmental theories.

Notes

1. Andrew Light, "Rereading Bookchin and Marcuse as Environmental Materialists," *Capitalism, Nature, Socialism,* Vol. 4, No. 1, March, 1993, pp. 69–98.

2. Murray Bookchin, "Response to Andrew Light's 'Bookchin and Marcuse as Environmental Materialists,' " *Capitalism, Nature, Socialism,* Vol. 4, No. 2, June, 1993, pp. 101–113. At the beginning of his article, Bookchin says that I contrast his materialist approach with my own "ontological bent" which in my distinction (see my chapter in this volume) means that my "bent" is the same as that of deep ecologists (p. 101). Nowhere in that original paper do I claim to be an environmental ontologist. Bookchin characterizes my description of deep ecology in the paper as "three pages of encomia" (p. 107). But I do not see how a mere description of how deep ecologists understand the structure of their theory could be considered formal praise. Finally, Bookchin says that he and Marcuse are similar in being rooted in revolutionary Marxism and neo-Hegelianism, and in having come to understand that the era of proletarian hegemony had come to an end. This, says Bookchin, was a "*very* widespread perception on the left. . . . and one that I personally developed in my experiences on a factory floor, not by reading Frankfurt School theorists" (p. 108). Since he cannot be referring to Marcuse in this last passage, because it wouldn't make sense to say that the only lessons that Marcuse learned about the proletariat were what he knew from reading himself, and since this article is about my paper, I can only conclude that I am the oblique target of this remark.

3. See, e.g., Murray Bookchin, "As if People Mattered," *The Nation,* October 10, 1988, p. 294.

4. Murray Bookchin, "Social Ecology versus Deep Ecology," *Socialist Review,* Vol. 18, No. 3, July–September, 1988, p. 13; emphasis added.

5. Bookchin, "Response to Andrew Light's 'Bookchin and Marcuse,' " p. 106.

6. See Michael E. Zimmerman, *Contesting Earth's Future: Radical Ecology and Postmodernity* (Berkeley and Los Angeles: University of California Press, 1994).

7. See Andrew Light, "Deep Socialism?: An Interview with Arne Naess," *Capitalism, Nature, Socialism,* Vol. 8, No. 1, March, 1997, p. 84.

8. This part of the Introduction is adapted from Alan Rudy and Andrew Light, "Social Ecology and Social Labor: A Consideration and Critique of Murray Bookchin," *Capitalism, Nature, Socialism,* Vol. 6, No. 2, June, 1995, pp. 75–106. This work is primarily Rudy's.

9. Murray Bookchin, *Our Synthetic Environment* (New York: Harper & Row, 1962), published in paperback under the author's real name in 1974.

10. Murray Bookchin, *Crisis in Our Cities* (Engelwood Cliffs, NJ: Prentice-Hall, 1965).

11. Many of Bookchin's early articles are published under pseudonyms including a series of articles by "M. S. Shiloh," "Harry Ludd," and "Robert Keller" that appeared primarily in the journal *Contemporary Issues: A Magazine for a Democracy of Content* between 1950 and 1958. A few other articles that would later appear in *Anarchos* were published in other places under the name "Lewis Herber" in the early 1960s, including the very influential and much republished "Ecology and Revolutionary Thought," in *Comment* in 1964.

12. Murray Bookchin, *Post-Scarcity Anarchism,* 2nd ed. (Montreal: Black Rose Books, 1986 [1971]).

13. Murray Bookchin, *Toward an Ecological Society* (Montreal: Black Rose Books, 1980).

14. Murray Bookchin, *The Modern Crisis* (Philadelphia: New Society Publishers, 1986).

15. Murray Bookchin, *The Ecology of Freedom* (Palo Alto, CA: Cheshire Books, 1982).

16. Murray Bookchin, *The Rise of Urbanization and the Decline of Citizenship* (San Francisco: Sierra Club Books, 1987); later republished as *Urbanization without Cities* (Montreal: Black Rose Books, 1992).

17. Murray Bookchin, *Remaking Society* (Montreal: Black Rose Books, 1989).

18. Murray Bookchin, *The Philosophy of Social Ecology* (Montreal: Black Rose Books, 1990). Several other books published since then have both defended Bookchin's views and segregated his work from other ecological and political theories. See Murray Bookchin, *Re-Enchanting Humanity* (London: Cassell, 1995), and Murray Bookchin, *Social Anarchism or Lifestyle Anarchism* (San Francisco and Edinburgh: AK Press, 1995).

19. Bookchin, *Ecology of Freedom,* p. 357.

20. Ibid., p. 358.

21. Ibid., p. 355.

22. Bookchin, *Toward an Ecological Society,* p. 59.

23. Bookchin, *Modern Crisis,* p. 11.

24. Ibid., pp. 42–43.

25. Ibid., p. 60.

26. Ibid.

27. Ibid., pp. 16–17.

28. Bookchin, *Ecology of Freedom,* p. 29.

29. Bookchin, *Modern Crisis,* p. 42.

30. Bookchin, *Rise of Urbanization,* p. 228.

31. Ibid., p. 245. By community Bookchin means "a municipal association of people reinforced by its own economic power, its own institutionalization of the grass roots, and the confederal support of nearby communities organized into a territorial network on the local and regional scale."

32. Ibid., p. 257.

33. Ibid., p. 257–258.

34. Ibid., p. 258.

35. Ibid., pp. 261–263.

36. For a commentary and brief account of the continued evolution of deep ecology into different schools of thought with varying degrees of continuing connection to the work of Arne Naess, see Andrew Light, "Deep Socialism?: An Interview with Arne Naess."

37. John Clark, "A Social Ecology," *Capitalism, Nature, Socialism,* Vol. 8, No. 3, September 1997, p. 3. For another attempt at revising social ecology, see David Watson, *Beyond Bookchin: Preface for a Future Social Ecology* (Brooklyn, NY, and Detroit, MI: Autonomedia, 1996). Bookchin has rejected both Watson's and Clark's attempts. See Murray Bookchin, "Whither Anarchism?: A Reply to Recent Anarchist Critics," available at http://www. pitzer.edu/~dward/Anarchist_Archives/bookchin/whither.html.

38. Clark, "A Social Ecology," p. 9.

39. Murray Bookchin, "Comments on the International Social Ecology Network Gathering and the 'Deep Social Ecology' of John Clark," *Democracy and Nature,* Vol. 3, No. 3, 1997, pp. 154–197; Murray Bookchin, "Advisory Board Resignation Letter," *Democracy and Nature,* Vol. 3, No. 3, 1997, pp. 198–201. Janet Biehl's letter of resignation from the editorial board was also published in the same issue. Bookchin's first piece can be read as a critique of an earlier version of the chapter by Clark in this volume.

40. Bookchin, "Advisory Board Resignation Letter," p. 200.

41. Editorial Board of *Democracy and Nature,* "Response to the Resignations of Murray Bookchin and Janet Biehl," *Democracy and Nature,* Vol. 3, No. 3, 1997, p. 205. Later, the editors point out that "in Britain . . . *Society and Nature* has been considered the organ of 'Bookchinism' " (p. 207).

42. Ibid., p. 204.

43. Ibid., p. 206.

44. Ibid., pp. 207–208.

45. Ibid., p. 211.

46. Ibid., p. 206. Something vaguely like a follow-up to this whole resignation affair has arisen at the Bookchin archive website over a clarification Bookchin has made concerning his role, or lack thereof, in the discontinuation of John Clark on the *Democracy and Nature* editorial board (after it restarted

from its earlier manifestation as *Society and Nature*). I had mistakenly raised a concern about this issue in my interview, "Deep Socialism?: An Interview with Arne Naess," p. 76, note 6. In a footnote to "Whither Anarchism?," which is posted at the site, Bookchin states that he had nothing to do with the removal of Clark from the board (note 72). In a reply to this account, Takis Fotopoulos, the editor of *Democracy and Nature*, says in "Comment on Bookchin's 'Whither Anarchism,' " May 1998 (http://www.pitzer.edu/~Edward/Anarchist_Archives/bookchin/fotopoulos.html) that while "Bookchin is right that he had nothing to do with the decision to 'drop' Clark" from the board, "he did try to influence the editorial board when, for instance, in an e-mail message to [Fotopoulos] dated 13/10/95 he stated categorically that 'I should advise you that I do not want to be on any editorial board that also contains this man's name' " (paragraph 3). Bookchin's reply to the issue of this e-mail (in "On Takis Fotopoulos's Complaint," May 1998; http://www.pitzer.edu/~Edward/Anarchist_Archives/bookchin/fotopoulosreply.html) is somewhat surprising: "If my regal authority is so great that Fotopoulos et al. could be influenced by my desire to dissociate myself from Clark, then they are probably too impressionable to run an independent journal, still less a radical one. If I overestimated their independence and radicalism, I apologize again" (paragraph 7).

It is hard to know if there are any lessons to draw from this last episode in relation to the larger point here concerning the exclusivity of Bookchin's interpretation of social ecology. On the one hand, again, Bookchin does have every right to not be on an editorial board with someone he is at odds with. It is certainly the case that the volleys between Bookchin and Clark have become increasingly vitriolic on both sides, to my mind to an unnecessary extent. Still, while the last comment above in "On Takis Fotopoulos's Complaint" is not directed at the published reply of Fotopoulos and the editorial board to Bookchin's resignation from the board of the journal (an act by Bookchin that could not have been precipitated by Clark's presence on the board because by then he was gone), one cannot help but wonder whether in some way this is Bookchin's response to the criticisms by the board of Bookchin's rationale for his resignation. But of course, this is just speculation. Nonetheless, Bookchin's last line in his response to Fotopoulos is very much at odds with an attempt to try to mend the fences of social ecology given his characterization of the radical-ness of the board: "So please, let us cool down and exhibit some common sense in this entire affair."

47. David Roberts assisted considerably in the preparation of this section, drafting the summaries of the chapters.

Part I

Dialectics and Ethics

Chapter 1

Negating Bookchin

JOEL KOVEL

Introduction

Murray Bookchin, originator of the doctrine known as "social ecology" and by any reckoning one of the principal figures of the ecological Left, poses an enigma to the critic. Of all the radical ecologies, Bookchin's has the greatest range of historical reference, the most considered grasp of political possibilities, and the most extended philosophical reflection into the interrelations between humanity and nature. Nor is Bookchin an armchair thinker or a mandarin removed from the world. Living simply and disdaining the comforts of academia, he has influenced a whole generation of libertarians with his focused rage against capitalism and fervent evocation of revolutionary goals. In an age of tepid nonbelievers and intellectual hacks, Murray Bookchin stands out as a figure of some genuine grandeur. His resolute atheism might resist the association, but there is something of the prophet in this ecoanarchist.

Or a fallen angel, for there is also something haunted, a sense of rancor and harsh discord. Struggle is the sine qua non of change, and the more radical the project the more ardent the struggle. And the greater, therefore, the potential for aggression to be turned destructively inward or toward comrades rather than against the actual adversary. For whatever reason, there is a legacy of what might be called "surplus frustration" in Bookchin's work. It is taken for granted that all radical projects face bleak prospects in this profoundly reactionary time. One learns, hopefully, to accept immediate defeat, or frustration, and to strive toward keeping some light flickering through the long darkness. But there are also the self-induced defeats stemming from misguided theories and divisive practices. These surplus frustrations need to be criticized all the more forcefully given the fragility of the Left today. It is in this spirit that I would like to examine some of Bookchin's ideas about society, nature, and the revolutionary project.[1] I am aware that doing so may open me as well to the charge of "aggression toward comrades." But there seems to be no way around this risk, only a hope that true critique works toward overcoming barriers. In any case, a certain combativeness in the sphere of ideas is, in my view, to be encouraged, so long as it is carried out creatively. Given the condition of the world, I see no alternative to what William Blake called "mental fight" as a way of advance.[2]

The Two Bookchins

In casting about for a way to articulate the unease I have felt with certain aspects of Bookchin's doctrine, I was struck by the thought that in reading his work I frequently alternate between attitudes: now feeling appreciation and admiration for a nobly stated ideal or a cogent insight, now feeling distress over something bombastic, or vague and confused, or violently polemical and distorted. From another angle, I could say that there is much about Bookchin's worldview with which I am profoundly in sympathy and that I even find inspiring. And yet, although a point *should* be made, it cannot truly be made because too much that is jarring gets in the way. After a while, it seemed to me that there was a pattern to these alternations; there seemed to be a consistency to the inconsistency. And after some further reflection, I decided that this consistent inconsistency could become the organizing principle of a useful critique.

One could, in other words, approach Bookchin as the author of multiple discourses, pasted together so as to appear unitary, but actually the workings out of a fundamental split. In what follows, then, I shall adopt the heuristic fiction that Bookchin's manifest work is an imbrication of two such discourses, which shall be called B1 and B2, as though they were different archeological scripts. Each discourse contains a coherent narrative. However, the narratives contradict one another, thus weakening the texts within which they are imbricated and giving rise to the bombast, vagueness, confusion, violent polemicism, and distortion I noted above.

The common theme is an expansive one, nothing less than the emergence of human society, its long, tangled history of hierarchy and domination, the role of hierarchy/domination in the ecological crisis, and the possibilities of liberation from both hierarchy/domination and ecodestruction, the two being regarded as different aspects of the same process. I should emphasize again that this is merely a heuristic device. There aren't actually two discourses in Bookchin, but there is, in my view, a split state of being. Thinking in terms of two discourses will help illuminate this split, after which the device, like a scaffolding that has served its purpose, may be disassembled and set aside.

The B1 Narrative

The B1 narrative, in which the substantive ideas of Bookchin's social ecology are advanced, treats the overall theme within the framework of the self-creativity of nature and the emergence of ethical possibilities as part of nature's bounty. The philosophical basis is resolutely naturalistic—hence Bookchin's choice of the term "dialectical naturalism" to signify his ontology and philosophical anthropology. In this worldview, reason is essential, but reason of a special kind, to be distinguished from mysticism and mechanism alike. Bookchin advances the notion of "dialectical reason, conceived as the logical expression of a wide-ranging form of developmental causality." More than a method, dialectical reason is also a system of causality: "it is ontological, objective and therefore naturalistic. It explicates how processes occur in the natural world as well as in the social world."[3] In this respect, Bookchin reveals himself to be a follower of Hegel and a defender of Max Horkheimer's idea of "objective reason," reason that resides in nature and which we embrace as we seek freedom from hierarchy/domination. "Reason . . . is not a matter of personal opinion or taste. It seems to inhere in objective reality itself—in a sturdy belief in a rational and meaningful universe that is independent of our needs and proclivities as individuals."[4]

B1 is very much a text promoting Enlightenment humanism and a universality in which the harmonization of humanity and nature is a real possibility. Since there is no essential contradiction between humanity and nature, to remove the historical contradictions of hierarchy/domination frees us for an inherent ethical development in which we care for nature, and so become able to heal the earth. In this narrative, freedom contends with unfreedom, that is, with hierarchy/domination, in the sphere of the mind as well as in the world. In the tradition of the Enlightenment, B1 identifies freedom with reason and unfreedom with myth and primitive backwardness. Thus Bookchin attacks the

> utterly arbitrary character of myth, its lack of any critical correction by reason, [which] delivers us to complete falsehoods. Viewed from a primitivistic viewpoint, "freedom" takes on the treacherous form of an absence of desire, activity and will—a condition so purposeless that humanity

ceases to be capable of reflecting upon itself rationally and thereby pre-
venting emerging ruling elites from completely dominating it. In such a
mythic—and mystified—world, there would be no basis for being guarded
against hierarchy or for resisting it.[5]

This passage, to which I will return below, also enables us to understand more fully a distinction drawn very forcefully by Bookchin,
the positive conception of freedom and its priority over notions of justice. Justice is merely the redistribution of scarcity, consisting of "corrective alterations in a basically irrational society."[6] Freedom, on the
other hand, is the participation in a fuller, more ethical life through the
overcoming of hierarchy/domination according to the adoption of dialectical naturalism. This is not merely a matter of increasing rationalization. To the contrary, freedom for Bookchin is also the enhancement
of pleasure and the exercise of sensuousness. It is the revolutionary motion as such.

Hierarchy/domination is the antithesis of freedom. The center of
the narrative of B1 becomes, then, the story of hierarchy and its overcoming. The subtitle of Bookchin's most important work, *The Ecology
of Freedom,* states this explicitly: the ecology of freedom is contained in
"The Emergence and Dissolution of Hierarchy." "There is a strong
theoretical need," we learn, "to contrast hierarchy with the more widespread use of the words class and State; careless use of these terms can
produce a dangerous simplification of social reality." A so-called class-
less, or even libertarian, society can still conceal hierarchical relationships and a sensibility that "even in the absence of economic exploitation or political coercion . . . would serve to perpetuate unfreedom."[7]

What is this "master" category? According to Bookchin, hierarchy
is the

> cultural, traditional and psychological system of obedience and command,
> not merely the economic and political systems to which the terms class
> and State most appropriately refer. . . . I refer to the domination of the
> young by the old, of women by men, of one ethnic group by another, of
> "masses" by bureaucrats who profess to speak in their "higher social in-
> terests," of countryside by town, and in a more subtle psychological
> sense, of body by mind, of spirit by a shallow instrumental rationality.

Hierarchy and domination therefore imply one another; for Bookchin, they are congruent sets of relations. All hierarchies are dominative, all dominations hierarchical; the terms can be linked as one: hier-
archy/domination. Bookchin goes on to summarize his view: hierarchy/
domination is a notion that may not be "encompassed by a formal defi-

nition. I view it historically and existentially as a complex system of command and obedience in which elites enjoy varying degrees of control over their subordinates without necessarily exploiting them." Here it is worth underscoring an idea that receives at least as much attention throughout Bookchin's work as any other, namely, the subsumption of class and economic relations (such as exploitation) to the more encompassing notion of hierarchy/domination. Hierarchy, for Bookchin, "although it includes Marx's definition of class and even gives rise to class society historically, goes beyond this limited meaning imputed to a largely economic form of stratification."[8] (It may be worth observing that class does not even appear in the above-quoted iteration of hierarchies.) The universalism of B1's political message is predicated on the ubiquity of hierarchy/domination, for something so pervasive and deeply rooted requires nothing less than a total revolutionary transformation.

A great deal of Bookchin's work consists of tracing the history of hierarchy/domination and its overcoming. Certain landmarks stand out in this narrative terrain: the origins of hierarchy/domination in the gerontocracies of primitive, or organic, societies (which according to Bookchin antedate the emergence of patriarchy); the origins of the ancient state as the coalescence of hierarchy with warrior elites; the emergence of the idea of freedom in the Athenian polis; the persistence of relatively liberated zones of freedom in the medieval commune, the guilds of early cities, and among the artisans who helped make the French Revolution; the reincarnation of the ideal of the polis in the New England town meeting; the early industrial utopias of Fourier and Owen; the later utopian vision of William Morris; the stirring example of Spanish anarchism, cruelly crushed in the Spanish Civil War of 1936–1939[9]; and shadowing it all, the cancerous growth of modern industrial society, capitalism on a grander scale than that envisioned by Karl Marx—a capitalism that can only be overcome through the emergence of an ecological society beyond all forms of hierarchy/domination. The downfall of this capitalism was prepared when it developed the means to overcome scarcity, hence providing the basis for freedom and universality. The New Left protests of the 1960s were fervently embraced by Bookchin, who saw in the student radicals and libertarian dropouts from industrial society the germ of a new and more universal revolutionary subject. Having published far-seeing critiques of the incipient ecological crisis as far back as the early 1960s, Bookchin envisioned the possibilities for a conjugation of the New Left's critique of domination with an emergent green consciousness and the anarchist tradition; in particular, the new radicalism would gain control of municipalities and neighborhoods from below, then gradually extend this "libertarian municipalism" over the whole of society, converting it into an ecological realm of freedom.[10] Social ecology is the name given by Bookchin to this synthesis, and B1 became its principal narrative.

The B2 Narrative

B1 is, so to speak, social ecology's public discourse, the manifest content of Bookchin's system he would pass on to the world and by which he would be known. If there is a critical judgment to be passed on Bookchin's philosophy, it will have to culminate in the critique of B1. But there is more to Bookchin. Another discourse lies embedded in his works, one never explicitly thematized yet inescapable once we attend to it. The shards of this narrative lie scattered about in the texts. Mostly they appear as fragments of thought, suggestive phrases that provide a kind of emotional tone, or mood, to the reasoned arguments of his philosophy. Consider two of the passages quoted above: If we are "careless" in using the terms "hierarchy," "class," and "the State," a "dangerous simplification of social reality" will result. Myth is seen as having an "utterly arbitrary character," the use of which "delivers us to complete falsehoods." Viewed so, freedom takes on a "treacherous form" in which emerging elites will succeed in "completely dominating" humanity. "In such a mythic—and mystified—world, there would be no basis for being guarded against hierarchy or for resisting it."

Overheated notions of this sort are neither accidental nor rare findings in Bookchin's writings. In themselves they scarcely constitute a logic or rise above the level of emotional coloring. However, they are not written in themselves, but as a set of gestures organized about another account of social bondage and transformation. Viewed from this perspective, what could be discounted as peccadilloes of a fiery temperament or lapses into theoretical hysteria emerges as another grand narrative, whose spirit is no longer that of the emergence of universals but of a defense against catastrophe. The adverbs—"utterly," "completely," and the rest—are suggestive of extreme states of being. The message is of clear and present danger, with an aroma of betrayal and skullduggery. Elites are out to dominate us; those who do not adequately defend against this (by means, it goes without saying, of social ecology) are not simply mistaken but guilty of treachery.

In this scheme of things the Enlightenment is dimly off in the future. The emergence of domination and hierarchy and its dissolution in the ecological society is now a retelling of the legend of the Fall and Redemption, the master mythos of the Judeo-Christian tradition. Humanity is now to be read as the Tribes of Israel, a pastoral people in bondage to Pharoah, or Rome, or Babylon—the cities of corruption.

These await their Redeemer. This figure is of, for, and by the people: for simplicity's sake, he can be called the Anarchist. The Anarchist is the prophetic instrument of higher powers—the self-creativity of nature—on whose behalf he seeks to liberate humanity. But the Anarchist is also fated to suffer and to endure persecution. He is not primarily set upon by the powers of Rome (Bookchin having little to say about specific modalities of state repression—the state, you will recall, is subordinated to domination-in-general in his system). The Anarchist, rather, is greatly burdened with false prophets from his own tribe. These Pharisees and Sadducees are the source of the "treachery" noted above, betraying the people of Israel to Rome. More, there seems to be a kind of "Great Satan" among them, an archbetrayer. This diabolic deceiver assumes now one form, now another, including, in the late 1980s, the shape of the Deep Ecologist.[11] However, if we look at Bookchin's work as a whole over the years, it would seem that the principal incarnation of Satan in the modern age has been the Marxist, and especially that progenitor of the devilish doctrine who took the name of Karl Marx.

A few remarks about the construction of B2 may be in order before we study it in greater detail. It should be clear that, though a finding of this sort may be interesting, it is in itself neither especially remarkable nor decisive for critique. In this case, it is not even controversial, since the observation is shared by Bookchin himself, who frankly admits of the "unabashed [sic] messianic character of [The Ecology of Freedom], a messianic character that is philosophical and ancestral."[12] Nor is it so unusual to find an alternative textual narrative embedded in the grand theory of a systematic thinker. In fact, it is difficult to imagine such a thinker without such a metanarrative. It may be that what makes a mind creative is the synthesis of internally negating lines of thought, some traces of which may be expected to persist in the final product. Nietzsche, Freud, and Marx himself are three thinkers who come immediately to mind as providing considerable material for what could be called N2, F2, or M2 texts—and in the case of Marx there would be no problem at all in demonstrating that M2 draws heavily upon the same mythos of Fall and Redemption employed by Bookchin. A thinker with large aspirations for humanity will find abundant thematic material in the Bible, from Genesis, to Deuteronomy, to Isaiah, to Daniel, to the Gospels, and finally, to Revelation, the last, grandest, and most apocalyptic rendition of all. Bookchin's appropriation of the legend of the Fall and Redemption puts him, therefore, in the splendid company of such as Dante, Milton, Goethe, Blake, and, among contemporary thinkers, Ernst Bloch, whom Bookchin rightly admires for his radical futurism, a trait essentially linked to the notion of redemption.

But these are just abstract considerations, whereas the effects of a metanarrative have to be decided concretely. Depending upon how it is appropriated, the mythos of the Fall and Redemption can turn into the vision of Blake's *Jerusalem* or the nightmare of *Mein Kampf*. It is a question of "spirit," that is, the manner in which being expands into new and more comprehensive syntheses.[13] This can be rephrased: a great deal depends on what kind of a messiah Murray Bookchin promises to be. There is in any case a latent contradiction between the democratic ideal of B1 and the messianic promise of B2, which will be realized or not according to Bookchin's treatment of his theme. B1, the ostensible narrative of social ecology, renders humanity, or "the People" (Bookchin's phrase chosen to avoid using class as a leading term), as the protagonist of history; B2, on the other hand, tends to focus on the Anarchist as redeemer. To the extent that this figure hogs the stage with his personal vendettas, so will democratic promise become fraudulent. Humanity is no longer the agent of its own transformation; it becomes, rather, a sign flashed by the redeemer to demonstrate his redemptive bona fides.

In this context, the running battles Bookchin wages with his presumed adversaries are more than irritating distractions. As it becomes evident that Bookchin is not merely dealing with the ontology of hope or the emergence of ethical being from nature, but is consumed, rather, with venom and rage against those he sees in the way of "freedom," the entire edifice of social ecology shows the strain. Intolerance and dogmatism are, to say the least, common human tendencies. But it is remarkable to find them in a thinker who announces his project as follows: "My definition of the term, 'libertarian,' is guided by my description of the ecosystem: the image of unity in diversity, spontaneity, and complementary relationships, free of all hierarchy and domination."[14] In a case of this sort, the contradiction between libertarian preaching and rancorous practice becomes stifling. Categories with at least latent explanatory or emancipatory potential—the very notions of hierarchy, domination, freedom itself—become drained of intellectual vitality and turn into rhetorical devices by means of which the Anarchist establishes redemptive authority.

There is indeed a kind of betrayal in Bookchin's texts. It is the betrayal of a professed Enlightenment rationality by vindictiveness of Old Testament proportion, and it severely distorts the possibilities of both B1 and B2 by splitting them into mutually repellent fragments. As a result, Bookchin's dialectic of reason withers and loses its claim on universality through a radical demythologization; the redemptive myth suffers spiritual disaster by being denied immanent rationality.

Consider the treatment of the Enlightenment in *Remaking Soci-*

ety.[15] Bookchin begins this two-page passage with a ringing defense of Enlightenment universality: "It fostered a clear-eyed secular view toward the dark mythic world that festered in feudalism, religion, and royal despotism." Enlightenment reason "tried to formulate a general human interest . . . and to establish the idea of a shared human nature that would rescue humanity as a whole from a folk-like, tribalistic, and nationalistic particularism." Bookchin is a great admirer of Hegel, whose rendition of Enlightenment rationality included a "dialectic of eductive development, a process that is best expressed by organic growth." Thus the true legacy of the Enlightenment is an expansive, dialectically inclusive rationality with the potential for the reharmonization of humanity and nature—if, that is, the burdens of tribalistic particularism can be overcome. Though the imagery of festering gives some pause for concern as to whether the "dark" world has been actually transcended, the timbre of this passage is largely consistent with the aims of enlightened reason.

Bookchin proceeds with a familiar—though essential—reminder that capitalism "warped these goals, reducing reason to a harsh industrial rationalism focused on efficiency rather than a high-minded intellectuality," and that it did so as part of a project of bourgeois domination. But he devotes barely more than six lines to this point, which serves less as a thesis concerning the dominant enemy of the Enlightenment than a launching pad for what he really wants to do, namely, assault his enemies. These emerge from the ranks of those who stand apart from the bourgeois synthesis, the Left and especially the ecological Left. The attack occupies a full page, and while it starts modestly enough with a reminder that the "trends that denigrate reason" are "perhaps understandable reactions" to the alienating conditions of late capitalism, momentum starts picking up in the second paragraph as the force of narrative B2 begins to kick in. Now the "perhaps" is brushed aside and the Anarchist takes off his gloves to go after the betrayers of humanity. However "understandable," these trends—which, roughly speaking, comprise all identitarian, deep-ecological, and spiritually driven political movements—may be, they are also "profoundly reactionary" because they dissolve general human interest "into gender parochialism . . . tribalistic folkdom . . . and a 'return to wilderness.' " They become "crudely atavistic," they "retreat into the mythic darkness of a tribalistic past." "Ecology's motifs of complementarity, mutualism, and nonhierarchical relationships are completely dishonoured." For "if the Enlightenment left us any single legacy that we might prize above all others, it is the belief that humanity in a *free* society must be conceived as a unity, a 'one' that is bathed in the light of reason and empathy." By now Bookchin is fairly shrieking:

Rarely in history have we been called upon to make a stronger stand for
this legacy than today, when the sludge of irrationality, mindless growth,
centralized power, ecological dislocation, and mystical retreats into quietism threaten to overwhelm the human achievements of past times. Rarely
before have we been called upon not only to contain this sludge but to
push it back into the depths of a demonic history from which it emerged.

This is certainly a fascinating passage, and a major, authentic eruption of B2. One is tempted to paraphrase it: rarely in history has a
thinker whose leading idea is the espousal of reason been so given over
to unreason. But the issue is not exaggeration or lapse; the argument,
bizarre as it is, expresses a logic. More, it has certain effects on the
whole, in this case, social ecology.

At first glance, the passage is ostensibly consistent: there are "atavistic" groups; they threaten to impair the Enlightenment project,
which is a good thing; therefore, they must be turned away. But here a
rupture occurs. Bookchin seems unable to recognize that the dialectical
and ecological sensibility he is espousing doesn't just turn things away,
or to use his vivid phrase, "push [them] . . . back into the depths of a
demonic history." That is to say, he does not recognize his self-
professed logic. For does he not himself assert in this very passage that
Enlightenment-oriented ecologic is "bathed in the light of reason," and
also, *"empathy"*—that is, fellow feeling with others, a condition that
would surely apply to those undergoing the painful struggle of contend-
ing with the ecological crisis? But who can feel with "sludge?"

Language, in this case excremental language, has real consequences; it is not just little boys talking "ca-ca." Bookchin's excremental metaphor sets up a universe of radical rejection and nonrecognition:
associating an alternative view with what is shit out of the body. To use
his apt evocation of Hegel, ecologic entails a "dialectic of eductive [i.e.,
drawing forth] development, a process that is best expressed by organic
growth." Exactly—and exactly what this apostle of a humanity harmo-
nized with nature rules out, because sludge can at best be further dedif-
ferentiated and broken down. But Bookchin cannot even tolerate this.
He wants the "sludge" of a spiritual worldview pushed back, buried,
even destroyed. The outburst is therefore a call for repression rather
than for dialectical unification. Since we are dealing here with human
beings and not just with ideas to be forgotten, the violent implications
of this pushing back cannot be overlooked. Not physical violence, but
the violation that comes from regarding another's existentially held position with contempt. Hegel's dialectic is a process of recognition, a
painfully won encounter between self and other. Bookchin will give the
requisite lip service to this notion, but when it counts, that is, when he

encounters another, a potential comrade no less, who may have areas of difference, then the other's position becomes sludge to be pushed away rather than a premonitory insight to be incorporated into a larger unity.

The premise of dialectic is central for Bookchin. But this is a dialectic of nonrecognition, of exclusion and splitting, which is to say that it is no dialectic at all. Nor is it ground for an ethic, nor any prescription for real democracy. Bookchin leaves those with "high-minded intellectuality," that is, the social ecological elect, on their pinnacle, with the remainder of humanity swarming below, lost in their pathetic spiritual delusions and parochial interests. This sectarianism in the midst of universalism has been paralleled in the actual record of Bookchin's political engagement, a trail littered with aborted or ruptured alliances between the high-minded intellectual and those not deemed good enough for him: solidarity groups, antinuclear affinity groups, labor groups, civic action groups, women's groups, cultural groups, even the Left-Green Network when it made honest demands for social justice. In other words, whoever stands for the fitful yet concrete emergence of humanity from domination in a way not sanctioned by purified social-ecological dogma is treated like toxic sludge. Would that Bookchin have applied to his practice his own theory: "A libertarian rationality raises natural ecology's tenet of unity in diversity to the level of reason itself; it evokes a logic of unity between the 'I' and the 'other' that recognizes the stabilizing and integrative function of diversity—of a cosmos of 'others' that can be comprehended and integrated symbiotically."[16]

Coping with Satan

The world, and especially the world of academia and the Left, is full of bad people in the eyes of Murray Bookchin. There are the postmodernists ("Yuppie nihilism"[17]); the deep ecologists ("well-to-do people who have been raised on a spiritual diet of Eastern cults mixed with Hollywood and Disneyland fantasies"[18]); the cultural ecofeminists ("mystified by contrived myths—an ensemble that is borne on a lucrative tidal wave of books, artifacts and bejeweled ornaments"[19]); and the whole "squalid ooze of atheistic religions, natural supernaturalisms, privatistic politics, and even liberal reactionaries."[20] But let there be no mistake that one big devil hangs over them all, Bookchin's bête noire, Karl Marx.

It is not that Marx is treated with more vituperation than the other false prophets. In fact, he often gets less; indeed, there are grudging expressions of praise scattered about the texts for Marx's insight into the capitalist economy or his enunciation of revolutionary goals. This may be a manifestation of the fact that Bookchin cannot afford to dismiss Marx as he can a goddess worshipper, with a few smears of the excremental brush. The scale of Marx's work forbids this approach; more importantly, its theoretical center stands squarely in Bookchin's path. If hierarchy/domination as Bookchin understands them are to become the centerpieces of radical ecological thought, then the central contributions of Marxism—class struggle, mode of production, and the like—have to be displaced. It is an unfortunate feature of the mantle of messianism that it can be worn by only one figure. Those applying for the position have to eliminate the competition. Bookchin has to wrestle with Marx and defeat him if his own messianic ambitions are to be fulfilled. And so there is a material reason why the figure of the Great Satan takes the shape of Karl Marx and figures so massively in Bookchin's texts.

A number of points need to be made before we can take up the substance of Bookchin's relationship to Marx. First, it should be emphasized that no serious assessment of Marx's relation to radical ecology can be made within the confines of this essay. Second, although my sympathies will be clear in what follows, nothing written here should be taken to mean that either Marx or the Marxist tradition are above reproach, especially from an ecological standpoint. An ecological critique of Marx is necessary in light of the economism, the fetishism of industrial growth, and the centralizing, antidemocratic record of actual Marxist projects. These may not have been truly Marxist, but they were not unrelated to Marxism, either. Further, although a distinction between Marx himself and the tradition that carries his name has to be made (he did, after all, say he was no Marxist), I do not believe that there is some "pure" essence of Marx that, if followed, could have avoided the counterecological and antidemocratic history of Marxisms.[21]

This is to say that a substantial portion of Bookchin's attack against the economism and centralism of the Marxist tradition is valid. It is certainly not original with him—which Bookchin strenuously implies; and it should be pointed out that similar criticism has come from within Marxism itself—which Bookchin strenuously ignores; but the critique remains important. On what grounds, then, does one question these texts? Simply this: that Bookchin on Marx all too often ceases to be critique. It deteriorates instead into grandiose posturing, propelled by roughly equivalent portions of hatred, ignorance, and lack of understanding.

For example, we are informed of Marx's "atrocious misreadings of 'savage' society"[22] in the course of a disputation of Marx's view that there is an essential "otherness" in our relation to nature. I will return to the more abstract point below, but it might be worth asking about this "atrocious misreading." No evidence is offered—a characteristic feature of Bookchin's writings; perhaps he refers to Marx's bias for town over country, or Engels's reading of primitive society, drawn from Morgan and presumably reflective of Marx's final views. Now there are difficulties with Engels's work, though it is scarcely "atrocious," but is it really reflective of Marx on primitive society? In fact, there was precious little in the standard Marxist oeuvre to indicate just what Marx really thought about this question. By 1980, however, when *The Ecology of Freedom* was being composed, serious students of this question finally had access to a remarkable document devoted just to this subject, the *Ethnological Notebooks*, a collection of notes and drafts on primitive society published in 1972 as the last of Marx's works. The *Notebooks* are doubly remarkable: for what they say about the vitality and genius of primitive society, and for what they reveal about Marx's continuing capacity, even in his fading last years, for intellectual renewal. Franklin Rosemont, a libertarian surrealist (if any label can be applied) deeply critical of established Marxism, finds the *Notebooks* quite wondrous, calling them a work that "still glows brightly with the colors of the future," yet that also demonstrates a return to the spirit of the *1844 Manuscripts*. Then there is the late Raya Dunayevskaya, a Marxist humanist whose theoretical ambitions could match those of Bookchin himself. Dunayevskaya's *Rosa Luxemburg, Women's Liberation, and Marx's Philosophy of Revolution*, published in 1981, a year before *The Ecology of Freedom*, was inspired by the *Ethnological Notebooks*, which revealed to her by its treatment of primitive society "how very deep must be the uprooting of class society and how broad the view of the forces of revolution."[23] Note that this is precisely Bookchin's own thesis. Do not expect him, however, to welcome the finding as evidence of convergence between radical social theories. There can, after all, be but one Messiah. And so, if there is an "atrocious misreading" here, it is Bookchin's—not the only instance, by the way, of the mechanism of psychological projection in his texts.

Common ground does not exist for Bookchin; there is one way the revolution must *not* go. Bookchin's hostility to socialism is adamant, all-consuming, and redolent with excremental vision: "A 'socialist' ecology, a 'socialist' feminism, and a 'socialist' community movement . . . are not only contradictions in terms; they infest the newly formed, living movements of the future with the maggots of cadavers from the past and need to be opposed unrelentingly."[24] To buttress this stunningly sectarian view, no distortion is too outlandish. Thus "the

Marxian revolutionary project reinforced the very degradation, deculturalization and depersonalization of the workers produced by the factory system. The worker was at his or her best as a good trade-unionist
or a devoted party functionary, not as a culturally sophisticated being
with wide human and moral concerns"[25]—a remarkable statement
from someone who grew up in the atmosphere of New York City communism of the 1930s, with its worker schools, libraries, and culturally
vibrant summer camps. Or perhaps Bookchin has never heard of Paul
Robeson, Diego Rivera, or Pablo Neruda. There were many things
wrong with the communist movement, but deculturalization of the
workers was not one of them.

The reader will be impressed to learn that all this dehumanization
started with the "remarkably insidious reduction of *human beings* to
objective forces of 'history' " by a certain nineteenth-century thinker,
who, like the serpent in the Garden, gives us "a mentality that is more
disconcerting than the most unfeeling form of 'anthropocentrism.' " No
wonder that the "contribution proletarian socialism made to the revolutionary project was minimal, at best, and largely economic in character."[26] This stems, in Bookchin's view, from Marx's objectification and
lack of moral vision: "Evil was not a word that Marx was wont to use
when he tried to turn the critique of capitalism into an 'objective' science, freed of all moral connotations."[27] Yet even within the terms of
objective science, Marx is second rate. He "mystifies"; he is impossibly
"vague"; his views on labor are apparently full of "innocence," but in
actuality "highly deceptive" and "riddled by ideology—an ideology
that is all the more deceptive because Marx himself is unaware of the
trap into which he has fallen. The trap lies precisely in the *abstraction*
that Marx imparts to the labor process, its ahistorical autonomy and
character as a strictly technical process."[28] And finally, from an earlier
broadside, Marx's "theories were still anchored in the realm of *survival,* not the realm of *life.*"[29] Bookchin sums it all up for the "remarkably insidious" Marx: "Tragically, Marxism virtually silenced all
earlier revolutionary voices for more than a century and held history itself in the icy grip of a remarkably bourgeois theory of development
based on the domination of nature and the centralization of power."[30]

What are we to make of these gross caricatures? This is not the
place to refute them in detail, to prove that no, Marx is not the figure
represented here, that he saw humans as the makers of history, that his
work is suffused with moral categories, that his is perhaps the most important contribution to the revolutionary project, that Marx's theory of
labor is anything but ahistorical and technicist—and finally, that his
ontology is anchored in the realm of life and not survival? There is no
point in doing so; this is not a study of Marx but of Bookchin, and the

convinced are not to be persuaded. But it is necessary to reflect on the significance of the fact that Bookchin's rendition of Marx is a carica-ture and not a critique, for it is precisely in this regard that we can ap-preciate where he loses his way.

To caricature some figure—and this would apply to Bookchin's treatment of the spiritual position as well as to Marx—it is necessary to reduce that figure, specifically, to take seemingly contradictory features that coexist in the living organism and split them apart, discarding what doesn't suit one's predilections and highlighting only what serves ideology. This is fine for political cartoons, but is a disaster for theory. Consider Marx for a moment. Anyone with reasonable familiarity with Marx's writings will find abstraction from the rich articulation of social existence—the feature singled out by Bookchin—but will also find the concrete wealth of that existence, as well as moral and objective cate-gories intertwining with each other.[31] Thus Bookchin caricatures Marx by reducing him and regarding him with single vision only.

But the alternative, namely, using twofold or multiple vision, is also how Marx surpasses the bourgeois worldview, for bourgeois thought is distinguished by the reduction of the universe to a monocu-lar perspective. Marx's capacity to range over multiple registers is a re-alization of the eternally fresh words of William Blake: "May God us keep / From Single vision & Newtons sleep."[32] Thus if we are to hurl epithets about, the bourgeois thought belongs to Bookchin for his re-ductionism—another instance of projection (which we also find in the ascription of vagueness to Marx). In his need to destroy Marx, Book-chin uses radical rhetoric while retreating theoretically into a one-sided and ultimately static materialism. It is as if he goes back from Marx to Feuerbach—except that Feuerbach couldn't have been expected to know better, nor was his thought tendentious and riddled with hateful projections.

We return to the point of dialectic. Bookchin professes himself a dialectical thinker, but dialectic and single vision are mutually exclu-sive. The sense of multiple possibility is the core of dialectic—that and the movement through contradiction. Dialectic requires recognition of the negative along with the affirmative. Such is the essential quality missing from Bookchin—the dwelling in negation, the capacity to hold together opposites so that the life immanent within their contradictori-ness can grow. Instead of negation, he sees "sludge," or something "in-sidious" or "deceptive," and expels it. Striving toward Enlightenment universality, he ends with repression. Dialectic, instead of unfolding, becomes static, frozen in an endless series of vendettas.

Dialectical stasis weakens ecological vision, because ecology itself is the recognition of dialectic in nature; it also undermines the theory of hierarchy/domination. To be more exact, Bookchin's undialectical notion of hierarchy/domination replaces theory with rhetoric, as, for example, when he proclaims that

> Hierarchy *as such*—be it in the form of ways of thinking, basic human relationships, social relations, and society's interaction with nature—could now be disentangled from the traditional nexus of class analysis that concealed it under a carpet of economic interpretations of society.[33]

Viewed as rhetoric, this passage makes sense. Its hidden claim is that Bookchin should be chosen over Marx for the role of Messiah, thanks to his all-encompassing view, which both includes and transcends that of poor old Marx. Promoting hierarchy/domination in general over all concrete determinations such as class or patriarchy gives the Bookchinite a trump card for all left debate, for if you disagree with Bookchin's view then logically you must be soft on domination—and if you agree, you are recognizing Bookchin's pre-"dominance" among radical ecologists.

Viewed as theory, however, the sense breaks down. Consider the axiomatic identity between hierarchy and domination. According to Bookchin in the above passage and throughout his work, hierarchy is an "as such," an essence that can be abstracted out of all concrete instances; it is an ontological flaw in B1, original sin in B2. Hence the assertion, repeatedly made, that the domination of nature is tightly linked to any and all manifestations of hierarchy, all of these being expressions of the same essence. Aside from the rhetorical claims, therefore, the anarchism of social-ecological politics rests upon the postulation of the absolute value of hierarchy, since anything less than eliminating all hierarchies (as, e.g., focusing on the empowering of workers, or of peasant women) will leave untouched the roots of the domination of nature.

One problem with this abstraction is that it remains abstract. Notwithstanding the enormous amount of attention given to the subject, at no point in Bookchin's work have I been able to find an account of what actually makes a hierarchy tick. He will say that "for reasons that involve complex evaluation" history was "diverted" from a cooperative to a hierarchical "direction,"[34] but we never learn what this complex evaluation entails. Hierarchy remains, therefore, a somewhat mysterious wrong turn to which human being is subject. What can have happened? Has nature mysteriously screwed up in the evolution of the human species—an impossibility by Bookchin's lights, who never ceases extolling the creative fecundity of nature. Or is humanity really in the grip of original sin, which would make impossible the optimism of social ecology?

Or could the question be rephrased? Could it be, in contrast to Bookchin's view, that we do not have to get rid of hierarchy as such, but need to attend, rather, to those hierarchies that degenerate into domination? In other words, the identity, hierarchy = domination, might be broken. Whether this is the case depends upon an essentially empirical issue that comes down to the following: Are hierarchies ubiquitous? Are there different dominative values attached to different hierarchies, so that some are less bad than others? Are some hierarchies even beneficial?

The answer is Yes to all these questions, which is to say, the category of hierarchy/domination as Bookchin develops it breaks down. Oddly enough, Bookchin himself recognizes this truth. He contradicts himself at the most vital point in his theory by recognizing that some hierarchies can be good, admitting tacitly that hierarchy is ubiquitous in human existence, and moreover—though he would surely deny that this is so—that class domination has priority within the set of hierarchies.

In a recent essay, "What Is Social Ecology?," Bookchin has the following to say:

> It is worth emphasizing that hierarchical domination, however coercive it may be, is not to be confused with class exploitation. Often the role of high-status individuals is very well-meaning, as in commands given by caring parents to their children, of concerned husbands and wives to each other, or of elderly people to younger ones. In tribal societies . . . the respect accorded to many chiefs is earned, not by hoarding surpluses as a means to power but by disposing of them as evidence of generosity.

There follows a discussion of the complex interrelations of class, hierarchies, and the domination of nature culminating in the observation that "until human beings cease to live in societies that are structured around hierarchies as well as economic classes, we shall never be free of domination, however much we try to dispel it with rituals, incantations, ecotheologies and adoption of seemingly 'natural' ways of life."[35]

There is a muddle here that is both revealing and puzzling. Logically, if the ecological society is to be without hierarchies, then it will have to dispense with the parent–child relationship, since this is a hierarchy according to Bookchin. Perhaps the little ones can be put in an autonomously run nursery, where they may learn the joys of freedom, preparing their own food and putting themselves to bed at night without anybody telling them what to do. If this is not to his taste and he would prefer, like other people, that small children be protected and cared for, then Bookchin must be willing to live with hierarchy, which is to say in his terms, domination. Where, then, is the theory of social ecology? Ah, but he says that the parent–child relationship is, or at least can be, a *good* hierarchy—a perfectly sensible idea that requires us, however, to detach hierarchy from domination. More specifically, we need then to concretely specify hierarchies to discern those that are harmful for humans and nature. And it also follows that there can be no necessary correlation between the domination of humans and that of nature, since such a correlation is grounded, according to Bookchin, in the generalization "hierarchy = domination."

In sum, we have to recognize that certain hierarchies can be both ubiquitous and at least potentially good. These would include the parent–child relation, but also that of teacher–student, or indeed any place in the cultural system where some people have something useful to impart to others and claim an authority to do so. Can we imagine a surgical operating room without hierarchy? An oceangoing ship? In other words, hierarchy is as inherent in the human situation as is childhood, culture, and the division of labor. Indeed, hierarchy is a concrete manifestation of human being. It is embedded in the very essentials of our species life, in the facts that we are born helpless and live through the creation and social transmission of a created world. In other words, if you want to eliminate hierarchies, you have to eliminate what is specifically human.

This distinction needs to be clearly drawn. There should be no doubt that any hierarchy contains the seeds of domination, and that therefore in every instance of hierarchy work has to be done to ensure that these dominative potentials are not realized, leading to the overcontrolling parent, the teacher who stifles the self-expression of the student, the surgical operating room that turns into an exhibit of patriarchy, and so on. In theoretical terms, what this means is that hierarchies need to evolve dialectically through a continuing sublation of their internal negativities.[36] All hierarchies have reciprocal character, which is to say, they are never unitary top-down structures except as an approximation in the extremes of domination. Children, to take the example of the most fundamental situation, reveal from earliest infancy a capacity to steer the parent–child hierarchy through the immanent logic of their emergent self-expression. As this grows, so does the parent –child relationship, with the parents growing—hopefully, but as experience confirms, all too rarely—along with it in a continuous evolution of roles and capacities.

But an evolving hierarchy is not necessarily a disappearing hierarchy. Reciprocity should not be confused with identity or the disappearance of structure. Certain differentiations can intrinsically remain and be given ordered, that is, hierarchical, status according to experience, wisdom, physical capacity, even symbolic need. We could call this the rational exercise of authority, so long as it is recognized that the form of reason involved is distinct from that established by the ruling order.

In any case, this reasoning removes the character of *as such* from hierarchy. It obliges us, rather, to think concretely about hierarchies in order to understand how they can be ecodestructive and murderous and to be aware of the forms they take as they become so. If there is no hierarchy as such then we are back to square one, where the job consists of looking closely at hierarchies and sorting them out. We might even find that the bad ones go under the names of Capitalism and Patriarchy, and decide on this basis to give priority to their immediate confrontation rather than going the route outlined by Bookchin. Admittedly, this might require bypassing the libertarian municipalist model of social transformation in certain instances, or to be more exact, downgrading it to the status of one pathway among many toward social-ecological transformation. If I am not mistaken, however, this would be on the whole quite a good thing. It would mean, at the least, that we would take the real world—and the rest of the world—a lot more seriously than Bookchin would allow. Imagine—seeing ecological possibilities in a Marxist–Leninist state, or in peasants who have never known the glories of the Enlightenment! These developments are off Bookchin's map; he would most likely warn us against the grave and insidious dangers that lay in store if we took them seriously or—Hegel forbid!—try to learn from them. But freed from Bookchin's rigid messianism, we could do just that.

Consider, for example, the substantial efforts made toward the development of organic agriculture in a Cuba driven to the wall by Soviet collapse and U.S. hostility, a project that Peter Rosset and Medea Benjamin have called "unprecedented, with potentially enormous implications for other countries suffering from the declining sustainability of conventional agricultural production."[37] This turn of events is by definition impossible according to the single vision of social ecology. However, if we attend to concrete and intermediate formations rather than to grand teleological abstractions such as the Ecological Society, we would be able to appreciate that there are degrees of freedom for restructuring agriculture afforded by the absence of agribusiness (i.e., capitalism)—and appreciate, too, the potentialities for growth persisting even in that despised entity, the socialist party-state. The lesson to be learned is the importance of getting capitalism off the back—not that Cuba is a good model for ecological transformation, which it cannot be, given all the deformations sedimented into Cuban society.

Then there is the spiritual/ecological radicalism emergent in places like the Indian subcontinent, made manifest in such phenomena as the tree-hugging Chipko movement. We might approach this through the work of Vandana Shiva, in whose person advanced science (Shiva having originally trained as a nuclear physicist), radical feminism, Marxism, and ancient spirituality come together in a working out of dialectic, with important consequences for the critique of science, of "development," and of modernism itself.[38] To do so commits neither to the fetishization of the East, nor to an ecofeminist "vanguard," nor to any particular spirituality as "the Way" to ecological transformation. As with the situation in Cuba, it simply consists of taking a more differentiated look at ecological politics.

The issue is no longer hierarchy as such, but hierarchy as it becomes domination—and domination as it is undone to become emancipation. Here a criterion is at hand in the notion of dialectic as the emergence of being through negation. Is this occluded or thwarted? Then we have an instance of domination. Is the occlusion undone so that negations emerge, proliferate, expand, and move toward universality? Then domination is to that degree overcome, while emancipation supervenes. Such an approach fosters concrete engagement with points of resistance and transformation as they spontaneously emerge. The abstract denunciation of hierarchy *as such* favors an equivalently abstract kind of politics, with the abstraction filling up with the localization of whoever enunciates it. Thus social ecology's municipalism, rigidly advanced by Bookchin, is a doctrine unable to be shared with or to learn from that 90% and more of the world population who do not share in the blessings of the Vermont town, German philosophy, or the emancipatory heritage of the white West. There is, in short, a kind of cryptoracism inherent in social ecology as Bookchin develops the notion, no matter how antiracist its individual practitioners may be.

A more dialectical ecological politics, by contrast, would be open to the emergence of multiple points of resistance and social transformation as these are conditioned in different settings by the forms of domination peculiar to the ecological crisis, namely, capitalism and patriarchy. Thus the ecological society is prefigured in the light of socialism and postpatriarchal social relations, for these are the specific negations of the prevailing domination of nature even if their positive content has not yet been adequately defined.

Questioning the hierarchy/domination nexus also leads to a rethinking of the freedom/justice nexus. Recall that Bookchin's view subordinates the latter to the former. This is not just a theoretical issue. Although he makes the customary obeisances to the victims of oppression in his writings, anyone familiar with Bookchin's political practice will know that he really does give a lesser role to struggles for a more just, egalitarian society as compared to his municipalism and the broad goals of social ecology. To Bookchin, justice is the mere rearrangement of inequities, while freedom is the positive good of overcoming hierarchy. Freedom is "not only the equality of unequals, but also the enlargement of our concepts of subjectivity, technics, science and ethics."[39] Justice, stained with the retrograde Marxian emphasis on economics, has actually become a bad thing:

> What is so stunning to the careful observer is that if justice never came to compensate but merely to reward, its spirit has finally become mean and its coinage small. Like every limited ideal, its history has always been greater than its present. But the future of justice threatens to betray even its claims to have upheld the "rights" of the individual and humanity. For as human inequality increases in fact, if not in theory, its ideology of equivalence assails the ideal of freedom with its cynical opportunism and a sleazy meliorism.[40]

Even after we clear away the rhetoric—"stunning," "betray," "assails," "cynical," "sleazy," and so on—there remains something insufferable about this way of thinking, which reduces the struggle against suffering and exploitation to an empty gesture, while the real work of overcoming "hierarchy" takes place so that the liberated social ecologist can frolic in the polis of the postscarcity utopia. The connoisseur of mental projections in Bookchin's work here finds one more specimen for his or her scrapbook, since meanness of spirit and smallness of coinage certainly apply to this callous passage. Callous and irrelevant too, for the notion of a postscarcity conjuncture that opens the way for "freedom" was based on a shallow reading of capitalism. The utopian hopes of the late 1960s now seem a hoax in a world where 850 million people are out of work and mass starvation looms.[41] The postscarcity project is now on the scrap heap of history, not because the means for overcoming scarcity are not at hand, but because it is more obvious than ever that capitalism is not about to let them be realized.

In any case, if there is no hierarchy/domination in general, then there can be no freedom in general as an undoing of same. And if the issue is those hierarchies that have turned into domination, then the approach to emancipation needs to pass through the overcoming of specific dominations. Another name for this is bringing justice to bear. A freedom detached from justice is, frankly, silly; fully developed, it turns into one of those New Age parodies Bookchin rightly despises, where the comfortable prance freely about their "growth center." In contrast, as anyone who has participated in a campaign for justice can testify, the struggle can generate an existential intensity that is the sine qua non for real social transformation. Here is where the revolutionary subject is forged, from Marx's youthful engagement with the wine growers of the Moselle to those who fight on for justice in places—of little importance to Bookchin's social ecology—like Haiti, Guatemala, and East Timor.[42] For justice is both individual, in that one particular case has to be addressed, and universalizing, in that the emancipation of one cannot fully take place until the emancipation of all is realized. This relation gives the various environmental justice movements a value they occupy in neither the theory nor the practice of Bookchinian social ecology.[43] As it is said, "No justice, no peace!," so may it be added, "No justice, no freedom!"

Toward a Negative Ecology

I have already observed that the self-contradictory character of Bookchin's libertarianism is reflected in his ontology and notion of dialectic. From another angle, this difficulty inheres in his conception of the human–nature boundary. Bookchin wants freedom, but he wants it in terms of the unfolding of nature into humanity. This is grounded in an Aristotelian notion of entelechy, or nature's self-generative capacity, which finds its self-expression in human being. Thus Bookchin sees humanity as directly continuous with nature: "There seems to be a kind of intentionality latent in nature, a graded development of self-organization that yields subjectivity and, finally, self-reflexivity in its highly developed human form."[44]

One implication of this is a teleological ethic in which what is and what ought to be converge toward identity. "Ethics is not merely a matter of personal taste and values; it is factually anchored in the world itself as an objective standard of self-realization."[45] This implies a definite, preset way toward which the development of things points. "Until [things and phenomena] are what they have been constituted to become, they exist in a dynamic tension."[46] Bookchin is at pains to point out that such a view only commits one to a general idea of growth and development of an ethical sensibility, not to any particular moral content, nor to a religious framework. "There is an 'end in view'—not preordained, to state this point from an ecological viewpoint rather than a theological one, but as the actualization of what is implicit in the potential."[47]

Nevertheless, Bookchin has a serious problem here, as would anyone who sought, for ecological reasons, to ground us in nature while also postulating freedom as a cardinal human goal. It is hard to see how things can be "not preordained," yet also "constituted to be" through an "objective standard of self-realization," something which is "implicit" and can be actualized. This is, after all, the way homo-phobes or racists tend to argue. In fact, it is difficult to see how a thor-oughly repressive naturalization can be avoided unless some principle is introduced that, so to speak, introduces plasticity and contingency into the schemata of nature where humans are concerned. That is, a chicken is pretty well locked in to becoming a chicken who behaves in a quite definite, "natural" way. If we found a hen, for example, that insisted on hanging out with dogs or showed an interest in Bartok, we would be justified in calling that chicken abnormal and looking for some wrinkle in the internal wiring or imprinting of said beast. Presumably we are not content to regard humans in this vein.[48] An individual who decides not to have children, or to undergo a sex-change operation, or to like or dislike Bartok, or to put a safety-pin in his lip, or to blow up a federal office building, or to become an ecological Marxist or a social ecologist, may be doing something good, bad, or indifferent, but in any case he or she is exercising a very basic aspect of human nature. This involves more than "subjectivity" or "self-reflexivity"; there is also what may be said to be a tendency to reject the given, whether this be defined as inherited social or natural convention. That is, humans are very much part of nature, but there is also something in us that is never content with nature. That is what the "self" in "self-reflexivity" is about—an entity that stands apart, or rather is engaged in a continuing dialectic of attachment and separateness in the expression of which na-ture is necessarily transformed. It seems very much impossible to think of a human being who does not function by transforming nature in such an expressive way. After all, we are the only species who cooks food or adorns the body, or buries the dead with ceremony, or thinks about the possible relations between humans and nature—all processes that involve the transformation of the given, that is, of nature.

Therefore, while it is essential to posit a *connectedness* between humanity and nature—any ecological way of being depends upon this—the thinker who wants to posit a direct *continuity* between hu-manity and nature has a burden to contend with. This is because in the very expression of the notion of continuity with nature one is also ex-pressing discontinuity with nature, and doing so in an entirely human-natural way. Thus I would go further: such a thinker is in deep concep-tual trouble unless she or he specifies some intermediate zone according to which the continuity/discontinuity between humans and nature can be understood.

Happily we have such a zone to conceptualize. Call it, if you will, the sphere of representation, or the imaginary, or of language, or of knowledge—whatever terms you think best to indicate the presence in human being of the possibility of a kind of dialectical space in which nature is affirmed/negated. Without such a notion, we either collapse human being into nature, or radically split the two apart in Cartesian dualism.

Unhappily, Bookchin does not seem to realize this, for there is no such zone conceptualized within his discourse. Or to the extent that he realizes it, he rejects it and flees from it. Remarkably, Bookchin is even anti-Kantian, in the sense that Kant put the critique of knowledge, or epistemology, in the foreground of inquiry. This, for Bookchin, is a graver philosophical problem than that of the Cartesian split that so aggravates ecological thinkers. Even some of the "best theorists" of ecology "commit an error. . . . They have dug their trenches poorly: they have defined themselves against Descartes rather than Kant. . . . As a result, philosophical theories of nature and the objective ecological ethics derived from them are still being created in the false light of the 'epistemological turn' that Kant ultimately gave to Western philosophy."[49]

The "epistemological turn" Bookchin so abhors contains among its components the notion that we should be humble about the limits of our knowledge and the kinds of propositions we make about the world. Whatever else Kant did, he taught us the virtue of skepticism about truth claims. Both modernism and postmodernism agree on this point—Marx, Nietzsche, Freud, Foucault, and Derrida alike. But not Murray Bookchin. He *knows*. Nature and history alike are open books to the social ecologist cum redeemer.

By denying a sphere of representation, Bookchin gains justification for his dogmatism, since whatever he decides becomes the one true way in which nature expresses itself. But he also locks himself into a highly objectivistic rendering of human nature. The varieties of interpretation, the play of language and the imagination, the exercise, in short, of our faculties of freedom—all this Enlightenment is denied as a false light. Such a maneuver may give rigid support to Bookchin's prejudices; but like any rigidity, the result is oppressive, stifling, and headed for collapse. The rejection of a concept of representation underlies his inability to give the category of hierarchy real substance, since what distinguishes dominative from nondominative hierarchies occupies the realm of the imaginary. And it enters into the single-visioned repressiveness that pervades his work. Without an intermediate sphere of representation, the negative can no longer be freely admitted into the real. Nature unfolds immediately into the human; it must be this way and no other.

The epistemological theme was not merely imagined by Kant. It is embedded, rather, in the actualities of human biology, in our dialectical relation to nature. Here, in the situation of childhood and the collective production of culture, arises the linguistic capacity and the ground of

its dialectical efflorescence. The condition of childhood and culture—where, as we have already seen, hierarchies arise "naturally" in
human existence—is also the condition of the imaginary in the configuration of language and desire. Here occurs the positing of an alternative, represented universe that cannot be collapsed into the given universe. Human being is shaped as the coexistence of subjective and
objective spheres, continually self-transforming. It is a state of differentiation from nature and will never be identical with nature—at least so
long as people cook their food, adorn their bodies, wonder about the
universe, and use speech. The differentiation from nature is paralleled
by the emergence of the human self as the internal universe of the person. This is, no doubt, the source of misery when it turns into greed,
the will to power or the endless varieties of self-aggrandizement—and
finds structures such as capitalism and patriarchy for realization. But it
is also the source of poetry, of music, of love, or indeed, of the ideas of
freedom and justice. And it is one of the places where the overcoming
of domination and the reconciliation with nature must occur.

The presence of a differentiated sphere of representation embeds
dialectic in human being and thereby raises it to a new level. For what
is noncollapsing is also mutually negating. It lives by a continual transform of absence/presence, a presentation, a re-presentation, a search for
the imaginary in the real, and a search for the realization of the imaginary. It is, in sum, the createdness/creativity of subject and object. The
young Marx recognized the implications of this and incorporated it
into the foundation for his worldview:

> Man is a directly natural being. . . . The fact that man is an embodied, liv-
> ing, real, sentient objective being means that he has real, sensuous objects
> as the objects of his life expression.

This is the vision that Bookchin claims is based on "the realm of *sur-*
vival, not the realm of *life*" (emphasis in original). But there is more to
Marx's philosophical anthropology.

> But man is not only a natural being, he is a human natural being. This
> means that he is a being that exists for himself[50] . . . that must confirm
> and exercise himself as such in his being and knowledge. Thus human ob-
> jects are not natural objects as they immediately present themselves nor is
> human sense, in its purely objective existence, human sensitivity and hu-
> man objectivity. Neither nature in its objective aspect nor in its subjective
> aspect is immediately adequate to the human being. And as everything
> natural must have an origin, so man too has his process of origin, history,
> which can, however be known by him and thus is a conscious process of
> origin that transcends itself. History is the true natural history of man.[51]

Humans are natural; they are other than natural. They are both subjective and objective, related through negation. They are rest-less—the given, nature, is not immediately adequate, and in order to express their being, they transform nature. They express this subjec-tively, through language, desire, and the imagination; and they express this objectively, by refashioning nature into objects of utility, beauty, or—increasingly under capitalism—exchange value. This is what labor is about—the labor that makes ideas as well as material things. Labor makes history and is given historical content according to its alienation, that is, domination, or its emancipation. Later generations—ours—have to expand the dialectical content in Marx by concern for what has been revealed in the intervening maturation of capitalism, in particular, eco-logical destruction on a world scale. Dialectical thought now needs to incorporate a greater awareness than Marx had of the inherent activity and worth of nature. It is perhaps the highest praise that can be given Marx to point out that his sense of dialectic was so sure that he built into his theory an openness to its own supercession.[52]

By the same token, Bookchin closes off his theory when he blocks the dialectical opening afforded by the imaginary. All these tendencies come together in his treatment of myth, to which I now return to round off this study. I have already commented on the following passage, where Book-chin attacks myth as the enemy of reason, lashing out at the

> utterly arbitrary character of myth, its lack of any critical correction by reason, [which] delivers us to complete falsehoods. Viewed from a primi-tivistic viewpoint, "freedom" takes on the treacherous form of an absence of desire, activity and will—a condition so purposeless that humanity ceases to be capable of reflecting upon itself rationally and thereby pre-venting emerging ruling elites from completely dominating it. In such a mythic—and mystified—world, there would be no basis for being guarded against hierarchy or for resisting it.[53]

Looked at more closely, this becomes less a defense of freedom and ra-tionality than a fear of receptivity. Openness to myth, that is to say, the collective imaginary, implies, for Bookchin, a destructive, terrifying re-gression to a state of archaic oneness with the universe, the "oceanic experience." The dangers of the mythic sensibility in general are repre-sented for Bookchin by one particular myth drawn from the *Odyssey*, that of the Lotus Eaters. In this legend, Odysseus's men lose their way on the island of the Lotus Eaters. They become drugged, forgetful, they lose their identity and live in a timeless immediacy. In short, they be-come babies. This is not my interpretation of Bookchin, but Bookchin's own reading of this legend:

Indeed, the island of the Lotus Eaters is a regressive myth of a return to infancy and passivity, when the newly born merely responds to caresses, a full breast, and is lulled into sedated receptivity by an ever-attentive mother. . . . [And thus] to retreat back into myth, today, is to lay the basis for a dangerous quietism that thrusts us beyond the threshold of history into the dim, often imagined, and largely atavistic world of prehistory. Such a retreat obliges us to forget history and the wealth of experience it has to offer. Personality dissolves into a vegetative state that antedates animal development and nature's evolutionary thrust toward greater sensibility and subjectivity. Thus even "first nature" is libeled, degraded, and denied its own rich dynamic in favor of a frozen and static image of the natural world where the richly coloured evolution of life is painted in washed-out pastels, bereft of form, activity, and self-directiveness.[54]

A bizarre statement, this, in which the germ of repressive intolerance is set forth. These are, first of all, the words of someone who thinks very literally. Bookchin starts interpreting, but then freezes in his tracks. He cannot seem to comprehend that one can move safely in and out of the sphere of representations. It is as if he does not realize that, barring psychosis or the abuse of drugs, the imagination and the perception of external reality do not collapse one into the other. That is, one can play with the imaginary and live to tell the tale. Indeed, one *must* play with the imaginary in order to fully develop as a human being. It is how we both strengthen and refresh ourselves. But to do so one must be able to feel for the possibilities of representation, a position closed off by "naturalism."

Second, these are the words of the typically masculinist position, fearing merger with the archaic mother and therefore establishing a kind of hypertrophic rationalism that sharply discriminates between subject and object. This position is, needless to add, depressingly familiar in the modern history of the West and specifically associated with the enemies of the ecological worldview.

Finally, this censorious attitude has a multiply repressive effect. It blocks not only the mythic worldview, with its immanent rationality, but closes off all radical spirituality. What implications does Bookchin's fear of regression have for Buddhist or Taoist conceptions of the Void? Are we to assume that these are simply infantilisms? That they lack the rationality of the clear-minded? That their lessons as to the reconciliation with nature and the critique of domination are meaningless?[55]

But perhaps enough has been said of the intellectual misadventures of Murray Bookchin. The great goal he has outlined—a free society in harmony with nature—remains the most fundamental project of all. In working to realize it, we should keep before us the famous aphorism of Terrence.[56] For we will not find our way home, this being the hidden text of ecology, until we are open to the entire universe, inner as well as outer, negative as well as affirmative, until it can be truly said that nothing human has been made alien to us.

Acknowledgments

I am indebted to Dee Dee Halleck, Mitchel Cohen, John Clark, Jonathan Stevens, and James O'Connor for constructive advice in preparing this essay.

Notes

1. The principle works considered here, with page references keyed in the notes are: *Toward an Ecological Society* (Montreal: Black Rose Books, 1980) [*TES*]; *Post-Scarcity Anarchism*, 2nd ed. (Montreal: Black Rose Books, 1986) [*PSA*]; *The Ecology of Freedom* (Palo Alto, CA: Cheshire Books, 1982) [*EF*], Bookchin's chef d'oeuvre; *Remaking Society* (Montreal: Black Rose Books, 1989) [*RS*]; *The Philosophy of Social Ecology* (Montreal: Black Rose Books, 1990) [*PSE*]; and "What Is Social Ecology," in Michael E. Zimmerman et al., eds., *Environmental Philosophy* (Englewood Cliffs, NJ: Prentice-Hall, 1993) [*WSE*].

2. As the great hymn at the beginning of *Milton* ends: "I will not cease from Mental Fight, / Nor shall my Sword sleep in my hand, / Till we have built Jerusalem / In England's green & pleasant Land" (William Blake, *Milton* [Boulder, CO: Shambala, 1978]), p. 62.

A brief personal note may be in order. For a while I to embraced social ecology and considered myself a comrade of Bookchin's. Having come into this area relatively late in my political and intellectual evolution, I was disposed to adopt the doctrine that seemed the most radical and comprehensive, and this appeared best exemplified by Bookchin's work. I made three appearances at the summer Institute for Social Ecology in Vermont, and belonged to the Left Green Network of the Green Party, both of which were heavily influenced by Bookchin. This essay represents the working out of a slowly growing conviction that arose on the basis of this experience that certain aspects of Bookchin's thought needed fundamental critique. It should also be noted that, although the Left Green Network is more or less defunct, perhaps for reasons having to do with the themes raised by this essay, I remain active in the Green Party and regard it as a potential force for radical change.

3. *PSE*, p. 29. Here and hereafter, quotes not followed by note numbers are from the same work and page of Bookchin's as the next quote that *is* attributed.

4. *EF*, p. 270.

5. *RS*, p. 102.

6. *RS*, p. 101.

7. *EF*, p. 3.

8. *EF,* p. 4.

9. Murray Bookchin, *The Spanish Anarchists* (New York: Harper & Row, 1977). This book is one of Bookchin's finest achievements.

10. In addition to the above, see, e.g., Lewis Herber (pseudonym), *Our Synthetic Environment* (New York: Knopf, 1962)—published the same year as Rachel Carson's *Silent Spring.* For the synthesis of Bookchin's municipalism, see *The Rise of Urbanization and the Decline of Citizenship* (San Francisco: Sierra Club, 1987). Bookchin also played a role in the emergence of the German Greens during the 1970s, and is virtually the only American ecological thinker taken seriously by them.

11. For a recapitulation of this quarrel, in which Bookchin backs away from his more strident denunciation while under a degree of public pressure, see Murray Bookchin and Dave Foreman, *Defending the Earth: A Dialogue between Murray Bookchin and Dave Foreman,* ed. Steve Chase (Boston: South End Press, 1991).

12. *EF,* p. 14.

13. See Joel Kovel, *History and Spirit* (Boston: Beacon Press, 1991) for an elaboration of this notion.

14. *EF,* p. 352.

15. *RS,* pp. 166–167.

16. *EF,* p. 306.

17. *RS,* p. 165.

18. *RS,* p. 11.

19. *RS,* p. 163.

20. *RS,* p. 162.

21. For a sense of the debate, see John Clark, "Marx's Inorganic Body," in Zimmerman, ed., *Environmental Philosophy*; John Bellamy Foster, "Marx and the Environment," *Monthly Review* 47, July–August, 1995. Clark, who was long associated with Bookchin, argues his anti-Marxism much more fairly and sophisticatedly than Bookchin himself, while Foster gives a vigorous defense of Marx.

22. *EF,* p. 304.

23. Lawrence Krader, trans. and ed., *The Ethnological Notebooks of Karl Marx* (Assen, The Netherlands: van Gorcum, 1972); Franklin Rosemont, "Karl Marx and the Iroquois," *Arsenal/Surrealist Subversion 4* (Chicago: Black Swan Press, 1989)—the quote is from p. 212; Raya Dunayevskaya, *Rosa Luxemburg, Women's Liberation, and Marx's Philosophy of Revolution* (Atlantic Highlands, NJ: Humanities Press, 1981), ix.

24. *TES,* p. 16.

25. *RS,* p. 136.

26. *RS,* pp. 136–137.

27. *RS,* p. 84.

28. *EF,* p. 225; *RS,* pp. 174, 191.

29. *PSA,* p. 232. The broadside in question was "Listen Marxist!," a pamphlet first distributed at a Students for a Democratic Society (SDS) convention in 1969 and reprinted in *Post-Scarcity Anarchism.* Two features about this work deserve some emphasis. First, the "Marxist" in the title immediately re-

fers to the neo-Stalinist tendencies such as Progressive Labor that were beginning to appear out of the frustrations of the New Left. Bookchin was trying to
steer the movement back into the libertarian–anarchist path; and in this effort,
the figures of Marx and Engels were not as demonized as they became in later
works. Thus despite its stridency, "Listen Marxist!" offers something close to a
real critique of Marxism, in contrast to the post-1980 works. Second, the foundation of this critique was similar to the point raised by Marcuse, that capitalism had solved the problem of scarcity, hence a revolution of needs could be
contemplated, rather than one confined to addressing the injustices of the proletariat. Obviously the crisis that has afflicted the capitalist world since the
early 1970s and produced ever-widening class differences, scrambles these assumptions badly. One might speculate that the increasing stridency of
Bookchin's later works reflects a reaction to the fact that the locomotive of history has pulled away from the station, leaving the postscarcity hypothesis behind. He attempts to deal with the problem in the Introduction to the second
edition of *Post-Scarcity Anarchism* by stating, not unreasonably, that we still
have to build on the realization that capitalism potentially can overcome scarcity. True, but this does take the steam from Bookchin's project. More, the notion was that of Marx, as well.

30. *RS,* p. 169.

31. For a treatment of abstraction in Marx's dialectic, see Bertell Ollman,
Dialectical Investigations (New York and London: Routledge, 1993).

32. From a letter of 1802. David Erdman, ed., *The Complete Poetry and
Prose of William Blake* (Garden City, NY: Doubleday, 1982), p. 722.

33. *RS,* p. 156; italics in text.

34. *PSE,* p. 178.

35. *WSE,* pp. 364–366.

36. Paolo Freire's theses on education remain the definitive study on how
dialectic should enter the educational process. His notions are primarily directed toward adult literacy campaigns—an instance in the real world where,
despite Bookchin, a good hierarchy needs to be introduced rather than
erased—but certainly apply to the education of children. Here the best teachers
remain definitely in command, yet continuously evoke the intrinsic unfolding
of the child's being in such a way that the outcome appears spontaneous. See
Freire, *Pedagogy of the Oppressed* (New York: Herder & Herder, 1971).

37. A category, by the way, that includes pretty much the entire world.
See Peter Rosset and Medea Benjamin, eds., *The Greening of the Revolution*
(Melbourne, Australia, and San Francisco: Ocean Press and Global Exchange,
1994), p. 8. See also Richard Levins, *The Struggle for Alternative Agriculture
in Cuba,* Red Balloon Pamphlet 1 (published 1991; for copies, write to 7652
Cropsey Ave. #74, Brooklyn, NY 11214).

38. Vandana Shiva, *Staying Alive* (London: Zed, 1989).

39. *EF,* p. 351.

40. *EF,* p. 166.

41. See, e.g., Lester Brown, *State of the World 1995* (New York: Norton,
1995), for the looming prospect of famine, especially in China.

42. As far as I know, the only time Bookchin ever intervened on behalf of

a struggle involving so-called third (or fourth) world people was to lend support to the Miskito Indians in their struggle with Nicaragua's Sandinistas. Here, however, his principle motive was clearly hatred of the presumed Marxism-Leninism of the FSLN, a sympathy he shared with the CIA. Eventually the main faction of the Miskitos and the Sandinistas made peace, no thanks to Bookchin, who had done his share in setting back the solidarity movement.

43. Richard Hofrichter, ed., *Toxic Struggles* (Philadelphia: New Society, 1993).

44. *EF,* pp. 353–354.

45. *PSE,* p. 35.

46. *PSE,* p. 30.

47. *PSE,* p. 171.

48. Unless, of course, we were modern practitioners of psychiatry, but that is another story.

49. *PSE,* p. 57.

50. In other words, for the self that is our natural legacy.

51. Karl Marx, "Economic and Philosophical Manuscripts," in David McClellan, ed., *Karl Marx: Selected Writings* (New York: Oxford University Press, 1977), pp. 104–105. It is worth noting that I cannot recall encountering a reference to this work in Bookchin's writings.

52. As argued by Jacques Derrida, in *Spectres of Marx,* trans. Peggy Kamuf (New York and London: Routledge, 1994).

53. *RS,* p. 102.

54. *RS,* p. 103.

55. For an argument, within anarchist theory, of the radical potentials of Taoism, see John Clark, "Master Lao and the Anarchist Prince," in *The Anarchist Moment* (Montreal: Black Rose Books, 1984).

56. Chosen by Marx (in a little quiz administered by his daughters) as his favorite motto.

Chapter 2

Divining Evolution
and Respecting Evolution

ROBYN ECKERSLEY

Divining Evolution: The Ecological Ethics of Murray Bookchin

In this essay I provide an exposition and critique of the ecological ethics of Murray Bookchin. First, I show how Bookchin draws on ecology
and evolutionary biology to produce a mutually constraining cluster of
ethical guidelines to underpin and justify his vision of a nonhierarchical
ecological society. I then critically examine Bookchin's method of justification and the normative consequences that flow from his position. I
argue that Bookchin's enticing promise that his ecological ethics offers
the widest realm of freedom to all life forms is undermined by the way
in which he distinguishes and privileges "second nature" (the human
realm) over "first nature" (the nonhuman realm). I conclude that
Bookchin's promise can only be delivered by a biocentric philosophy
(which he rejects) rather than by his own ecological ethics.

Introduction

It has become unfashionable for social and political theorists to ground
political norms in biological theory or a philosophy of nature. Many of
the more famous past attempts to explain, "naturalize," or justify a certain political state of affairs by recourse to "the way things are" have
been dismissed in more modern times as mere apologies for different
forms of political domination, whether it be oligarchy (Plato), slavery

(Aristotle), hierarchy (Aquinas), or capitalism (Spencer).[1] One of the legacies of such critiques has been a general detachment from the life sciences on the part of social and political theorists that has served to reinforce the conceptual cleavage in Western thought between culture and nature, between human history and natural history, and between the human and the nonhuman. This, in turn, has reinforced the view that humans are somehow "above" nature, that they are technically able and morally obliged, indeed destined, to adapt nature to human ends. While this widespread view is now being increasingly challenged on many fronts by environmental philosophers in the wake of the environmental crisis, social and political theorists have been slow to question the human domination of nature and to search for ways of transcending the nature/culture schism in Western thought.

A notable exception is Murray Bookchin. Over the past three decades, Bookchin's numerous publications on "social ecology" have sought to undermine the cleavage between the social and the natural and to restore a sense of continuity between human society and the creative process of natural evolution as the basis for the reconstruction of a communitarian or anarchist politics.[2] Bookchin describes his thought as carrying forward the "Western organismic tradition" represented by scholars such as Aristotle, Hegel, and, more recently, Hans Jonas—a tradition that is process-oriented and concerned to elicit the "logic" of evolution.[3] According to Bookchin, the role of an ecological ethics is "to help us distinguish which of our actions serve the thrust of natural evolution and which of them impede it."[4] For Bookchin, evolution is developmental and dialectical, moving from the simple to the complex, from the abstract and homogeneous to the particular and differentiated, ultimately toward greater individuation and freedom or selfhood.[5] Social ecology—a communitarian anarchism rooted in an organismic philosophy of nature—is presented as the "natural" political philosophy for the green movement because it has grasped the true grain of nature and can promise the widest realm of freedom for both nature and society.[6]

Bookchin's ideas are diffuse and wide-ranging. Indeed, the sheer breadth of subjects he covers in his writings is both his source of strength and his weakness, making assessment of his work a challenging and difficult task. In this essay, I will focus on only one aspect of his work (albeit a very important one), namely, the ecological ethics derived from his philosophy of nature. Specifically, I will show how Bookchin draws on biological theory (particularly ecology and evolutionary biology) to develop an ecological ethics that provides the basis of, and the justification for, his vision of a nonhierarchical, ecological society.[7] The central question I examine is whether his synthesis of biological and political ideas is a coherent and desirable one from both a methodological and normative standpoint. As to method, has Bookchin managed to transcend some of the criticisms that have discredited previous attempts to ground a political philosophy in this way? As to ethics, how far does he go in challenging the human domination of nature? This question is of particular interest to those who have sought to comprehend the basis of Bookchin's recent scathing attacks on the biocentric philosophy of deep ecology.[8]

I show that Bookchin's claim that his ethics is *objectively* anchored in evolution arises from his reworking of the Hegelian dialectic in ecological terms. However, this idiosyncratic use of the term *objective* cannot serve to justify his ethics in the way that he would like. Social ecology is best understood and described as an ecologically inspired utopianism since it seeks to foster certain potential evolutionary pathways that are felt to be desirable and attainable but are by no means "objectively right" in the ordinary sense of the term, that is, independently verifiable as the "true" path. Moreover, while Bookchin's organismic philosophy may have transcended the nature/culture schism, it has only partially undermined the idea that we should dominate nature. In particular, his promising claim that his ecological ethics offers the widest realm of freedom to all life forms is undercut by the way he develops his distinction between first and second nature (corresponding to the nonhuman and human realms, respectively). Indeed, I argue that there is a certain arrogance in his claim that humans have now discerned the course of evolution, which they have an obligation to further, on the ground that it ultimately favors human attributes over the attributes of other life forms and therefore cannot deliver his central promise of freedom or self-directedness writ large.

The Dialectics of Nature

In his magnum opus, *The Ecology of Freedom*, Bookchin has developed a philosophy of nature that is presented as being objectively anchored in natural history insofar as it recognizes the reality and thrust of evolution. All we need to do, Bookchin argues, is to permit nature "to open itself to us ethically on its own terms."[9] If we allow this, then we will find that nature, far from being amoral,

> exhibits a self-evolving patterning, a "grain," so to speak, that is implicitly ethical. Mutualism, freedom, and subjectivity are not strictly human values or concerns. They appear, however germinally, in larger cosmic and organic processes that require no Aristotelian God to motivate them, no Hegelian Spirit to vitalize them.[10]

Bookchin holds that truth and meaning lie in a self-organizing nature rather than in a transcendent God or Spirit. Although arguing that humans are simply part of "nature rendered self-conscious," he does not deny "the high degree of nisus, of self-organization and self-creation, inherent in nonhuman phenomena," all of which are "immanently and creatively constituted."[11] According to Bookchin, human consciousness and reason emerged out of a wider realm of *subjectivity* that inheres in *all* phenomena: "The gradual emergence of mind in the natural history of life is part of the larger landscape of subjectivity itself . . . a common bond of primal subjectivity inheres in the very organization of matter itself."[12] This larger landscape of subjectivity or mentality "inheres in nature as a whole—specifically in the long development of increasingly complex forms of substance over the course of natural history."[13]

Bookchin argues that we can only understand an organism by looking at its history, by making sense of the way it has developed over time, both as an entity and in relation to its environment. His exploration of natural history and his reading of recent developments in the life sciences has led him to a particular teleological[14] interpretation of nature that combines the insights of ecology (which focuses more on the spatial orientation of phenomena and the interconnections between organisms within an ecosystem) and evolutionary biology (which focuses more on patterns of change and the ways in which natural phenomena develop over time). In Bookchin's view, nature does not develop blindly or randomly, but rather dialectically and with a general directionality that is encapsulated in the ecological principles of diversity (or unity in diversity), complexity, mutualism (or complementarity—capturing the interdependent and nonhierarchical character of the ecosystem), and spontaneity. (Bookchin often speaks of fecundity and creativity, but these descriptions appear to be collective attributes of the principles already mentioned.[15]) Moreover, unlike Leopold's land ethic, which endorses those actions that tend to preserve the "integrity, stability, and beauty of the biotic community," Bookchin's ecological ethics place a greater emphasis on novelty and change. Distilled into the pithy Leopoldian format, Bookchin's ecological ethics would run as follows: "A thing is right when it tends to foster the diversity, complexity, complementarity, and spontaneity of the ecosystem. It is wrong when it tends otherwise." While each of these qualities is heralded as an important end in itself,[16] taken together they are also presented as tending toward still higher norms, what might be called more advanced forms of subjectivity—freedom, reason, and selfhood—the ultimate desideratum of evolution. In Bookchin's own words, "The striving of life toward a greater complexity of selfhood—a striving that yields increasing degrees of subjectivity—constitutes the internal or immanent impulse of evolution toward growing self-awareness."[17]

There are certain parallels here between Bookchin's organismic philosophy and the interdisciplinary philosophy of the French theologian Teilhard de Chardin, although they should not be pressed too far. Both thinkers understand the evolutionary process in terms of advancing subjectivity that has reached its most developed form in humans, who have become "nature rendered self-conscious," at the helm of evolution. However, unlike Teilhard de Chardin, who sees the evolutionary process dissolving in an omega point of organic oneness, Bookchin sees evolution as continuing to unfold toward ever increasing degrees of subjectivity.

Nature and the Grounding of Ethics

The above principles provide the foundation of Bookchin's entire social and political philosophy. While Bookchin presents them as mutually reinforcing ends in themselves, he gives diversity a special status as a guarantor of ongoing freedom, an aspect of Bookchin's ecological ethics that has been further developed in essays published after *The Ecology of Freedom*:

> Diversity may be regarded as a source not only of greater ecocommunity stability; it may also be regarded in a very fundamental sense as an ever-expanding, albeit nascent, source of *freedom* within nature, a medium for objectively anchoring varying degrees of choice, self-directedness, and participation by life-forms in their own evolution.[18]

Accordingly, says Bookchin, we ought to foster natural diversity rather than monocultures, for rendering "nature more fecund, varied, whole, and integrated may well constitute the hidden desiderata of natural evolution."[19]

In his more recent book, *The Modern Crisis*, Bookchin has reiterated the basic themes set out in *The Ecology of Freedom* in terms of two mutually reinforcing concepts: participation and differentiation.[20] The former seeks to capture the symbiotic, mutualistic character of interactions between plants and animals within ecocommunities, while the latter underscores the stability and nascent freedom engendered by variety and complexity: "The greater the differentiation, the wider is the degree of participation in elaborating the world of life. An ecological ethics not only affirms life, it also focuses on the *creativity* of life."[21] In short, the more participatory and differentiated an ecocommunity, the more its human and nonhuman inhabitants are seen to be able to realize their individual and collective forms of "selfhood."

Bookchin thus reaffirms his central claim that his ecological ethics promises the widest realm of freedom for both nature and society. At the same time, Bookchin presents his ethics as unifying his natural and social philosophy by showing that what is authentic and good for humanity is authentic and good for nature (and vice versa). Such fitting ethical symmetry is presented as flowing naturally from social ecology and as giving rise to clear political imperatives such as developing ecotechnologies and ecocommunities that foster diversity, participation, and self-management. Herein lies the justification for Bookchin's vision of a self-directed, anarchist society in which human inhabitants are fully aware of their special evolutionary responsibilities. In Bookchin's view,

> It is eminently *natural* for humanity to create a second nature from its evolution in first nature. By second nature, I refer to humanity's development of a uniquely human culture, a wide variety of institutionalized human communities, an effective human technics, a richly symbolic language, and a carefully managed source of nutriment.[22]

The important issue, argues Bookchin, is to show how second nature is derived from first nature and to discern the libertarian pathways that are open to it. In Bookchin's words, "We cannot hope to find humanity's 'place in nature' without knowing how it *emerged* from nature with all its problems and possibilities. Our result yields a creative paradox: second nature in an ecological society would be the actualization of first nature's potentiality to achieve mind and truth."[23]

According to Bookchin, since there is no road back from second nature to first nature, there must be "a radical integration of second nature with first nature along far-reaching ecological lines."[24] Bookchin thus writes an active human presence into nature, enabling it to "act upon itself rationally, defined mainly by co-ordinates created by nature's potential for freedom and conceptual thought."[25] Minimally, Bookchin's ecological ethics demands human stewardship of the planet, which involves "new ecocommunities, ecotechnologies, and an abiding ecological sensibility that embodies nature's thrust toward self-reflexivity."[26]

Ethical Naturalism

Bookchin describes his organismic philosophy of nature as transcending both ethical subjectivism and relativism because it is *objectively* anchored in nature. The word *objectively* is used here in a neo-Hegelian sense to mean "based on a mode of dialectical reasoning that is projective and sharply critical."[27] When Bookchin claims that his ethics is objective, he means that it is based on potentialities that are actually latent in nature, potentialities that really exist as concrete possibilities standing beyond the present:

> That possibilities, i.e., the actualizations of very existential potentialities, could be regarded as "objective" is quite as valid as the notion that an oak tree objectively inheres in an acorn. Ethically, this is a highly illuminating approach. It establishes a standard of fulfillment—an objective "good," as it were—that literally informs the existential with a goal of objective fulfillment, just as we say in everyday life that an individual who does not "live up" to his or her capabilities is an unfulfilled person and, in a sense, less than a "real" person.[28]

The same analogy is extended further in *The Modern Crisis,* in which Bookchin argues that "what is potential in an acorn that yields an oak tree or in a human embryo that yields a mature, creative adult is equivalent to what is potential in nature that yields society and what is potential in society that yields freedom, selfhood and consciousness."[29]

This analogy is both telling and problematic. In drawing a parallel between the developmental path of an acorn, a human embryo, nature, and finally society—as if all have an equally discernible objective standard of fulfillment—Bookchin is collapsing ontogenetic development (i.e., the sequence of events involved in the development of an individual organism) into phylogenetic evolution (i.e., the sequence of events involved in the evolution of the human species, including its culture). Even at the level of ontogenetic development, it is a very confusing analogy, since the similarities between an acorn and a human embryo are essentially confined to the growth patterns of the physical organism; they tell us nothing about consciousness or about what humans may properly value or do with their hands and tools. The analogy thus begs the question as to what characterizes a mature and fulfilled adult psychologically, intellectually, and ethically. More importantly, there are very real limits to the extent to which the "objective" developmental path of an acorn can be reasonably compared with that of the human species as a whole and, in particular, with that immensely more complex and open-ended phenomenon we call human society. Indeed, how can it be said that there exists some objective standard of fulfillment latent within human society itself, urging it toward mind and truth? Why are not *all* of the myriad potential paths of human development also objective and desirable ones in Bookchin's sense? What is it about

Bookchin's evolutionary path of mutuality, diversity, and "advancing subjectivity" that makes it the good and true path as compared to, say, Herbert Spencer's struggle of the fittest?

In his effort to advance his principle of mutualism or complementarity in nature, Bookchin has done more than take sides with Peter Kropotkin and his theory of "mutual aid": he not only emphasizes the prevalence of symbiosis as part of the implicit *moral* fabric of nature, he downplays and at times *redefines* in more benign terms what would ordinarily pass as competitive or aggressive behavior by nonhuman animals (e.g., such behavior is defined as episodic, instinctual chainlike links between individuals rather than as the result of *institutionalized* hierarchy).[30] (The latter, incidentally, is seen as peculiar to human society and identified as the root of our social and ecological problems.) Bookchin is right to point out that, too often, our descriptions of nature *are* anthropomorphic projections that are then used for social ends (e.g., justifying social stratification). In this respect, he is keen to dissociate social ecology from what he sees as previous reactionary attempts to appeal to the authority of nature to justify political domination (e.g., fascism, scientific socialism, and sociobiology). But isn't Bookchin caught in his own criticism? Isn't his "mutualistic" nature one more anthropomorphic projection? How can Bookchin's claims that nature is neither hierarchical nor egalitarian, neither evil nor virtuous,[31] be reconciled with his claim (quoted earlier) that the patterns in nature (embodied in his ecological principles) are "implicitly ethical"? Bookchin has argued that nature is merely the *ground* of his ethics rather than the paradigm or model. Yet this is hard to grasp alongside his statement that "My definition of the term 'libertarian' is guided by my description of the ecosystem: the image of unity in diversity, spontaneity, and the complementary relationships, free of hierarchy and domination."[32] If this is not a paradigm or model, then what is? Moreover, merely asserting that these principles constitute an "objective" statement of what is implicit in nature does nothing to elevate such principles over other potential pathways of evolution.

Bookchin claims that his ecological principles and the ultimate norms they serve are verified in the processes of real life.[33] However, his case rests on intuitive reasoning and ingenious rhetorical questions rather than on testable hypotheses. For example, he scorns any attempt to explain the development of natural phenomena as a purely chance event:

> To invoke mere fortuity as the *deus ex machina* of a sweeping, superbly organized development that lends itself to concise mathematical explanations is to use the accidental as a tomb for the explanatory.[34]

Bookchin's philosophy of nature is predicated on the intuition that there *must* be a telos (in the sense of a general directionality as distinct from a fixed end) in nature by virtue of the wondrous patterns it reveals:

> From the ever-greater complexity and variety that raises sub-atomic particles through the course of evolution to those conscious, self-reflective life-forms called human beings, we cannot help but speculate about the existence of a broadly conceived *telos* and latent subjectivity in substance itself that eventually yields intellectuality.[35]

Yet Bookchin's derivation of this latent subjectivity or immanent striving from the outward *results* of evolution is presented more or less as a fait accompli, as something that is intuitively self-evident once it is pointed out. In a characteristically rhetorical mode, he asks why is it that the onus of proof should lie on the shoulders of those who argue that nature does have a telos rather than on the shoulders of the likes of Bertrand Russell, with his "image of humanity as an accidental spark in an empty meaningless void."[36] This is indeed a good question, but it cannot thereby serve to raise Bookchin's own presuppositions and theoretical constructs to a privileged status simply by reversing the onus of proof.

It would seem less confusing and more to the point for Bookchin to focus on the desirability rather than the supposed objectivity of the ends that his ecological ethics serves, for his claim that his ethics has discerned the true telos of nature is highly contentious and cannot perform the work of shoring up his ethics in the way that he would like. That it is contentious is apparent from his very method of reasoning. Bookchin's ethical naturalism rejects the sharp fact/value divide of hypothetico-deductive logic by insisting that ethics *is* part of, or at least implied by, the natural world. Judgments about the "goodness" of an action are seen to be factual or objective judgments about what is conducive to the fulfillment of certain natural ends or tendencies. Evolutionary biology and ecology are treated as both objective and normative sciences in that they not only tell us what reality is like, but also provide a "deep wisdom" as to what we ought to value. On this point, Bookchin is untroubled by the logical difficulties associated with deriving an "ought" from an "is"—an impasse that Callicott has described as the "Achilles heel" of academic environmental ethics.[37] This Humean dichotomy, argues Bookchin, is a "positivistic mousetrap" that

> is not [so much] a problem in logic as it is a problem in ethics and the right of the ethical "ought" to enjoy an objective status. . . . Speculative philosophy is by definition a claim by reason to extend itself beyond the

given state of affairs. . . . To remain within the "is" in the name of logical consistency is to deny reason the right to assert goals, values, and social relationships that provide a voice to the claims of ecology as a social discipline.[38]

Bookchin is critical of hypothetico-deductive logic and argues instead for a mode of thinking that is structured around what he calls *eduction,* that is, a phased process that renders "the latent possibilities of phenomena fully manifest and articulated."[39] In Bookchin's own words, such an approach is creative because it ceaselessly contrasts "the free, rational, and moral actuality of 'what-could-be' that inheres in nature's thrust toward self-reflexivity with the existential reality of 'what-is.' "[40]

How, then, does Bookchin manage to persuade others as to the rightness of his ecological ethics? It is not enough simply to invoke the authority of nature as known by science, since even if Bookchin is right in his argument that there is a telos in nature, this discovery does not in itself tell us why we ought to further it. Yet Bookchin's discussion (particularly in the Epilogue to the *Ecology of Freedom*) of recent developments in science, which are claimed to corroborate his teleological view of nature and hence his naturalistic ethics, comes very close to this form of argument. He thus joins that long line of ethical naturalists, referred to by Donald Worster, who, "critical of scientific thought [i.e., its mechanistic and materialist outlook], have nonetheless found their way back to science, finding its authority indispensable."[41] As Worster has observed, the argument is that "the 'Is' of nature must become the 'Ought' of humanity. Ever since Immanual Kant peremptorily severed the two, men have been trying to stitch them back together, most recently through ecology."[42] But as Worster has noted, the "Ought" appears to be shaping the "Is":

> In the case of the ecological ethic, for instance, one might say that its proponents picked out their values first and only afterwards came to science for its stamp of approval. It might have been the better part of honesty if they had come out and announced that, for some reason or by some personal standard or value, they were constrained to promote a deeper sense of integration between man and nature, a more-than-economic relatedness—and to let all the appended scientific arguments go. "Ought" might then be its own justification, its own defense, its own persuasion, regardless of what "Is."[43]

Seizing the Helm of Evolution

What, then, are the *normative* arguments that underpin Bookchin's ecological ethics? It is here that the reader is invited to *imagine* the alternative world proffered by Bookchin and to decide whether it might not be more authentic and rewarding than the ecological and social monoculture that is increasingly becoming the "price" of modern society. Diversity, argues Bookchin, is a particularly desirable goal because it promises relief from a culturally impoverished society and a denuded landscape; from a longer term perspective, it also provides a potential or embryonic form of future freedom by opening up a greater number of evolutionary pathways, ultimately enabling species to play a more active and creative role in their own evolution. What is more, not only will our future be assured if we foster diversity, but we and other life forms will also be able to enjoy a greater capacity for freedom and creativity.

So far so good. But can Bookchin's ethics really deliver the kind of freedom promised, that is, freedom in nature writ large? If we foster diversity and mutualism along the lines urged by Bookchin, will the potential for self-directedness in nonhuman life forms also expand along with that of humans? I argue that a close examination of Bookchin's philosophy reveals serious cross-purposes that tend to undermine his central promise that his ecological ethics guarantees the widest realm of freedom to all life forms. I argue that these problems largely stem from the way in which he distinguishes and privileges second nature over first nature and from his presumptuous conclusions concerning the state of human understanding of ecological and evolutionary processes.

While Bookchin takes pains to point out that ecocommunities are interdependent and nonhierarchical, he nonetheless sees humans as occupying a very special role within ecocommunities, for it is through humans as second nature that first nature is now able to act upon itself *rationally.* That is, we are nature rendered self-conscious, and provided that we act rationally in Bookchin's terms, we will be able to further "first nature's potentiality to achieve mind and truth."[44] This creative role assigned to humans in fostering nature's evolution is the essential basis upon which Bookchin rejects asceticism, stoicism, biocentrism, or any worldview that he interprets as involving the "quietistic surrender" or resignation by humans to the natural order. Bookchin interprets such approaches (quite wrongly, as will be seen, in the case of deep ecology) as idolizing and reifying nature, setting it apart from a "fallen humanity"—an approach that Bookchin claims is an insult to humanity by denying us our creative role in evolution.[45] There must be an infusion of human values into nature, he argues, because humans are the fulfillment of a major tendency in *natural* evolution. Indeed, Bookchin claims that our uniqueness cannot be emphasized too strongly, for "it is in this very *human* rationality that nature ultimately actualizes its own evolution of subjectivity over long aeons of neural and sensory development."[46]

The clear message of Bookchin's ethics, then, is that humanity, as a self-conscious "moment" in nature's dialectic, has a responsibility to rationally direct the evolutionary process, which in Bookchin's terms means fostering a more diverse, complex, and fecund biosphere. Indeed, we may "create more fecund gardens than Eden itself."[47]

What is troubling about this stance is that Bookchin's vision of human stewardship does not qualify how and to what extent our responsibility is to be discharged. Indeed, Bookchin seems to be urging us to take active steps to speed up the evolutionary process and become "algenists" in Rifkin's sense of the term, albeit in the name of ecological freedom rather than in the name of power and profit. But where is the line to be drawn? Should we enlist the aid of computers and the latest biotechnology and step up the selective breeding of plants and animals so as to foster the development of more complex ecosystems and more intelligent species? Take, for example, Bookchin's central principle of diversity. One can envisage the greening of deserts to enhance ecosystem diversity, the speeding up of the international trade in seeds and sperm, the growth of "gene banks" and gene splicing, and the proliferation of "exotic" species of flora and fauna in native ecosystems. These are all troubling scenarios for those concerned with the preservation of native ecosystems, for whom the fostering of more diverse ecosystems (in terms of number of species or habitat) is not necessarily more "valuable" than simpler ecosystems in which the result is the displacement of local species by opportunistic weeds and feral animals. Ecologists have shown that attempts to "manage" ecosystems to ensure maximum diversity can often wind up increasing the number of marginal or nonindigenous species in a given area.[48]

There is, of course, another approach to ensuring biological diversity, and that is to do so in situ through the preservation of large tracts of wilderness (an approach strongly favored by deep ecologists). Bookchin, however, has said very little on the subject of wilderness preservation as compared to, say, organic agriculture practices.[49] Indeed, as a measure to promote diversity in terms of Bookchin's ecological ethics, the preservation of wilderness seems somehow too Leopoldian, too concerned with the stability and the *maintenance* (as distinct from the *cultivation*) of diversity, too prepared to allow the process of succession and genetic change to take place in historical and evolutionary time, and thus too *passive*.[50]

Whatever Bookchin's intentions, the above scenarios are presented as implications that can reasonably be drawn from his theoretical position, particularly that elaborated in his post-*Ecology of Freedom* writings. Indeed, the diversity example serves to bring into sharper relief the major difference between Bookchin's ethics and a biocentric approach. As we have seen, Bookchin has rejected the latter, as imposing needless restraints on human conduct, as denying us our creative role in evolution. To be sure, Bookchin has been quick to draw attention to some of the logical difficulties associated with some formulations of biocentrism. For example, he argues that "if all organisms in the biosphere are equally worthy of a right to life and organic fulfillment, as many biocentrists believe, *then* human beings have no right, *given the full logic of this proposition,* to stamp out malarial and yellow-fever mosquitoes."[51] According to Bookchin, the choice is between recognizing the fact that we are unique moral agents in the biosphere who are capable of practicing a rational, ecological stewardship *or* surrendering our creativity to a passive quietism that posits humans and viruses as "equal 'citizens' in a 'biospheric democracy' "[52]—a position sometimes cited by those who are opposed to giving humanitarian aid to starving Africans on the grounds that nature should be left to take its own course.[53]

These contrasting choices as outlined by Bookchin are, however, overdrawn. The "equal rights" kind of biocentrism logically impaled by Bookchin is not the only kind of biocentrism; there are other biocentric approaches that avoid these logical problems *and* acknowledge the uniqueness of human life. Indeed, overly literal interpretations of the deep ecology principle of "biospherical egalitarianism," a phrase coined by Arne Naess,[54] have been shown to be *misinterpretations* of the central message of deep ecology. Warwick Fox has shown that deep ecology is not a formal axiological (i.e., value theory) or rights-based approach (i.e., it does not adopt an objectivist values-in-nature position), but rather is an approach that seeks the cultivation of a certain *ideal state of being* for humans, namely, one that sustains the widest possible identification with other life forms and entities.[55] The basic thrust of this kind of biocentrism or ecocentrism—which is found not only in the empathic ecological sensibility that has been extensively elaborated by deep ecologists, but also in Leopold's land ethic—is a defense of a *prima facie orientation* of nonfavoritism, of live and let live. Moral considerability is not apportioned on the basis of the possession of characteristics that are suggestive of respected human attributes, such as reason or sentience, since that would denigrate the unique mode of being and integrity of nonhuman life forms and entities; rather, respect is extended to all organisms, populations, species, and ecosystems as well as to the interlinked whole—the biotic community. This biocentric orientation, however, does not imply the passive surrender of humans to the natural order, as Bookchin has claimed, since humans, like any other organism, are recognized as special in their own unique way and

are entitled to modify the ecosystems in which they live in order to survive and blossom in a way that is simple in means and rich in ends.[56] In
the context of this orientation, it is not inconsistent for humans to act
in their own self-defense by keeping in check or eradicating life-
threatening organisms *where there is no alternative* (and when the action is taken with reluctance). However, it is inconsistent when the activities of humans wantonly or needlessly interfere with or threaten the
existence or integrity of other life forms, when humans no longer see
themselves as plain members of the biotic community, but assume instead the role of planetary directors who have discerned the true path
of evolution and hence determined what should be the destinies (or
nondestinies!) of other life forms and entities.

To be sure, Bookchin himself has made many early and important
inroads into anthropocentrism. He has repeatedly emphasized his rejection of *environmentalism,* a term he employs to denote "a mechanistic,
instrumental outlook that sees nature as a passive habitat composed of
'objects' such as animals, plants and minerals, and the like that must
merely be rendered more serviceable to human use."[57] Indeed, this emphasis follows from Bookchin's organismic philosophy, which recognizes subjectivity as present, however germinally, in all phenomena, not
just in humans. For Bookchin, it is crassly instrumental to reduce the
richly textured ecocommunities (he prefers this term to *ecosystem* because of the latter's mechanistic connotations) of nonhuman nature to a
mere storage bin of raw materials for human use. That much is quite
clear. On the other hand, while subjectivity does not exclusively reside
in humans, it is nevertheless preeminent in humans. It is by virtue of
our advanced consciousness that we must assume the responsibility of
evolutionary stewards and further the advancement of yet higher forms
of subjectivity.

Bookchin's anthropocentrism, then, is not of the more familiar
kind. Indeed, his philosophy of nature is in part a critique of mechanistic materialism and instrumentalism along with the idea that humans
must dominate and control nature so as to adapt it to *human* ends.
Rather, human activity must be guided by overarching evolutionary and
ecological processes, not the instrumental needs of humans, an approach that seeks to reconnect human social activity with the natural
realm. This is indeed a promising approach that reaches beyond the human realm. The problem arises, however, in Bookchin's claim that we
now know enough about these processes to *foster and accelerate* them.
But are we really *that* enlightened? Can we really be sure that the thrust
of evolution, as intuited by Bookchin, is one of advancing subjectivity?
In particular, is there not something self-serving and arrogant in the

(unverifiable) claim that first nature is striving to achieve something that has presently reached its most developed form in us—second nature? A more impartial, biocentric approach would be simply to acknowledge that *our* special capabilities (e.g., a highly developed consciousness, language, and tool-making capability) are simply *one* form of excellence alongside the myriad others (e.g., the navigational skills of birds, the sonar capability and playfulness of dolphins, and the intense sociality of ants) rather than *the* form of excellence thrown up by evolution. Ecologists and evolutionary biologists have repeatedly stressed our profound ignorance of nature's processes. Indeed, the present scale and depth of the environmental crisis is testimony to how *little* we know about nature; nor can we afford to dismiss the possibility that nature is more complex than we *can* know.

These arguments should not be construed as meaning that we should therefore abandon all responsibility for the planet, avoid ecological restoration work, and do nothing. As Walter Truett Anderson has emphasized, there is no escaping the fact that *whatever we do* has implications for future ecological and evolutionary processes, and that we have been influencing these processes ever since our arrival on the evolutionary scene.[58] Yet Bookchin treats this undoubted fact as not only demanding that we become more conscious of our ecological and evolutionary interventions (a move that I fully endorse), but also as conferring on us a mandate to seize the helm of evolution on the grounds that we have grasped the direction of evolution and are now ready and able to give it a helping hand. But surely our mistakes tell us that the real issue is one of slowing down and revising the pace and scale of human interventions in ecocommunities so that we can keep abreast of the consequences—not only for our own sake, but also for the sake of other life forms. In short, an ecological ethics must be commensurate with our ecological (and evolutionary) understanding. Our empathy toward other beings should therefore naturally lead us to practice humility in the face of complexity and to acknowledge how little we know of our rapidly changing and crisis-ridden world. The wisest course of action—and this is not a passive surrender, but a deliberate and reasoned choice—is, wherever practicable, simply to let beings be.

Conclusion

My conclusion, then, rests on an important irony. Bookchin's enticing promise of the widest realm of freedom to all life forms is best delivered not by his own ecological ethics but by a biocentric philosophy. However, in view of Bookchin's scathing criticisms of biocentrism, particularly that expressed by deep ecology, the prospect of any kind of philosophical revision or reconciliation in this direction on Bookchin's part is, sadly, unlikely.

Acknowledgments

Parts of this essay were presented at the First National Ecopolitics Conference, held at Griffith University, Brisbane, Queensland, Australia, August 30 and 31, 1986. I wish to thank Warwick Fox, Peter Hay, and the referees of *Environmental Ethics* for helpful comments on earlier versions of this essay. Permission to reprint this essay, which originally appeared as "Divining Evolution: The Ecological Ethics of Murray Bookchin," *Environmental Ethics*, Vol. 11, 1989, 99–116, is gratefully acknowledged.

Notes

1. The society of "mutual aid" defended by the Russian prince Peter Kropotkin in *Mutual Aid: A Factor of Evolution* (Boston: Porter Sargent, n.d.) is considered to be one of the few nonreactionary examples of this kind of reasoning.

2. Bookchin's publications are too numerous to list exhaustively here. His major books include *Our Synthetic Environment* (New York: Knopf, 1962), which was published under the pseudonym Lewis Herber and appeared six months before Rachel Carson's influential *Silent Spring* (New York: Fawcett Crest, 1962); *Post-Scarcity Anarchism* (Berkeley, CA: Ramparts Press, 1971); *Toward an Ecological Society* (Montreal: Black Rose Books, 1980); *The Ecology of Freedom* (Palo Alto, CA: Cheshire Books, 1982); and *The Modern Crisis* (Philadelphia: New Society Publishers, 1986). Along with the three last-mentioned books, the major articles by Bookchin relevant to the present essay are "Toward a Philosophy of Nature—The Bases for an Ecological Ethic," in Michael Tobias, ed., *Deep Ecology* (San Francisco: Avant Books, 1984), pp. 213–235; "Freedom and Necessity in Nature: A Problem in Ecological Ethics," *Alternatives*, Vol. 13, No. 4, 1986, 29–38; and "Thinking Ecologically: A Dialectical Approach," *Our Generation*, Vol. 118, No. 2, 1987, 3–40.

3. Bookchin, "Thinking Ecologically," p. 4.

4. Bookchin, *Ecology of Freedom*, p. 342.

5. These terms are nowhere specifically defined by Bookchin, although they appear throughout his writings as ultimate norms and the desideratum of evolution. They are best encapsulated in the notion of *self-directedness*, which is central to both Bookchin's organismic nature philosophy and his social philosophy of anarchism—both of which have strong Hegelian overtones. Indeed, this is openly acknowledged by Bookchin with the qualification that his ecological dialectic differs from the Hegelian dialectic insofar as (1) it is more concerned with the existential details of nature rather than with the idea of nature; (2) it does not terminate in an Absolute; (3) it leans more toward differentia-

tion rather than toward conflict; and (4) it would "redefine progress to emphasize the role of social elaboration rather than social competition." See "Thinking Ecologically," pp. 26–27.

6. Bookchin has argued that an anarchist society is "a precondition for the practice of ecological principles" (*Post-Scarcity Anarchism*, p. 70). Indeed, Bookchin's contribution to anarchist thought has earned him the title of "the grand old man of American anarchism." See the Editor's Preface to "The Concept of Social Ecology," adapted by Peter Berg from the *Ecology of Freedom* for *Co-evolution Quarterly,* Vol. 32, 1981, p. 14.

7. I do not intend to discuss Bookchin's controversial thesis that the domination of nature by humans stems from the domination of humans by humans. Suffice it to say that I agree with Warwick Fox's conclusion that there is no necessary relationship between these two forms of domination. See Fox, "The Deep Ecology–Ecofeminism Debate and Its Parallels: A Defence of Deep Ecology's Concern with Anthropocentrism," *Environmental Ethics,* Vol. 11, 1989, pp. 5–25.

8. Murray Bookchin, "Social Ecology versus 'Deep Ecology': A Challenge for the Ecology Movement," *Green Perspectives,* Nos. 4–5, 1987, 1–23, and "Social Ecology vs. Deep Ecology," *Resurgence,* Vol. 127, 1988, p. 46. Many of these criticisms had already been expressed (albeit in considerably more tempered terms) in "Thinking Ecologically: A Dialectical Approach."

9. Bookchin, *Ecology of Freedom,* p. 364.

10. Ibid., p. 365.

11. Ibid., p. 275.

12. Ibid., p. 276.

13. Ibid., p. 279. Since Bookchin's notion of subjectivity is rather elusive, some historical comparisons may help to illustrate what I understand Bookchin to be getting at (although they do not perform the task of arguing his case). His notion of subjectivity in nature appears to stand midway between the vitalism of Bergson and the teleological view of nature espoused by Aristotle. Bergson rejected the mechanistic model of life in favor of a view that depicts organisms as constituted not just by physical matter, but also by a special life force or *elan vital*. But, whereas for Bergson the vital force has no aim or goal, it simply pushes in every direction—"a vast extemporization" (R. G. Collingwood, *The Idea of Nature* [Oxford, UK: Clarendon Press, 1945], p. 138)—for Bookchin, nature is pushing in particular directions and is exhibiting particular tendencies. Nevertheless, Bookchin is not calling forth any transcendental spirit to animate the living. Rather, his claim is that subjectivity inheres in the very organization of matter itself; it does not exist over and above it. Bookchin thus shares Aristotle's view of nature as "self-moving," as exhibiting *nisus* (i.e., an impulse toward or striving after a goal). Yet, whereas for Aristotle the change was cyclical, for Bookchin the change is evolutionary, representing a process that moves via creative advances toward increasingly complex forms. This model is clearly less deterministic than the Aristotelian one. In Aristotle's teleology, with its notion of "final causes," the end point of change is fixed in eternal forms such that the process leading to that end must necessarily move along predetermined lines. In contrast, Bookchin regards self-organization in

nature as an evolving process, which entails a degree of open-endedness, thus leaving room for creativity and spontaneity.

14. I am using the term *teleological* here in the broad sense to refer to an approach that seeks to uncover the grand design, end, or purpose in nature. Such an approach obviously assumes that such a design (Bookchin would prefer the word *directionality*) is there and that we are able to discern it—a problem that will be discussed below. For the moment, I am merely concerned to clarify Bookchin's particular approach, which must be understood as more open-ended than classical accounts. As Bookchin has put it: "If there is an 'end' it is the actualization of what is implicit in the potential—potential that has its ancestry in the dialectical process that precedes it" ("Thinking Dialectically," pp. 28–29).

15. In the course of developing his ethics, Bookchin has sometimes varied the relative weight and emphasis given to these different principles. For example, in *Toward an Ecological Society,* he described the wisdom of ecology in terms of three principles—diversity, spontaneity, and complimentarity (p. 60)—whereas, more recently, he has distilled his ethics into two principles—participation and differentiation (see *Modern Crisis*).

16. Bookchin, *Ecology of Freedom,* p. 365.

17. Bookchin, "Thinking Ecologically," p. 30.

18. Bookchin, "Freedom and Necessity in Nature," p. 5; emphasis in original.

19. Bookchin, *Ecology of Freedom,* p. 342.

20. Bookchin, "Rethinking Ethics, Nature, and Society," in *Modern Crisis,* p. 25.

21. Ibid., p. 26.

22. Bookchin, "Thinking Ecologically," p. 21.

23. Ibid., p. 35.

24. Ibid., p. 32.

25. Ibid., p. 33.

26. Ibid., p. 32.

27. Bookchin claims that "an ecological interpretation of dialectic, in effect, opens the way to an ethics that is rooted in the objectivity of the potential, not in the commandments of a deity [or the waywardness of the opinion poll]. . . . Hence, the 'should be' or 'could be' acquires not only objectivity, but it forms the objective critique of the given reality" ("Thinking Ecologically," p. 32).

28. Ibid., p. 35, n. 22.

29. Bookchin, *Modern Crisis,* p. 13.

30. Bookchin, *Ecology of Freedom,* pp. 26–29.

31. Bookchin, "Thinking Ecologically," p. 22, n. 14; *Ecology of Freedom,* p. 278.

32. Bookchin, *Ecology of Freedom,* p. 352.

33. On the issue of verification, Bookchin has argued that "in truly dialectical thinking, an empirical test must explore whether a given *process* in its theoretical form explains a given *process* in real life" (*Modern Crisis,* p. 15, footnote). Elsewhere, Bookchin has explained that "common sense demands

only inference, consistency, and the verification provided by ordinary sense experience" ("Thinking Ecologically," p. 23).

34. Bookchin, *Ecology of Freedom,* p. 354.

35. Ibid., p. 364.

36. Ibid., p. 355.

37. J. Baird Callicott, "Hume's *Is/Ought* Dichotomy and the Relation of Ecology to Leopold's Land Ethic," *Environmental Ethics,* Vol. 4, 1982, p. 164.

38. Bookchin, "Thinking Ecologically," p. 17.

39. Ibid., p. 24.

40. Ibid., p. 31.

41. Donald Worster, *Nature's Economy: The Roots of Ecology* (San Francisco: Sierra Club Books, 1977), p. 336.

42. Ibid., pp. 335–336.

43. Ibid., pp. 336–337.

44. Bookchin, "Thinking Ecologically," p. 35.

45. The central concern of such deep ecologists as Arne Naess, Bill Devall, George Sessions, Warwick Fox, and Alan Drengson is to cultivate a sense of identification with nature, a sense of interdependence and interconnectedness, in short, a sense of empathy for the fate of other life forms. This can hardly be interpreted as an approach that "reifies" nature and sets it apart from humanity.

46. Bookchin, "Thinking Ecologically," p. 20.

47. Bookchin, *Ecology of Freedom,* p. 343.

48. For an insightful discussion of the problems that can result from the promotion of diversity per se, see Reed Noss, "Do We Really Want Diversity?." *E.F.!,* June 21, 1986, p. 21. For a related discussion of the problems associated with what is usually thought of as an enlightened form of organic agriculture that promotes diversity, namely, permaculture, see John M. Robin, "Permaculture—An Ecological Nightmare," *Larrikin,* Spring, 1985, pp. 22–23. Robin, a Tasmanian botanist, has argued that permaculture (as designed by fellow Tasmanians Dave Holmgren and Bill Mollison) is a considerable advance over Monoculture, but that it is nonetheless anthropocentric (in concentrating only on species that are useful to humans) and indiscriminate (in selecting species that have a high propensity to become environmental weeds and in transferring these "useful" self-seeding species to different geographic regions with similar soils and climate). The result, according to Robin, is that "permaculture is condemning whole floras and faunas to irreversible change, even extinction" (p. 23).

49. The one reference to wilderness that I have come across Bookchin's work is consistent with the scenario I have sketched. Bookchin writes: "In advocating human stewardship of the earth, I do not believe it has to consist of such accommodating measures as James Lovelock's establishment of ecological wilderness zones. . . . What it should mean is a radical integration of second nature with first nature along far-reaching ecological lines" ("Thinking Ecologically," p. 32).

50. James A. Heffernan has argued that "when Leopold talks of preserving the 'integrity, stability, and beauty of the biotic community' he is referring

to preserving the characteristic structure of an ecosystem and its capacity to withstand change or stress. Moreover, maintaining the characteristic structure of the ecosystem, its objective beauty, is the key to preserving its stability." Heffernan notes that current ecological science has shown that an increase in the diversity of an ecosystem does not always increase its stability when the introduced species are exotic. He therefore suggests that Leopold's land ethic be rephrased to read: "A thing is right when it tends to preserve the *characteristic diversity* and stability of an ecosystem (or the biosphere). It is wrong when it tends otherwise" (emphasis added). By "characteristic diversity," Heffernan means the preexisting abundance and distribution of indigenous species. See Heffernan, "The Land Ethic: A Critical Appraisal," *Environmental Ethics,* Vol. 4, 1982, pp. 237, 247. On this important question of diversity, J. Baird Callicott has argued that the point of the land ethic is not to "enshrine the ecological status quo and devalue the dynamic dimension on nature," but rather to condemn the increasing rate and scale of human intervention in ecosystems, including the indiscriminate introduction of exotic species, to ensure, among other things, that speciation outpaces extinction. See J. Baird Callicott, "The Conceptual Foundations of the Land Ethic," in J. Baird Callicott, ed., *Companion to "A Sand County Almanac": Interpretive and Critical Essays* (Madison: University of Wisconsin Press, 1987), p. 204.

51. Bookchin, "Thinking Ecologically," p. 36; emphasis added.

52. Ibid., p. 37.

53. Fortunately, this stance has very few defenders. Dave Foreman, editor of *Earth First!,* is one such person who has gained notoriety for his biocentric militancy as a result of his interview (conducted by Bill Devall) with the Australian magazine *Simply Living.* See "A Spanner in the Woods," *Simply Living,* Vol. 2, No. 2, n.d., pp. 40–43.

54. Arne Naess, "The Shallow and the Deep, Long-Range Ecology Movement: A Summary," *Inquiry,* Vol. 16, 1973, p. 95. It should be noted that Naess expressed the phrase as "biospherical egalitarianism—in principle" in order to acknowledge that "any realistic praxis necessitates some killing, exploitation and suppression" (ibid.).

55. In his earlier work, Fox had criticized biospherical egalitarianism for advocating a "Procrustean ethics" that "attempts to fit all organisms to the same dimensions of intrinsic value," thus losing sight of the diversity of organisms and their varying capacities for richness of experience. See Warwick Fox, "Deep Ecology: A New Philosophy of Our Time?," *Ecologist,* Vol. 14, 1984, p. 199. However, Fox has since argued that his interpretation of biospherical egalitarianism (i.e., positing intrinsic value as evenly spread throughout nature) was a misinterpretation of deep ecology and he has provided a thoroughgoing critique of this kind of value theory interpretation along with an extensive elaboration and defense of the experiential, wider-identification approach. See Fox's *Approaching Deep Ecology: A Response to Richard Sylvan's "Critique of Deep Ecology,"* Occasional Paper No. 20 (Hobart, Australia: University of Tasmania, 1986), esp. p. 38.

56. This approach is encapsulated in the maxim "Live simply that others may simply live."

57. Bookchin, *Ecology of Freedom*, p. 21.

58. Walter Truett Anderson, *To Govern Evolution: Further Adventures of the Political Animal* (San Diego, CA: Harcourt Brace Jovanovich, 1987).

Respecting Evolution: A Rejoinder to Bookchin

In "Divining Evolution: The Ecological Ethics of Murray Bookchin," originally published in *Environmental Ethics* in 1989, I provided both a methodological and a normative critique of Bookchin's ecological philosophy from a biocentric perspective.[1] In "Recovering Evolution: A Reply to Eckersley and Fox," published in *Environmental Ethics* in 1990, Bookchin took strong exception to my characterization of his position, provided a spirited defense of the dialectical philosophy of social ecology, and reiterated his now well-known aversion for deep ecology.[2] In this rejoinder, I offer a belated response to the key points raised by Bookchin and offer some further thoughts on the methodological and normative differences between social ecology and what I now prefer to call an ecocentric philosophical perspective.

Before addressing Bookchin's reply, it is perhaps salutary to remind readers of the common ground between social ecology and ecocentrism.[3] Both social ecology and ecocentrism are critical of the purely instrumental orientation toward nonhuman nature that has informed the dominant political ideologies of modern times. Moreover, social ecology and ecocentrism find common cause in defending biological and cultural diversity, and both seek a radical ecological rapprochement between human society and the rest of nature. Given this general agreement, one should not be surprised to find agreement between social ecologists and those of an ecocentric persuasion in relation to a broad spectrum of environmental issues, ranging from the prevention of logging of old-growth forests to the protection of the rights of indigenous peoples to their traditional homelands. Indeed, prior to Bookchin's much publicized broadside against deep ecology at the National Greens conference held in Amherst, Massachusetts, in July 1987,[4] many ecological activists and theorists regarded deep ecology and social ecology as loosely overlapping, complementary ecophilosophies.[5] To the extent that ethical and political differences could be found, they were seen as minor border skirmishes rather than serious or irreconcilable conflicts when set against the broader canvass of environmental ideologies.

In the nine years that have elapsed since the publication of "Divining Evolution" in 1989, the differences between social ecology and eco-centrism (particularly deep ecology) have been closely scrutinized and have become much more sharply etched.[6] What lies behind these growing differences? To what extent may they be attributed to Bookchin's feisty approach toward his detractors, perhaps partly stemming from his concern to vindicate his pioneering credentials as a radical ecological theorist against newcomers (after all, Bookchin's vast oeuvre dates from the early 1950s, while deep ecology did not find its way to the United States until the late 1970s and did not become popular until the 1980s)? Or can the sharpening of differences simply be put down to unapologetic political aspirations on the part of both protagonists to capture the intellectual high ground, and the hearts and minds, of the growing green movement? While both of these possible reasons may have some bearing on the current debate, I wish to reiterate instead what has emerged as a more fundamental set of metaethical differences that stem from different "epistemologies of nature." These differences cast humans in somewhat different roles in the evolutionary drama. In terms of the exchange between Bookchin and myself, whether we should be cautious about "divining evolution" (and therefore respect evolution) or actively seek to "recover evolution" remains the serious moot point. Set in this context, semantic differences over how we should understand, say, the meaning and application of the principle of ecological diversity, are indicative of much more fundamental differences concerning how the necessary ecological rapprochement between humanity and the rest of nature should be conducted. Behind Bookchin's sometimes barbed prose and occasional ad hominem attacks lies a spirited defense of a very different way of thinking about the relationship between ethics and nature.

In "Recovering Evolution," Bookchin raises a host of objections to my critique in "Divining Evolution." For convenience, these objections and associated arguments may be stripped down and sorted into three broad categories (this precis is for the benefit of readers who are unfamiliar with Bookchin's reply; Bookchin does not present his argument in this order and I will not necessarily respond in this order). The first category concerns what he regards as my mischaracterization of his ideas, the second comprises a counterattack against deep ecology, and the third relates to the more fundamental methodological and metaethical differences between social ecology and ecocentrism.

In the first set of objections, Bookchin has rejected my characterization of his argument that the ecological ethics of social ecology provide a justification for humans seizing the helm of evolution, or "taking wanton command of nature." A close reading of his work, he insists, should make it clear that he has always emphasized the importance of allowing spontaneity and diversity in nature. In any event, a truly ecological society would be a nonhierarchical society, one based on an ecological ethic that would "add the dimension of freedom to nature" (p. 259). However, this does not mean making nature sacrosanct and beyond the reach of prudent human intervention (which is implicit in the quietistic "romantic wilderness cult"). Rather, it means developing "a sophisticated and nuanced ecological synthesis between nonhuman and human needs" (p. 272). On this point, Bookchin wonders why I should favorably cite a passage from Walter Truett Anderson's *To Govern Evolution* when Anderson is so critical of a "hands off" management philosophy and so enamored of the idea of "remaking nature."

To flesh out his case, Bookchin offers some illustrations of appropriate intervention. For example, he argues that while it might be anthropocentric for today's growth-oriented, egoistic society to turn the Canadian barrens into an area supporting a rich biota, it would be a legitimate activity if undertaken by an ecological society informed by the principles of social ecology. Nor would it be wrong to prevent massive asteroids from pounding the earth (which might exterminate highly evolved life forms) or to actively engage in ecological restoration of damaged ecosystems. In short, "nature" does not always "know best" when left to her own devices. On the basis of these arguments, Bookchin is nonplussed (to put it mildly) as to why I might wish to impugn the ecologically enlightened practice of permaculture.

Finally, he argues that I (along with Fox) have confused his rendition of the links between the domination of people and the domination of nonhuman nature. According to Bookchin, he is not simply arguing, as an empirical claim, that an egalitarian society will necessarily be nondomineering toward the nonhuman world, or the converse. Both of these claims, he concedes, can be easily refuted. Rather, he is concerned with making links between actual damage caused by particular societies and broad cultural mentalities or "epistemologies of rule," notably, those societies that ideologically equate human progress with the domination of nature. Such an ideological orientation would have no place in an ecological society, where both hierarchical thinking and hierarchical social relations would be eliminated.

The second set of arguments that can be teased out of Bookchin's reply bear upon his objection to deep ecology. Here he reiterates his now familiar arguments that deep ecology is both asocial and misanthropic. The third set of arguments comprises an attack on what he calls my Humean skepticism, some reflections on the use of developmental analogies, and the relationship between science and ethics—all by way of a defense and further elaboration of what it means to think dialectically.

In order to properly address the substantive points raised by Bookchin, I find it helpful to begin with the deeper methodological and metaethical differences between social ecology and ecocentrism, since they inform and underpin the social and ethical differences.

Bookchin thinks "generatively" by employing a speculative, organismic theory of knowledge that seeks to discern the dialectical logic or directionality of evolution. It is this dialectical unfolding that provides he "objective" ground of an ecological ethics. That is, from nature's unfolding, Bookchin believes it is possible for ("rational"?) humans to discern and grasp the principles of diversity, complexity, complementarity (or nonhierarchical, symbiotic relationships), and spontaneity. Moreover, Bookchin's method of dialectical reasoning regards determinable potentialities (what could/should be) as no less real and "objective" than what actually exists. Indeed, the "ought" may be regarded as more real and rational than the "is" if it is more in accord with nature's logic of development (p. 270). (The resonances with Hegel's philosophy are palpable, although Bookchin strongly denies that he is Hegelian or neo-Hegelian.) According to Bookchin, "The dialectical philosopher focuses on the *transitions* of a developmental phenomenon, on its passing from an 'is' into an 'other' " (p. 268). The task is to discriminate between those developments that further the logic of development of an entity and those that may "deflect" an entity from such logic (p. 268). According to Bookchin, the dialectic "tries to hold together past, present, and future in a unified thought" (p. 269).

It is one thing, however, to structure our experience and understanding of history in terms of intelligible principles or tendencies, but altogether different to project these principles and tendencies into the future. Set against Bookchin's "epistemological speculation," my version of ecocentrism is indeed based upon an "epistemological skepticism" (as Bookchin puts it) insofar as it declines to make any claims about the thrust or directionality of nature's unfolding. However, contrary to Bookchin's charge, this is not an atomistic, Humean empiricism. Rather, my approach proceeds on the basis of a more contingent, agnostic, and open-minded understanding of nature's unfolding or "logic of development" in terms of what we may properly claim to know. It also proceeds on the basis of a more modest metaethical understanding of how an "ecological ethics" might be defended. Like social ecology, ecocentrism defends general principles such as diversity, complexity, complementarity, and spontaneity. However, they are justified as *desirable norms* without recourse to claims that they represent the logic of evolution. To argue, as Bookchin does, that certain tendencies are "objectively" present in nature insofar as they are real (as distinct from fanciful) possibilities and that their very presence/potential in nature makes them normatively compelling is ultimately question-begging. Given that there are many real/potential pathways of evolution, the important issue is surely which pathways are *desirable and defensible* rather than which pathways are *objective or natural*.

Bookchin is right to observe that his particular form of dialectical reasoning vexes me (p. 269). It vexes me because he slides backward and forward between the "what could be" and the "what should be." So, one may find cryptic passages where he declares: "The *telos* that is imputed to dialectic is not what it *becomes*, but what it *should* become, given its potentiality to be so realized or fulfilled" (p. 269). Bookchin argues that things become what they are constituted to become by the logic of their own development (p. 267). In what respect is such a claim ethically meaningful as distinct from tautological?

Take my claim (vehemently resisted by Bookchin) that it is an illegitimate move to treat ontogenetic and phylogenetic development as analogous. While I would agree that one will always encounter limits when using analogies to explain ideas, I would still maintain that there are *serious* limits in the extent to which we may draw analogies between the development of individual organisms and entire species, or between physical and ethical developments. To be sure, it seems to be reasonable enough to say that an acorn, under the right conditions, is destined to become an oak tree, not a rose bush (although physical developmental paths are by no means fixed in the very long term; oak trees may evolve over time into thorny plants as a result of unpredictable genetic mutations, mutations that are essential to the process of speciation). In this sense, it is possible to talk about a "logic of development" and to identify "deviations" from that logic. However, the analogy becomes much more tenuous in relation to the development of entire species, and even more strained in relation to the psychological and ethical developments of individuals and communities. Can mind really "divine" an *ethical* logic in nature's dialectic? If so, how should we distinguish between rational or aberrant mutations, using *nature* (as distinct from human moral argument) as our guide?

One might agree with Bookchin that ethics "emerged" from the so-called natural world in the unfolding of evolutionary time. In this respect, I do not dispute the way in which Bookchin emphasizes the graded nature of natural and social evolution. However, my point is that the pathways of physical and social/ethical development are not analogous. As Bookchin knows, the latter are considerably less predictable than the former. What are humans constituted to become ethically? While human embryos will develop, in the ordinary course of things, into human adults, such embryos will, in the ordinary course of things, develop into adults with a wide variety of psychological and ethical orientations. Yet if we are to apply Bookchin's mode of reasoning without further qualification, the actualization of Hitler's potential is no less "real" and "objective" than that of Mother Theresa's. To repeat my point: in the move from "what could be" to "what ought to be," the putative "objective" status of any given set of potentialities is of much less interest than their *desirability* at any given historical juncture.

Bookchin would probably reply that there is an endless number of hypothetical factors that might "deflect" an entity from the logic of its development (p. 268). That is, while Hitler's moral development may be understood dialectically, this does not mean that his behavior is rational/ethical. Indeed, it may be shown to be irrational when set against the "logic of nature's unfolding," which is toward greater diversity and mutuality. Yet I still find such qualifications unsatisfying and question-begging. By what yardstick might we collectively agree with Bookchin's particular rendering of nature's unfolded past and future potential, given that there exists a range of alternative interpretations/speculations? While the social ecology principles discerned by Bookchin may serve to discriminate between domineering (irrational) and mutualistic (rational) relationships, his *method* may be employed with equal effect by social Darwinists to discriminate between competitive (rational) and cooperative (irrational) relationships. In this respect, it should be clear that I am much less troubled by Bookchin's principles than by his method of normative justification.

For his part, Bookchin claims that my version of ecocentrism is "positivistic" in insisting that an "ought" cannot be derived from an "is" and, accordingly, that ethics cannot be derived from nature. However, much turns on what is meant by "derived." An ecological ethic may be informed and inspired by "what happens in nature." However, formally speaking, such an ethic cannot be logically derived from nature simply because a value cannot be logically derived from a fact (or a contingent set of "facts," since our knowledge of nature is imperfect and uncertain). Premises, assumptions, and other values must necessarily form part of the moral reasoning process.

Moreover, I believe that it is quite possible to make this "is"/"ought" separation as a general heuristic or analytical aid while also subscribing to a post-Kuhnian philosophy of science (which rejects a *rigid* distinction between facts and values) and a poststructuralist account of knowledge (which would not herald science as the "one true story"). It is possible because my argument is not that "is" and "ought" are unrelated. Rather it is simply that *the "is" of nature (whether past, present, or future) cannot stand alone as an argument for the "ought" of human individuals and the communities to which they belong.* Murray Bookchin doesn't like hierarchy, so he looks to nature to find an "objective ground" for his defense of mutuality. It would have been much more straightforward if he were simply to have argued that hierarchy is bad and that cooperative forms of sociality are good, and then left it to his readers to evaluate his case.

The task of ethical justification requires that we advance normative (rather than "naturalistic") arguments concerning the *appropriateness* of certain practices or forms of thinking in relation to particular problems or situations. Normative arguments cannot be proved; they can only be made more or less compelling. This requires creativity, clarity, general consistency, overall coherence, and general plausibility in terms of contemporary understandings of the world. To this end, natural and social history, personal experience, imagination, and logic are all relevant. Moreover, what has happened in the natural and social worlds (whether understood by science, history, or literature) may be enlisted to *inform and support* arguments concerning the desirability of either existing or potential human orientations toward the rest of nature. Such understandings can also be used to refute alternatives that seem out of touch with contemporary understanding. For example, many people find creationism uncompelling because it conflicts with evolutionary theory. However, particular understandings of nature cannot be enlisted to *prove the rightness or goodness* of certain ethical principles simply by demonstrating that certain principles are "objective" or "natural." The justification of any ecological ethics must rest on the ability of its advocates to invite others to think imaginatively and critically beyond the strictures of the present and consider whether they might be happier, more authentic, freer, more connected, or otherwise more fulfilled (as individuals *and* community members) by the principles and vision of an ecological society that are being offered. (One need not be a follower of Jürgen Habermas to accept the general idea that "valid" norms emerge from an intersubjective agreement reached by public spirited and "uncoerced" dialogue based on mutual recognition and egalitarian reciprocity. Under such circumstances, the "force of the better argument" is able to prevail.) The crucial questions ultimately boil down to: What kind of humans do we wish to become? and What kind of world do we wish to inhabit? In response to these questions, an ecocentric ethics seeks to cultivate open-minded, tolerant, and empathetic citizens who affirm and celebrate the experience of belonging to an ecologically and culturally diverse world. The general appeal of such an ethic to me is that it seeks to maximize the relative autonomy (or "options to unfold") of *both* human and nonhuman life forms. I find this conducive because it is *inclusive* and because it softens the divide between self and other without denying difference.

In "Divining Evolution" I suggested that social ecology aspired to a broadly similar ideal, but failed to deliver because of the way in which Bookchin consistently privileges "first nature" over "second nature." However, Bookchin claims that it is a misrepresentation of social ecology to say that it seeks "the *widest* realm of freedom for *all* life forms." Rather, social ecology merely seeks to "add the *dimension* of freedom, reason and ethics to first nature" (p. 259). I must confess some difficulty in grasping Bookchin's argument here. If the dimension of freedom is "added" to first nature such that *both* first and second nature have freedom, then surely the general realm of freedom is widened. Moreover, Bookchin's rendering of my argument that an ethic that promises the widest realm of freedom to all life forms would confer freedom on humans to exploit nature seems strange and misplaced. Surely such an ethic asserts the very contrary: if we are serious about respecting the autonomy of *both* human and nonhuman nature, then our own autonomy must necessarily be exercised in ways that are compatible with the relative autonomy of others, both human and nonhuman. A's freedom should not become B's "unfreedom." Rather, we should seek ways to mutually enlarge A's and B's freedom, or where that is not possible, at least try to ensure a "fair distribution" of freedom, as it were. This, at any rate, is what I meant to convey.

Moreover, an ecocentric philosophy cannot preach a simple norm of human noninterference. Such a posture is undesirable and in any event impossible if it is accepted that all organisms are partly constituted by their ecological relationships. For humans, as moral agents, a general affirmation and celebration of the *relative autonomy* of other species must necessarily be qualified by principles of human self-defense and self-preservation, which are incorporated into the deep ecology notion of "vital needs" (after all, humans, like the other innumerable life forms, also seek to unfold).[7] The mutual enlargement or sharing of freedom need not entail a rigid "biospherical egalitarianism" (a confusing and much maligned deep ecology notion), nor does it place the unique characteristics of humans on a par with those of other species. Rather, an inclusive, ecocentric perspective would highlight the incommensurability and "radical otherness" of the characteristics of many nonhuman species, some of which may not readily conform to Bookchin's ideals of symbiosis and mutualistic harmony in nature.

When one turns from the realms of epistemology and metaethics to the realm of practical ethics and politics, the differences between social ecology and ecocentrism are less marked. Nonetheless, important differences still remain, some of which are a direct spillover from the metaethical debate while others stem from different domains of inquiry and different theoretical preoccupations. In "Divining Evolution" my primary focus was on the *ecological* component of social ecology and deep ecology, or more particularly, the human–nature (rather than human –human) relationship. In defending deep ecology over social ecology in this respect, I do not wish my comments to be taken as defending deep ecology in *every* respect. In particular, I have recently argued that deep ecology should be understood primarily as an ecophilosophy and an ecopsychology, not as a well-developed political theory that addresses questions such as power and authority.[8]

In contrast to deep ecology, social ecology may be understood as both an ecophilosophy and a political theory insofar as questions of power, hierarchy, authority, and democracy have always been its major theoretical preoccupations. This partially explains why Bookchin misconstrues deep ecology as misanthropic in appearing to attribute blame to a monolithic humanity rather than paying theoretical attention (as distinct from personal, ad hoc concern) to the problem of social hierarchy. There is, of course, much more that has been said, and could be said, by way of claim and counterclaim in the wide-ranging debate between deep ecology and social ecology. However, "Divining Evolution"—and this reply—take issue with only one dimension of this debate. While I would maintain that deep ecology requires considerable development as a political theory, these apparent failings (which may be attributed to deep ecology's primary focus on ontological and psychological questions rather than political questions) do not detract from the general normative argument I wish to advance here concerning the relevance of evolutionary and ecological theory to ethics.

Bookchin places considerable emphasis on the active and creative role played by humans in ecological and evolutionary processes. Humans ("second nature") are seen as the most complete expression thus far of a major tendency in natural evolution ("first nature"). Bookchin has also characterized humanity as a self-conscious "moment" in nature's dialectic, arguing that we have both the rationality and the ethical responsibility to foster (as distinct from merely to maintain) the evolutionary process. "Recovering evolution," then, means not merely defending the characteristic diversity, distribution, and abundance of existing species (the primary concern of ecocentrism). Rather, following the logic of evolution, the task is to create (or engineer?) a *more* diverse, complex, and fecund biosphere than the one we presently have. Indeed, Bookchin suggests that we may even "create more fecund gardens than Eden itself" (*Ecology of Freedom*, p. 343).

This idea of not simply maintaining but increasing diversity and fecundity in nature is an important component in Bookchin's philosophy. To take his telling example, should we ecologically aware stewards take up Bookchin's suggestion of turning the apparently lifeless "Canadian barrens" into a "richer" and more diverse ecosystem? My ecocentric response would be to err against lending an ecological helping hand in precisely this kind of circumstance, in the name of protecting not diversity per se, but *characteristic* diversity. That is, paradoxically, increasing the richness of the biota in *every* region of the globe may lead to an overall *loss* in global diversity as the more simple and fragile ecosystems (of, say, polar or desert regions) are replaced by more resilient, human-cultivated ecosystems made up of species selected from elsewhere. (A very loose parallel may be made here in relation to human cultural diversity. Replacing each distinct culture in the world with a mixture of cultures is likely to lead to global monoculture rather than greater cultural diversity as the more resilient, flexible, and/or powerful cultures establish dominance.) The case for protecting representative ecological diversity is not an argument against agriculture per se or ecological restoration, both of which are consistent with ecocentric principles. Rather, it is an argument against ecological colonialism—the total human appropriation of the process of species selection.

Set against Bookchin's philosophy of nature, ecocentric philosophies—particularly those inspired by Leopold—have been cast as relatively static, passive and "pictorial." Moreover, the idea of "protecting nature" by "fencing off" wilderness appears somewhat naive and misconceived. Indeed, the very idea that large areas of "first nature" should be cordoned off as generally out of bounds to "second nature" (except for very low impact uses or indigenous human lifestyles) is, in Bookchin's schema, as "unnatural" as the idea that the body should be segregated from the brain. Such a step would deny or curtail the opportunity for humans to perform their creative stewardship role in the evolutionary drama. As Bookchin has explained, "In advocating human stewardship of the earth, I do not believe it has to consist of such accommodating measures as James Lovelock's establishment of ecological wilderness zones. . . . What it should mean is a radical integration of second nature with first nature along far-reaching ecological lines."[9] Indeed, Bookchin argues that our uniqueness cannot be overemphasized, "for it is in this very human rationality that nature ultimately actualizes its own evolution of subjectivity over long aeons of neural and sensory development."[10]

When compared to social ecology, I think it is fair to say that my version of ecocentrism (which has since been explicated in *Environmentalism and Political Theory* and more recent work) is more concerned with the biological "here and now" than with a potentially richer future. Indeed, it is precisely this concern with the biological here and now that would look dimly on genetic engineering as "playing God," or act to deflect comets hurtling toward the earth—even though both events may open up new evolutionary pathways.

Nonetheless, Bookchin's criticisms of naïve preservationism are well made and form part of a growing chorus of critiques against a simplistic "hands-off" management philosophy in relation to wildlife and wildlands. I agree that defending the earth's existing biodiversity should not entail a "postcard ecology" that seeks literally to freeze-frame wilderness areas along with the biological status quo, for that would rule out the important practice of ecological restoration that is necessary to repair degraded ecosystems and protect threatened and endangered species. Moreover, protecting the characteristic diversity, distribution, and abundance of the earth's species requires much more than ecosystem reservations. Indeed, ecosystem reservations may be seen as a last-ditch, lamentably necessary effort to protect what is a rapidly disappearing asset. Large ecosystem reservations would not be necessary if human patterns of settlement and economic development were compatible with the maintenance of biodiversity and ecosystem resilience. *If* this is what Bookchin means by the "radical integration of second nature with first nature along far-reaching ecological lines," then I am in full agreement. However, so long as we fall short of such an ecological society, the preservation of large tracts of representative ecosystems (along the lines of the U.S. Wildlands Project) is now the most reliable and appropriate means of protecting in situ threatened and endangered species and ecosystems from the insatiable demands of capitalist economic development. In the meantime, of course, it would be self-defeating to focus exclusively on protecting biodiversity while ignoring problems of poverty, pollution, and agricultural land degradation, since these problems are likely to have detrimental impacts upon the relatively wilder areas. There are also other compelling human welfare reasons for engaging in ecological reforms in the urban and agricultural environment.

A strategy of protecting biodiversity in situ—which is central to ecocentrism—is clearly not a passive exercise. Ecosystem reserves and national parks require active human management. In some cases, for example, this may require fire prevention; in other cases, it may require regulated burning, depending on the history and characteristics of the ecosystem and the vulnerability of the species in question. There can be no doubting, then, that humans are active, creative agents in ecological and evolutionary processes. However, Bookchin has mistakenly taken my quoting of a passage from Walter Truett Anderson's book *To Govern Evolution* as an indication that I favor Anderson's position over his own. But to quote approvingly a particular passage from a work does not carry with it an endorsement of the general argument in the work. Bookchin's assumption that I endorse Anderson's general position (which is far more supportive of humans actively "remaking nature") is simply unwarranted in the face of my general argument in favor of humility and prudence. The differences in orientation between ecocentrism and Anderson's position should have been obvious enough not to warrant a general disclaimer.

The crucial question in relation to human creative agency is not whether we should assume a "hands-off" posture or play an active role (since a "hands-off" posture carries ecological consequences —this was the point of the passage I quoted from *To Govern Evolution*). Rather, if we are to aspire to become self-conscious and reflexive in relation to our creative agency, which I believe we must, then the crucial question is *What kind of "creative agents" do we wish to be?* With what claims to knowledge and what sensibilities should we "facilitate" human and nonhuman nature's unfolding? In short, how should we coevolve?

One of my primary concerns in "Divining Evolution" was to argue that we should approach the task of biodiversity protection and ecological repair and reconstruction with an appreciation of the limitations of our ecological and evolutionary knowledge and with an attitude of humility rather than one of arrogance. My main problem with Bookchin's philosophy of nature is that it tends to overemphasize human knowledge and prowess at a time when we should be applying the precautionary principle. My comments about revisions to Permaculture were intended to illustrate the contingent character of our ecological knowledge and the need for ongoing adjustments in our understanding of ecological interactions. Far from wishing to impugn permaculture (and I certainly did not claim that it was anthropocentric, as Bookchin suggests), my comments were intended to illustrate the fact that even the more promising and ecologically aware forms of landscape design and cultivation can sometimes pay insufficient heed to the impact of introduced species on indigenous species. To their credit, the authors of *Permaculture III* have revised some of their recommendations to take into account constructive criticism of this kind.

Bookchin has taken exception to my suggestion that he wishes humanity to "seize the helm of evolution," and he would no doubt wish to reiterate his charge that ecocentric philosophies are misanthropic, passive, and quietistic. In his defense, he has reiterated his principles of spontaneity and mutuality, which presumably serve to check any move toward, say, the total human domestication of the planet. According to Bookchin, social ecology seeks a relationship of symbiosis rather than domination, a relationship that allows for "a fuller level of mutualistic harmony" (p. 258).

Many of my critical comments in "Divining Evolution" were directed to Bookchin's post–*Ecology of Freedom* writings. It would seem reasonable to assume (bearing in mind that ideas develop and change over time) that if different emphases can be found in an author's writings over time, then the later writings should be given more prominence.[11] Accordingly, although I noted Bookchin's earlier emphasis on spontaneity, I have interpreted the more recent elaborations of his evolutionary stewardship thesis as a further development and qualification of his earlier writings. Bookchin cannot have it both ways. In any event, I would still maintain that it is an expression of human chauvinism to claim that we have grasped the direction of evolution, irrespective of whether Bookchin delineates this direction in appealing terms and personally dissociates himself from practices such as genetic engineering. Again, my central argument is that while we have some knowledge of evolution and ecology, we cannot "divine" a telos from nature's unfolding for the purposes of developing an ecological ethics. We must treat our knowledge as contingent rather than settled if we are to adapt flexibly and with minimum disruption to changing circumstances. And throughout this adaption process, our ecological ethic must remain commensurate with that ecological (and evolutionary) knowledge.

Notes

1. *Environmental Ethics*, Vol. 11, 1989, pp. 99–116. Reprinted as the earlier part of this chapter.

2. *Environmental Ethics*, Vol. 12, 1990, pp. 253–274.

3. For an explication of my version of ecocentrism, see Eckersley, *Environmentalism and Political Theory: Toward an Ecocentric Approach* (Albany: State University of New York Press, 1992); and "Beyond Human Racism," *Environmental Values*, Vol. 7, No. 2, 1998, pp. 165–182.

4. See notably "Social Ecology" versus "Deep Ecology," *Green Perspectives*, Summer 1987, p. 2.

5. See Devall and Sessions, *Deep Ecology: Living as if Nature Mattered* (Layton, UT: Gibbs M. Smith, 1986), p. 259.

6. This remains so notwithstanding an apparent reconciliation between Bookchin and Dave Foreman; see Murray Bookchin and Dave Foreman, *Defending the Earth: A Dialogue between Murray Bookchin and Dave Foreman*, ed. Steve Chase (Montreal: Black Rose Books, 1991).

7. For a fuller discussion of the general deep ecology idea of "vital needs," see Eckersley, "Beyond Human Racism."

8. See Eckersley, "The Political Challenge of Left–Green Reconciliation:

A Response to Roger Gottlieb," *Capitalism, Nature, Socialism*, Vol. 6, No. 3, September, 1995, pp. 21–25.

9. "Thinking Ecologically: A Dialectical Approach," *Our Generation*, Vol. 18, No. 2, 1987, pp. 3–40; quotation from p. 32.

10. Ibid., p. 20.

11. In *Toward an Ecological Society* (Montreal: Black Rose Books, 1980), Bookchin encapsulated his ecological ethics into three principles—diversity, spontaneity, and complementarity (p. 60)—whereas in *The Modern Crisis* (Philadelphia: New Society Publishers, 1986) he has distilled his ethics into two guiding principles: participation and differentiation.

Chapter 3

Ethics and Directionality in Nature

GLENN A. ALBRECHT

Introduction

It is commonly thought that the Darwinian theory of evolution entails the view that evolution produces progress in the living forms of nature from lower to higher, and furthermore that this *progress* can provide a natural foundation for ethics. Darwin himself seemed ambivalent on this issue. We find in the *Origin* evidence that he saw natural selection as a mechanism in nature giving "progress toward perfection."[1] However, there is also evidence that he cautioned himself not to use the words *higher* and *lower* in an evaluative way.[2] Despite this evidence of caution about the normative implications of the theory of evolution in Darwin's own work, there has been no shortage of theorists prepared to see some notion of ethical progress in evolution. Even before the publication of Darwin's *Origin* in 1859, Herbert Spencer had in 1851 in his *Social Statics*[3] understood natural evolution in terms of the "survival of the fittest," which he saw as a "law of nature." Life in human society, as in nature, was for Spencer a competitive contest where *success* was achieved at the expense of "less fit" humans. Such an ethic stressed the need to let social inequality take its course and let individuals find their own level in social life. Social Darwinists and neo-Malthusians have followed Spencer in reading, directly from nature, individualistic and competitive ethics.

Following the anarchist Peter Kropotkin and his conception of *mutual aid,* contemporary ecoanarchists, among others, have examined ecoevolution and seen a different direction in its history. Murray Bookchin, for example, in developing his philosophy of social ecology argues for a view of nature as exhibiting a "self-evolving patterning, a 'grain,'

92

so to speak, that is implicitly ethical."[4] This *directionality* is the potential and freedom manifested in life to evolve toward ever-increasing interrelated diversity and complexity.

Rather than seeing competition and struggle as the major features of evolution, Bookchin identifies cooperation and symbiotic interrelationship as fostering its most important outcomes. The social correlate of this natural order would be a human community that complements in its social design the structure and processes of nature. From the perspective of social ecology, human society has, for the greater part of its own evolution, created communities that have been consistent with this natural order. The scope, scale, and organization of society has been inspired by an understanding of the *spontaneous natural organization* from which it arose and in which it remains embedded. As with natural ecologies, social ecologies are defined by the organic interactions that occur at the local and wider levels, with spontaneously generated order achieved without the need for outside (authoritarian) regulation.

Bookchin's use of directionality in establishing an objective, naturalistic, and environmental ethic as the foundation for society is one of the elements that makes his philosophy of social ecology such a distinctive contribution to the contemporary ethics and politics of the human–nature relationship. This essay shall outline Bookchin's thesis on directionality and examine some of the major objections to it from within the current literature.

In defending and expanding the directionality thesis, Bookchin's position shall be connected to relevant material in the contemporary philosophical and scientific literature that deals with this issue. In what has become known as "complexity theory," the directionality thesis has been taken to new levels and in new ways that possibly strengthen the foundations of social ecology. It may also be possible to provide promising new pathways for the application of social ecology to areas such as design, technology, and politics. The implications of these new directions for social ecology shall be explored at the conclusion of this essay.

Bookchin's Position on Directionality

Bookchin's attempt to provide an objective foundation for ethics arises out of his analysis of the relevant scientific literature and his Aristotelian and Hegelian philosophical background. In *The Ecology of Freedom* (1982), he reviews the then current bioevolutionary literature, relying in particular on Cairns-Smith (1974), Trager (1970), Margulis (1981), and Lewin (1980) to reach the conclusion that within evolution there is an ethically significant sense of the interdependence of all life and the way this interdependence promotes the continuity of life. Bookchin summarizes:

> Contemporary biology leaves us with a picture of organic interdependencies that far and away prove to be more important in shaping life forms than either a Darwin, a Huxley, or the formulators of the Modern Synthesis could ever have anticipated. Life is necessary not only for its own self-maintenance but for its own self-formation.[5]

The three attributes of organic life that are of primary ethical significance are what Bookchin calls "mutualism," "freedom," and "subjectivity." Mutualism depicts the relations of mutual dependence in time and space that are the essence of ecological relationships. According to Bookchin, the geographer and anarchist Kropotkin[6] was on the correct path when he saw mutual aid within and between species as a factor in evolution. This pioneering insight has been expanded in the twentieth century by ecological advances in understanding the complexities of *symbiosis*[7] and types of living relationships formed by mutual dependence. In addition, mutualism assists in the creation of new features of the biosphere where new varieties of organic life have the potential to emerge. "Freedom" is best described as the active effort by organic beings to assert their identity and presence and to preserve themselves, while "subjectivity" is Bookchin's term for the ability of matter to self-organize toward consciousness (sentience) and finally self-consciousness in humans.

Bookchin's Aristotelian–Hegelian philosophical background is made manifest in his explanation of the emergence of increasing degrees of subjectivity in the unfolding of life from the simple to the complex. Life, in spontaneously producing itself, also produces a material substance that "eventually yields mind and intellectuality."[8] This ability of nature to become self-conscious was a life potential that was latent until the evolution of the ancestors of humans.

Both Aristotle and Hegel saw life in terms of inner teleology, or the unfolding in a spontaneous way of potentialities that lay dormant in life from its outset. Bookchin sees inner teleology as an entirely natural phenomenon. He argues:

> Hence our study of nature . . . exhibits a self-evolving patterning, a "grain," so to speak, that is implicitly ethical. Mutualism, freedom and subjectivity are not strictly human values or concerns. They appear, however germinally, in larger cosmic and organic processes that require no Aristotelian God to motivate them, no Hegelian Spirit to vitalize them.[9]

Bookchin sees the values of mutualism, freedom, and subjectivity as "implicit in nature"; they must be made explicit by humans, who can act as the "self-reflexive voice of nature" in the production of an ecologically inspired rationality and ethic.[10] Humans are in a position to do this because their sense organs, their language, and their intelligence are themselves the products of natural evolution and have the potential to reflect the structure and processes of nature in conceptual form. Bookchin believes that we must find concrete social expressions of these naturalistic ethical values. They will be made manifest in ideas and action that assist the "grain of nature" in that they support the movement of "self-organizing reality toward ever-greater complexity and rationality"[11] and fight against homogeneity and antirational forces.

Although humans have created social ecologies that are substantially out of balance with the natural ecologies that once sustained them, as natural agents with natural intelligence and rationality humans can manage their own affairs in such a way that could enable the "remaking" of society along lines that restore the harmony between the natural and the social. The ethical impetus to do so arises out of a realist understanding of the way life is structured and how organic processes work. Bookchin argues that the social values that arise out of a naturalistic ethic—unity in diversity, spontaneity, and nonhierarchical relations—are objectively grounded in this understanding. He suggests that they are the "elements of an ethical *ontology,* not rules of a game that can be changed to suit one's personal needs."[12] By making this claim, Bookchin is advocating a type of ethical realism in which value exists independently of any human valuer (objectivity) but can be discovered alongside other facts about the nature of reality.

As expressions of social action that would support or complement the direction or grain of nature, Bookchin forwards technologies that are based on an array of renewable energy resources (on a human scale and in harmony with the environment), direct democracy, decentralized urban communities, and organic food production. All these strategies would have to be artistically and intimately integrated to create *ecocommunities.*

Bookchin maintains that an ecological vision is compatible with the kind of social structure and relationships that communitarian anarchist theorists have championed since the late nineteenth and early twentieth centuries.[13] Although not supportive of the term "bioregionalism," Bookchin does see important linkages between ecocommunities developing on confederal lines. Such confederal relationships between communities would recognize cultural as well as ecological links. In addition, Bookchin, in his recent work *Urbanization without Cities,* has suggested that it is still possible to create a social ecocommunity in the form of a modern city if the "cancerous phenomenon" of urbanization is overcome by a revival of participatory democracy and grassroots institutions.[14]

Critiques of Bookchin's Position

The Naturalistic Fallacy and the Is/Ought Problem

As I indicated above, there is a long tradition in ethics that has warned philosophers and others not to derive values from facts. The tradition argues that the good that ethics describes has no relationship to the natural and that identifying value in the facts of nature, or committing the "naturalistic fallacy,"[15] is one of the worst errors that can be perpetrated in moral philosophy. T. H. Huxley, Darwin's "bulldog," avoided this problem by suggesting that the theory of evolution was amoral and that a humanistic ethic was needed in society to counter the caprice, contingency, and complete lack of ethics in nature.

Contemporary philosophers such as Paul Taylor[16] and Robyn Eckersley[17] have criticized any attempt to found ethics on ecological and evolutionary facts because, as the traditional argument goes, factual knowledge is logically different from normative judgment. In addition, the facts about evolution provide contradictory advice for normative judgments because evolutionary science can generate facts that support competition and conflict as well as facts that support cooperation and harmony in nature. Bookchin's attempt to found an objective, naturalistic ethic is thought to founder right at the outset because of the fact/value, is/ought problem.

The long and tortuous history of debate on the is/ought problem in Western philosophy stands as a warning to those who wish to enter it; however, it is useful to note that in many ways Bookchin's project runs counter to the central elements of the Humean, Kantian, and Mooreian traditions that create the logic and structure of the problem. Directionality in evolution (nature) is an empirical claim that stands or falls on scientific evidence. Although certain facts about particulars in nature might seem to contradict the general claim, the historical totality of life and the process that generates it are what the claim is about. Particular instances in evolutionary and ecological relationships that seem to contradict the thesis can be overcome by a more powerful general and long-term tendency of increasing complexity and diversity. Not even occasional catastrophes in nature that produce, for example, mass extinction of species can negate the long-term direction. When evolution recommences, it can be shown that it again moves in the direction of increasing complexity and diversity.

Bookchin's ethical project is holistic in that it views the human–nature relationship as a collective or an organic unity and rejects any attempt to found ethics, be it social or natural, on atomistic and individualistic foundations. Philosophers such as Bookchin who are working within the organicist Hegelian tradition do not find the is/ought problem intimidating.

Ecologically inspired environmental philosophers such as John Rodman have also become impatient with those who continue to apply the logic of atomic relationships and individualism to the domain of dynamic, organic totalities. Rodman, writing in the 1970s, suggested that an ecological understanding of life provided a new factual basis for human values and that the naturalistic fallacy should no longer have such a pervasive grip on debate. According to Rodman, the naturalistic fallacy has, among other things, reduced the quest for ethics "to prattle about 'values' taken in abstraction from the facts of experience" and because of this, "the notion of an ethics as an organic ethos, a way of life, remains lost to us."[18]

Rodman's suggestion that ethics only makes sense when it is linked to an organic ethos or way of life is indirectly supported by Alasdair Macintyre in *After Virtue*. Macintyre criticizes much that passes as contemporary moral philosophy because it has become divorced from the social and historical context in which it once had conceptual and practical relevance. The notion of a *right,* for example, was generated in the seventeenth century "to serve one set of purposes as part of the social invention of the autonomous moral agent."[19] The *autonomous moral agent* was to possess rights and a universally valid justification was supposed to defend them. Despite, according to Macintyre, the failure of the Enlightenment project to create an "objective and impersonal criterion" for moral rightness, the concept of the autonomous moral agent and the individualistic language of rights continue to dominate ethical discourse in the late twentieth century. Such a situation has produced an "incoherent" conceptual scheme for ethical thought where there is now no relevant context or ethos where our ethical concepts can meaningfully be put to work. Resolution of conflict generated by, for example, competing rights occurs through "manipulative modes of relationship." In general, ethics is in a state of despair, with relativism and nihilism the prevailing postmodernist moods.

It seems that the possibility of an objective foundation for ethics arising, in part, out of eco-evolutionary evidence may be one way of avoiding the relativism and nihilism inherent in Macintyre's impasse and of returning ethics to a meaningful organic ethos or way of life. Such a foundation for ethics would provide a base rationale for the choice of certain sets of values over others and the projection of those values into other domains of human enterprise such as policy and planning.

Moreover, given the continuing disputes within other traditions in ethics about the viability and foundations of an environmental ethic, it seems that these traditions will only exacerbate the conflict over policy and praxis that already prevails. The extension of rights from the human to nonhuman domains has already generated such conflict, where human rights and the so-called rights of animals and/or ecosystems are thought to be mutually exclusive. The right of nations to develop their "natural capital" and benefit materially from that development also opens arenas of conflict between economic development and the rights of the biotic world. The notion of *intrinsic value* proffered by the deep ecologists[20] is also unhelpful in resolving the clashes of interests that occur as different, supposedly equally valuable species attempt to "live and flourish" according to their potential. The role of humans in a world governed by such a biocentric and egalitarian ethic remains contentious since, as a natural species with creative intelligence, humans will inevitably negatively impact on other species and elements of ecosystems. Fear of the violation of another's intrinsic value could lead to complete inertia as individuals attempt to make choices about their survival.

For Murray Bookchin, an ethic based on contractual self-interest and a subjectively founded sense of how we would wish the planet to be treated has the potential to give us ecoideologies ranging from misanthropism and ecofascism to an anthropocentrism that "confers on the privileged few the right to plunder the world of life, including human life."[21] As far as Bookchin is concerned, without an objective foundation for ecological ethics we are left with "an ethical relativism that is subject to the waywardness of the opinion poll."[22] Environmental philosophies that fail to provide a foundation for an ethics that is objective in the sense that its truth is independent of subjective experience and instrumental value, leave open the possibility that self-orientated and relativistic positions will emerge that promote environmental destruction and other negative outcomes.[23]

Evolutionary-Based Critiques of Bookchin's Thesis

Darwin's ultimate rejection of progress in the direction of evolution suggests that he, or more correctly, his supporters, like T. H. Huxley, could not see any reason why it was inevitable that life evolved from the simple to the complex. Even though the fossil record may have shown a general increase in complexity over time, there was no guarantee that this direction was the inevitable outcome of evolutionary forces. In Darwinian theory, evolution could also produce persistence of species and even possible retrogression from complex to simple organisms. The mechanism of evolution is blind and random, so its outcomes reflect no immanent drive to some end state like high-level complexity. Such a conclusion applies even to human consciousness, which is often held up as the climax of evolutionary forces moving toward increasing complexity. The human brain, according to classical Darwinian theory, is nothing over and above an adaptive advantage that enables Homo sapiens to successfully exploit a given environment. As Stephen J. Gould argues, the fact that a "tiny and accidental evolutionary twig called *Homo sapiens*" now "reflects back on its own production and evolution"[24] is of no ethical significance in itself. Indeed Gould, on the basis of his reading of the palaeontological evidence, suggests that we cannot even perceive a direction in evolution of gradual, increasing complexity. The idea that after the Cambrian era life has evolved in a series of rises and falls in complexity and variety, a theory known as "punctuated equilibrium,"[25] has challenged the gradualist "great chain of being" orthodoxy in evolutionary thought. Gould's position has been summarized by Lewin:

> The fact of historical contingency, which Stephen Jay Gould has championed ever more strongly in recent years, means that the world we inhabit is simply one of a virtual infinity of worlds. Run the tape again, he says, and even the most modest turn on the long road of history translates into a dramatic effect a hundred million years or so later. Multiply such excursions of fate a million-fold, and the end result is a world unrecognizable to our eyes.[26]

The hardheaded conclusion that arises out of the collapse of the gradualist interpretation of evolution is that complexity and diversity are things that may be desirable from an anthropocentric point of view, but they are not the inevitable products of natural forces. The level of complexity we see around us in the present may not be as high or as productive of variety as previous peaks in evolution, and future peaks may also be more productive of variety. Indeed, by enacting wholesale change on present ecosystems, humans may be providing the conditions favorable for new forms of life and hence new types of complexity to emerge. Once gradualist assumptions are removed, there is no reason to conserve or assist the present type of complexity we see manifested around us.

From a Darwinian understanding of the evolution of life, no naturalistic ethic is present that could link human ethics and action to the maintenance of present levels of biodiversity. There are a number of possible pathways that are compatible with ecoevolutionary theory. Commitment to the "diversity and complexity pathway" is an aesthetic and ethical choice that might arise out of human needs and interests. Equally, humanity might choose a path of progress that is inconsistent with the current level of biodiversity, and this too could be justified within contemporary evolutionary theory.

Bookchin's account of evolution is thought to be incapable of addressing the major objection to an ethics based on the direction of evolution, namely, that there are numerous, equally viable directions, not all of them consistent with a complexity that features unity (mutualism) and diversity (freedom). The extension of an inadequately argued scientific basis for unity and diversity into advocacy of "nonhierarchical" relations in nature and society and increasing degrees of "subjectivity" further compounds problems with Bookchin's lack of an adequate conceptual basis for directionality.

It shall be argued below that it is possible to strengthen Bookchin's position by supplementing it with new material from complexity theory that takes evolution well away from its Darwinian roots and provides new evidence for the thesis that life is a complex adaptive system that self-organizes or moves in the direction of increasing unity and diversity—and that included in this naturally evolving complexity is human consciousness.

Another potential criticism of Bookchin's general thesis is that it depends on life being self-generated and directionality being spontaneously achieved. Standard accounts in evolutionary theory stress that an information transfer system (cybernetics) is responsible for the biological order that we see in the evolutionary record. Genetics is required to explain life and the evolution of complex structures, and it is a major weakness in Bookchin's work that he does not provide any account of how complex structures emerge and are reproduced over time. The issue of self-generated increasing complexity and diversity shall be examined in greater detail below through contemporary developments in evolutionary theory that stress the nongenetic basis of order in the biological realm.

Further problems with Bookchin's thesis have been identified by those like Simon, who focuses on the inherent difficulty in defining what "variety and complexity mean."[27] He cites the work of Levins and Lewontin (1985) in providing what he sees as examples from nature that work against the direction (end) of increasing complexity and variety. Simon asks rhetorically: "Forest fires, which seem to decrease complexity and variety, naturally occur. Are these teleological developments? Are we to cooperate in this development?"[28] Another critic, Eckersley, argues that Bookchin "projects" his own set of social values into an idiosyncratic reading of the direction of evolution and that this position amounts to support for the direct intervention by humans in evolutionary processes. She suggests that Bookchin's naturalistic ethic founders because "his case rests on intuitive reasoning and ingenious rhetorical questions rather than testable hypotheses."[29] She is also critical of Bookchin's organismic philosophy when it claims that there can be objective knowledge of the movement from potentiality to actuality in evolution.

Bookchin sees all development in individual organisms, ecosystems, and society as exemplifying the same type of underlying dynamic of immanent or self-generated movement from potentiality to actuality. Eckersley counterargues that in doing this Bookchin is in part "collapsing ontogenetic development . . . into phylogenetic evolution."[30] The debate between Eckersley and Bookchin continues in this collection of essays and Bookchin has produced a lengthy reply to Eckersley's critique[31] that focuses on the issue of human control of evolution. I do not intend to go over this ground again.

On the basis of these numerous criticisms, it is evident that Bookchin does not provide adequate detail about the foundations of, and crucial terms in, his directionality thesis. Despite the references to scientific literature, Bookchin's position ultimately rests on the conviction that directionality is in some way self-evident. He suggests that the "*directive development of nature toward complexity* . . . requires no greater intellectual justification than the fact of Being itself."[32] The critics are correct in seeing such a justification as inadequate. A more detailed and current explanation of evolutionary theory and its key terms is required to defend and expand the basic thesis. The status of these terms within contemporary evolutionary theory and the relationship between dynamic events in nature like fire and flood, diversity and complexity, and human agency shall also be examined below.

Directionality in Nature

There is continuing general support from within science for the view that there is directionality in nature toward the increasingly diverse and complex. In any contemporary text on biological and evolutionary science it is possible to view the evidence based on the fossil record of increasing diversity and complexity of species and ecosystems over time.

On the issue of complexity, the fossil record shows that there has been a sequence of developments in life from the simple to the complex. The Monera, or single-celled organisms, have evolved into other, more complex kingdoms such as the Fungi, Plantae, and Animalia over great time. From the Triassic through to the Pleistocene epochs, life has

manifested a tendency to produce more complex organisms. The genetic and anatomical structure of mammals represents the most complex form of life that has been known; humans, while they share 98%
of their genes with their closest relatives, the great apes, are at the leading edge of the evolution of the brain into new dimensions of complexity.

On the question of diversity, according to our best information at
the present time, there are between 5 and 100 million species existing
on the planet. Over time, despite periods of mass extinction, there has
been an increase in the rate of speciation as compared to the rate of extinction; we can therefore conclude that biodiversity has increased over
evolutionary history. According to Wilson, "Today the diversity of life
is greater than it was one hundred million years ago—and far greater
than 500 million years before that."[33] Increasing complexity and diversity in nature is not simply a philosophical assertion; it is supported by
the best scientific evidence available.

Over the last two decades the issue of how the complexity and diversity of life has arisen has led to some revolutionary new theories in
evolution and related fields of inquiry. The self-organization of complex
order in the world has been a constant theme in this literature. It is related to increasing speculation from within science about the existence
of deep and fundamental principles within complex systems creating an
emergent order that acts on evolution and ecosystems long before natural selection and genetics have an influence.[34] The field of complexity
theory is relatively new and is a continuing source of fertile new directions for research and its applications. The application of the field to
Bookchin's thesis on directionality in nature and the future development of social ecology can only be tentative at this stage. However, in
the remainder of this contribution, I shall argue that there is indeed a
potentially fertile union of complexity theory and social ecology that
begins to counter the major objections to Bookchin's work outlined
above.

Under the influence of pioneering researchers such as Prigogine and Stengers (1984),[35] there has been an attempt within contemporary science to
apply new ideas about how open systems, be they biological or nonbio-
logical, are potentially capable of self-organization in the face of forces
that might cause them to degenerate. Prigogine has used the term "dissipative structure" to describe how it is that living and nonliving systems
can postpone the effects of the second law of thermodynamics which describes the inevitable loss of directly accessible energy that occurs in the
universe (entropy). Dissipative structures take in energy from their environment and create new forms of order. As Davies explains:

This is the key to the remarkable self-organizing abilities of far-from-equilibrium systems. Organized activity in a closed system inevitably decays in accordance with the second law of thermodynamics. But a dissipative structure evades the degenerative effects of the second law by exporting entropy into its environment. In this way, although the total entropy of the universe continually rises, the dissipative structure maintains its coherence and order, and may even increase it.[36]

The relevance of nonequilibrium thermodynamics (NET) to ecology, evolution, and ultimately ethics has provided a fertile meeting point for science and philosophy, where we now have a universally applicable theory that attempts to explain "the generative capabilities in nature."[37] Such an investigation has provided new insight into the evolutionary process and may even answer some of Simon's complaints about the lack of detail in Bookchin's use of terms like *complexity, variety,* and *potentiality.* as well as addressing some of Eckersley's difficulties with Bookchin's immanent or internal teleology.

Those who have applied the new insights generated by nonequilibrium thermodynamics to evolutionary theory have, in an admittedly controversial area, done a great deal to assist the defense of a naturalistic ethic along Bookchinist lines. The principles of NET are themselves evolving within complexity theory; however, there is an emerging perspective that dissipative structures develop in an irreversible way through self-organization to states of increased complexity. As Depew and Weber argue:

> It is an essential property, as we have seen, of dissipative structures, when proper kinetic pathways are available, to self-organise and, when initial boundary conditions are specified, to evolve toward greater complexity. . . . Thus if we grant that biological systems are constrained by the same physical laws that made their emergence possible, we can expect that such systems—organisms, populations, species, clades, ecosystems, and the biosphere as a whole—will evolve, and evolve toward greater complexity.[38]

In contrast to a Darwinian evolutionary perspective, which is based on Newtonian closed systems, evolution can be understood in terms of dissipative structures. This means that biological systems are engaged in an endogenous (modern science) or immanent (Hegelian nature philosophy) movement toward complexity and that this feature of life leads, unlike in Darwinian theory, to the idea that the direction of evolution is irreversible. That is, an evolving system cannot, within its own dynamic, return to an earlier and simpler state. As argued by Depew and Weber, it is a feature of systems explicated in NET terms to be irreversible not just in fact, but in principle, because, lacking inertial states to which they would tend to return when forces are removed, the entities in the system are defined historically—in terms of the entire sequence of their interactions over a series of irreversible changes.[39]

Once biological systems are conceptualized as dissipative structures, the self-generated movement from simplicity to complexity in living systems provides a measure of support for Bookchin's philosophical and evolutionary arguments for directionality in nature. The order that arises out of increasing complexity in biological systems is not then something that can be treated in atomistic and reductionistic terms. The creation of order-in-complexity, when understood in terms of eco-evolution and NET, is the outcome of the interaction of energy systems over very long periods of time. Although periods of stability may occur in geological time, changes that create new levels of complexity will be the result of fluctuations or perturbations in that system. The stability we have now in biogeochemical systems is itself the product of irreversible movement from simplicity to complexity over deep time. The search for the fundamental principles of order in the creation of life has seen the suggestion that "attractors," or states toward which a dynamic system eventually settles, play an important role in the maintenance and creation of diversity. In complexity theory there is a recurrent theme that very complex and seemingly chaotic systems can give rise to regularity and order. However, the order that is achieved is not fixed, but rather delicately poised between order and instability. As Goodwin summarizes:

> For complex non-linear dynamic systems with rich networks of interacting elements, there is an attractor that lies between a region of chaotic behaviour and one that is "frozen" in the ordered regime, with little spontaneous activity. Then any such system, be it a developing organism, a brain, an insect colony, or an ecosystem, will tend to settle dynamically at the edge of chaos. If it moves into the chaotic regime it will come out again of its own accord; and if it strays too far into the ordered regime it will tend to "melt" back into dynamic fluidity where there is rich but labile order, one that is inherently unstable and open to change.[40]

The application of complexity theory to all types of complex, dynamic systems has produced renewed interest in a project that unifies all natural phenomena under common natural laws. Biological and physical entities are subject to the same natural forces, and different levels of biological organization are subject to common principles of organization. Such a perspective begins to explain how it is possible for organic form to emerge in the first place. The process of morphogenesis, as Goodwin points out, describes how development occurs from an immature organism (embryo) to a mature adult; however, Goodwin further argues:

> During morphogenesis, emergent order is generated by distinctive types of dynamic process in which genes play a significant but limited role. Morphogenesis is the source of emergent evolutionary properties, and it is the absence of a theory of organisms that includes this basic generative process that has resulted in both the disappearance of organisms from Darwinism and the failure to account for the origin of the emergent characteristics that identify species.[41]

The morphogenesis hypothesis can be used to explain the similarities that exist between species like the Tasmanian wolf, a doglike carnivorous marsupial with stripes on its back, and placental mammals such as the wolf. Unrelated species, evolving in different epochs on different continents have very similar shapes, anatomical features, and behaviors.[42]

In the light of the argument that there are "generic forms that characterize organismic morphology,"[43] it becomes more difficult to maintain the traditional definitions of ontogeny and phylogeny. As Hull summarizes with respect to the biological realm, "Today the clear distinction between ontogeny and phylogeny is once again being brought into question."[44]

The Darwinian tradition that emphasizes the contingency of evolution is also under revision. Indeed, Goodwin argues, in contrast to Gould, that if we run the tape of evolution over again, it is possible that we would get a biological world very similar to the one we have now.

It is clear, then, that within contemporary biology there are theoretical movements that are consistent with Bookchin's Hegelian account of the emergence of organic order. Given the embryonic stage of Kauffman's attempt to find the fundamental principles of self-organization and Goodwin's work on the types of attractors that define morphological possibilities, Eckersley's concerns are not without foundation. However, it is clear that critics of the directionality thesis must examine this new area in contemporary theory before any comprehensive dismissal of Bookchin's position is attempted.

Bookchin's claim that human consciousness (subjectivity) is a vital part of the evolution of complexity and diversity is consistent with the idea within complexity theory that consciousness is "the climax of one kind of progress (in complex adaptive systems), that of information processing."[45] Similarly, the application of complexity theory and NET

to events in nature like bushfires that seem, in line with Simon's interpretation, completely random and productive of decreased complexity
in ecosystems, produces a completely different perspective. As Johnson
maintains, ecosystems display "natural cyclic processes in the vicinity
of the climax"; he shows that fire plays an important role in perpetuating such cycles. He summarizes:

> Long-term cyclic effects result from intermittent fires in the Northern
> Woodlands of the United States. A mosaic of cells, each at a different
> stage within the cycle, is thus formed allowing the whole to remain in a
> dynamic steady state. The formation of cells is, in itself, a damping
> mechanism since complete synchronisation of oscillations would lead to
> violent and possibly lethal fluctuations.[46]

Thus fire does not necessarily lead to a decrease in the complexity in
nature; it may in fact be integral to the maintenance of the biodiversity
present in different stages or successions in an ecological system. Modern biologists are now focusing on the nonequilibrium determinants of
biological communities as a factor in explaining species diversity. As
Reice argues:

> In some systems the return frequency of disturbance is so long that the
> impression of equilibrium conditions develops. This is what underlies the
> traditional idea of climax communities. However, careful observation re-
> veals that disturbance is ubiquitous and frequent relative to the life spans
> of the dominant taxa. Thus communities are always recovering from the
> last disturbance. It is the process of that recovery that produces the high
> diversity we find in disturbed systems.[47]

Arguments of a similar nature could be made about other events in
nature like floods and drought that seem to decrease complexity and
diversity. When humans try to control or manage water so as to remove
the uncertainty of flooding, they attempt to create conditions of quasi-
equilibrium in which the stability of the system seems assured over long
periods of time. However, if conditions arise in which change in the
system can take place, such change is likely to totally alter the system.
Hence, flood mitigation projects may lead to worse floods than would
have occurred if such attempts to engineer a solution were not implemented. In NET terms, closed systems are in jeopardy by virtue of the
large changes that they are likely to suffer. Although it seems counterintuitive, open systems in a far from equilibrium situation are likely to
experience more controlled and smaller changes through spontaneous
self-organization within the system.

Conclusion: New Directions for Social Ecology

Bookchin, until 1990, did not fully endorse a systems approach to the understanding of eco-evolution as support for his naturalistically derived ethic. He comments with regard to Prigoginian systems theory that

> I feel obliged to note that a system of positive feedback allows for no concept of potentiality. We know only from Prigoginian "fluctuations" that when a system approaches a "far from equilibrium" situation . . . there is no way to determine whether the system will simply fall apart into "chaos" or assume an immanently predictable form. . . . Prigogine's emphasis on the irreversibility of time, appropriate as it may be in exorcising a mechanistic dynamics based on time's irreversibility, is not congruent with process and evolution.[48]

Such fluctuations could produce the postmodernist version of chaos theory, where nothing is stable and all can be deconstructed into a Nietzschean nihilism with no objective points of reference.[49] Clearly, such a prospect is anathema to the totality of Bookchin's work. He is also critical of systems theory for its "mechanistic mentality," which leads to the reduction of organismic thinking to mathematical abstraction. However, since he concedes that "randomness is subject to a directive ordering principle,"[50] Bookchin does seem to be prepared for the possibility that there can be greater understanding of such ordering principles. This is precisely what complexity theorists have been attempting to do; their achievements so far have brought into question some of the fundamental assumptions of Darwinian evolution and reductionistic genetic and cellular biology.

The picture that emerges from these new developments in NET and complexity theory is one in which the mechanistic, Newtonian certainty of a closed-system model of the world is no longer tenable. The uncertainty of far-from-equilibrium situations that Bookchin identifies is a part of the new picture; however, their irreversibility, if "appropriate boundary conditions are maintained,"[51] is particularly relevant for biological systems understood in terms of NET. The periods of apparent stability with "near equilibrium" systems are predicated upon the operation of NET in eco-evolution over very long periods of time. Furthermore, the order present in life on earth at all levels of complexity is now being explicated by complexity theorists in terms of fundamental ordering principles that cut across all levels of complexity. At the very least, within this new domain we have a number of very important hypotheses that can be explored in any context that is explicable in terms of the flow of energy and the spontaneous production of order. Such a conclusion may not be as strong as that desired by Bookchin, but it does explicitly connect a naturalistic ethic to an emerging understanding of the nature of reality. If humans impose change on dynamic systems in such a way as to create wild fluctuations, then clearly such fluctuations could bring about major ecological disturbances and consequent catastrophes for life on earth. Human action, when mediated by an understanding of complexity theory and NET, will work within design parameters that ensure that our impacts on ecosystems are within the tolerances of dissipative structures. We can continue to export our socially produced entropy into environmental "sinks" only if such sinks are capable of absorbing such "waste" and can continue to do so indefinitely. The ability of sinks to do this work may, in turn, depend on the ongoing level of biodiversity in, for example, wetlands, oceans, and forests, since we know that the current relationship between plants, organisms, and the physical components of the planet is one that is capable of maintaining long-term self-regulatory ecosystem relationships.

As humans denude biodiversity on the planet and move to create vast monocultures that are susceptible to dangerous change, we risk destroying complexity and diversity that it has taken the whole of evolutionary history to achieve. We can choose to support the ongoing evolution of self-organized systems or work to impose structures that destroy complexity and diversity.

Humans are no longer constrained by evolution to always work within limits imposed by nature. That we can prevent the free unfolding of evolutionary potential toward increasing complexity and diversity is manifestly obvious; however, in doing so we sever our connection to past and future *natural* evolution of the planet. As argued by Slocombe, to be truly sustainable, human society can no longer attempt to impose an artificial equilibrium on social or ecological systems. On the contrary, he maintains that "success at adapting, at becoming part of the evolving system and capitalising on its particular structure, organisation, and dynamics is a more appropriate strategy."[52] Bookchin agrees with such a strategy in that he sees human involvement in such systems, as Charles Elton had observed, as "more like steering a boat" in a complex stream than applying rigid rules to a fixed system. Bookchin suggests that such sensitive management of human activity requires detailed knowledge and that "what ecology, both natural and social, can hope to teach us is the way to find the current and understand the direction of the stream."[53] As Bookchin might be willing to agree, understanding complexity theory and NET and their relationship to biodiversity provides a framework to "help us distinguish which of our actions serve the thrust of evolution and which of them impede them."[54]

Understanding the direction of evolution does not mean that humans must return to a hunting-and-gathering lifestyle or commit two-thirds of the planet to "wilderness." As products of evolution and with an understanding of its direction, humans can, as Bookchin puts it, evolve socially in a way that complements this direction. An ethics of complementarity could see humans open up new and innovative ways of linking their needs to a foundation of social life based on renewable energy pathways and technology that is organically connected to it. Such innovation includes a complete change in the design principles that inform all of our technologies; it also provides social ecologists with a framework to critically and objectively evaluate the homogeneity and monocultures that are currently the outcome of the globalization of capitalist culture.[55]

A challenge for social ecologists now is to apply the insights of NET and complexity theory to the remaking of contemporary society. It is clear that our current system of production and consumption is unsustainable since it relies on nonrenewable energy and creates social structures such as cities and their urban complexes that do not operate as viable, long-term dissipative structures. The vision from within social ecology for an energy and material base for social life that is consistent with self-organized directionality has received considerable support from those who have applied the new science of complexity to economic and social systems. Dyke, for example, has encouraged social theorists to apply "entropy bookkeeping" rather than "economic bookkeeping" to understand how a city might be conceptualized as a dissipative structure.[56]

Such a perspective counters the attempt by planners and engineers to force all parts of the social system in diverse parts of the planet to conform to universal standards. The origin of such mentalities lies in institutional and political forces that use their economic power to impose fixed standards on what are locally defined dynamic systems. The dangers of such high-risk engineered design in physical systems such as rivers (e.g., the Mississippi) is now well understood and provides theorists such as Bookchin who have long championed the cause of social and cultural diversity a new foundation on which to make and defend such claims. As an integral part of the critique of globalization, its entropy-maximizing tendencies, and the homogenization of social systems, social ecologists can help to draw the biophysical limits of human settlements and institutions, at specific places on earth, in terms of entropy-minimizing dissipative structures and dynamic, self-organizing systems.[57]

Social ecologists and ecoanarchists have long argued that an ecological understanding of life provides foundations for and guidance about how human communities and their institutions can be made to harmonize with natural systems. Ecosystems have been highlighted as examples of self-organizing systems that exhibit high degrees of order without external design or hierarchical power structures. As such, the understanding of ecosystems has provided considerable support to the idea that human social, economic, and political systems could be organized in a similar way. Complexity theorists are now identifying more fundamental sources of order that underpin the biological order found within ecosystems. The combined insights of Bookchin's dialectical naturalism, eco-evolutionary science, and complexity theory provide an expanded foundation upon which to build a more theoretically sophisticated social ecology.

The application of this expanded foundation to the future direction of our cultural evolution is a major challenge. Human ethics, aesthetics, and values that apply the principle of maintaining or increasing the complexity and diversity of social systems while at the same time maintaining or increasing the complexity and diversity of natural systems will drive the creation of whole new ways of doing economics, politics, technology, education, art, and architecture. The advocacy by social ecologists of ecocommunities run by participatory democracies structured by renewable energy and ecologically sustainable technologies can be defended as the most appropriate response humans can make to the new understanding we have of complex, dynamic systems and the way they can self-organize to increased states of complexity and diversity.

We directly experience the tragic loss of potential to mature toward fuller complexity and richer diversity when confronted with the unexpected death of a child. The same sense of tragedy is confronting us with the premature destruction of complexity and diversity of the planet. A sense of the greater directionality of nature and an ethics of complementarity just might help in the building of self-generated human communities that are organically connected to the local, regional, and continental physical systems that provide the energy, resources, and order they require for continuity.

Acknowledgments

I wish to thank Andrew Light and the anonymous referees who have helped in the evolution of this essay from one that lacked an overall direction to something that begins to systematically address the main issues in what is an emerging field of inquiry.

Notes

1. Charles Darwin, *On the Origin of Species,* Facsimile of the first edition (Cambridge MA: Harvard University Press), p. 489.

2. In his copy of the *Vestiges of the Natural History of Creation,* Darwin wrote a note to himself saying "never use the words 'higher' and 'lower' " (in F. Darwin and A. C. Seward, *More Letters of Charles Darwin* [London: J. Murray, 1903], Vol. 1, p. 114, n.).

3. See H. Spencer, *The Man versus the State* (Harmondsworth, UK: Penguin, 1969), pp. 139–141, where he discusses how the material published in his Social Statics (1851) on "survival of the fittest" is supported by Darwin's work on natural selection.

4. Murray Bookchin, *The Ecology of Freedom* (Palo Alto, CA: Cheshire Books, 1982), p. 365.

5. Ibid., pp. 362–363.

6. Peter Kropotkin, *Mutual Aid: A Factor of Evolution* (London: Freedom Press, 1914). It is interesting to note that Stephen Jay Gould, in his essay "Kropotkin Was No Crackpot," supports the concept of mutual aid as a valuable counterbalance to the Malthusian–Darwinian view of competitive struggle in nature; see Stephen J. Gould, *Bully for Brontosaurus: Reflections in Natural History* (London: Radius, 1997), pp. 338–339.

7. Jan Sapp, *Evolution by Association: A History of Symbiosis* (New York: Oxford University Press, 1994).

8. Bookchin, *Ecology of Freedom,* p. 364.

9. Ibid., p. 365.

10. See his article "Recovering Evolution: A Reply to Eckersley and Fox," *Environmental Ethics,* Vol. 12, 1990, pp. 255–256.

11. Bookchin, *Ecology of Freedom,* p. 365.

12. Ibid., p. 365; emphasis in original.

13. The communitarian theories of the Russian anarchist Kropotkin are seen by Bookchin as a foundation for his own theory of social ecology.

14. Murray Bookchin, *Urbanization without Cities: The Rise and Decline of Citizenship* (Montreal: Black Rose Books, 1992).

15. The naturalistic fallacy can be traced from Hume's warning that we cannot derive an "ought" from an "is" to G. E. Moore's concerns, in *Principia Ethica* (Cambridge: Cambridge University Press, 1971 [1903]), about identifying "the natural" with "the good."

16. Paul Taylor, *Respect for Nature* (Princeton, NJ: Princeton University Press, 1986).

17. Robyn Eckersley, "Divining Evolution: The Ecological Ethics of Murray Bookchin," *Environmental Ethics,* Vol. 11, 1989, pp. 99–116 (reprinted as

the first part of Chapter 2, this volume). Eckersley continues the debate with Bookchin on this and other issues in "Respecting Evolution: A Rejoinder to Bookchin," the second part of Chapter 2, this volume.

18. John Rodman, "The Liberation of Nature," *Inquiry,* Vol. 20, 1977, pp. 83–131.

19. Alasdair Macintyre, *After Virtue,* 2d ed. (South Bend, IN: University of Notre Dame Press, 1981), p. 68.

20. See, e.g., Arne Naess, *Ecology, Community, and Lifestyle,* trans. and ed. D. Rothenberg (Cambridge: Cambridge University Press, 1989), and Bill Devall and George Sessions, *Deep Ecology: Living as if Nature Mattered* (Layton, UT: Peregrine Smith, 1985).

21. Murray Bookchin, "Social Ecology versus Deep Ecology," *Socialist Review,* Vol. 18, 1988, p. 27.

22. Murray Bookchin, *The Philosophy of Social Ecology* (Montreal: Black Rose Books, 1990), p. 176.

23. See G. A. Albrecht, "Social Ecology and Ecological Ethics," in *Changing Directions* (Adelaide, Australia: University of Adelaide, 1990), for a critique of the relativist tendencies in the ethical position arising out of deep ecology.

24. Stephen J. Gould, *Bully for Brontosaurus: Reflections in Natural History* (London: Radius, 1991), p. 13.

25. N. Eldredge and S. J. Gould, "Punctuated Equilibria: An Alternative to Phyletic Gradualism," in T. J. M. Schopf, ed., *Models in Palaeobiology* (San Francisco: Freeman Cooper, 1972).

26. Roger Lewin, *Complexity* (London: Phoenix, 1993), p. 72.

27. Thomas Simon, "Varieties of Ecological Dialectics," *Environmental Ethics,* Vol. 12, 1990, p. 223.

28. Ibid., p. 223.

29. Eckersley, "Divining Evolution," p. 108 (Chapter 2, p. 65, this volume).

30. Ibid., pp. 106–107.

31. Bookchin, "Recovering Evolution: A Reply," pp. 253–274.

32. Bookchin, *Philosophy of Social Ecology,* p. 84; emphasis in original.

33. E. O. Wilson, *The Diversity of Life* (London: Penguin Books, 1992), p. 328.

34. See Stuart Kauffman, *The Origins of Order: Self-Organization and Selection in Evolution* (New York: Oxford University Press, 1993), for a detailed account of these and other related claims.

35. Ilya Prigogine and Isabelle Stengers, *Order Out of Chaos* (New York: Bantam, 1984).

36. Paul Davies, *The Cosmic Blueprint* (London: Unwin, 1989), p. 85.

37. Ibid., p. 85.

38. David Depew and Bruce Weber, "Consequences of Nonequilibrium Thermodynamics for the Darwinian Tradition," in B. Weber, D. Depew, and J. Smith, eds., *Entropy, Information, and Evolution: New Perspectives on Physical and Biological Evolution* (Cambridge, MA: MIT Press, 1988), pp. 337–338.

39. Ibid., p. 333.

40. Brian Goodwin, *How the Leopard Changed Its Spots: The Evolution of Complexity* (London: Phoenix Giants, 1995), p. 169.

41. Goodwin, *How the Leopard Changed Its Spots,* p. xiii.

42. See Roger Lewin, *Complexity: Life on the Edge of Chaos* (London, Phoenix, 1993), pp. 72–74. The Tasmanian wolf is now extinct, having been exterminated by bounty hunters earlier this century, and the wolf is in danger of suffering a similar fate.

43. Goodwin, *How the Leopard Changed Its Spots,* p. 173.

44. David Hull, "Introduction," in Weber, Depew, and Smith, eds., *Entropy, Information, and Evolution,* p. 5.

45. Lewin, *Complexity,* p. 170.

46. Johnson, "Thermodynamic Origin of Ecosystems," p. 92.

47. Seth Reice, "Nonequilibrium Determinants of Biological Community Structure," *American Scientist,* Vol. 82, September–October, 1994, pp. 424–435.

48. Bookchin, *Philosophy of Social Ecology,* p. 192.

49. See N. Katherine Hayles, ed., *Chaos and Order: Complex Dynamics in Literature and Science* (Chicago: University of Chicago Press, 1991), for an account of the parallels between chaos theory and postmodern theories.

50. Bookchin, *Ecology of Freedom,* p. 84.

51. Depew and Weber, "Consequences of Nonequilibrium Thermodynamics," p. 333.

52. Scott Slocombe, "Assessing Transformation and Sustainability in the Great Lakes Basin," *GeoJournal,* Vol. 21, No. 3, 1990, p. 269.

53. Bookchin, *Ecology of Freedom,* p. 25.

54. Ibid., p. 342.

55. See Glenn Albrecht, "Social Ecology and Organic Environmental Design," in J. Birkeland, ed., *Rethinking the Built Environment: Proceedings of Catalyst '95* (Canberra, Australia: University of Canberra, 1995).

56. See C. Dyke, "Cities as Dissipative Structures," in Weber, Depew, and Smith, eds., *Entropy, Information, and Evolution,* pp. 355–368.

57. Bookchin, in *Urbanization without Cities,* argues that cities can be "ecocommunities" but that the process of urbanization has effectively separated social and natural evolution. He suggests that the city can be seen as a "uniquely human, ethical, and ecological community that often lived in balance with nature and created institutional forms that sharpened human awareness of their sense of natural place as well of social place"; see Bookchin, "Preface," in *Urbanization without Cities,* p. x. It remains to be seen how much political and economic compatibility lies between Bookchin's ecoanarchist view of the city as human-scaled places of enrichment and the idea of cities as dissipative structures.

Chapter 4

Social Ecology and Reproductive Freedom
A Feminist Perspective

REGINA COCHRANE

Introduction

In "Feminism and Ecology: On the Domination of Nature," Patricia Jagentowicz Mills undertakes a critical evaluation of the ecofeminist theory of Ynestra King. Acknowledging that ecofeminism addresses valid concerns that do not receive sufficient attention in any other approach to feminism, Mills refers to King's work as "groundbreaking." However, besides finding that work inadequate on the theoretical level, Mills takes King to task for what she terms an "abstract pronature stance" that calls for a reconciliation between humanity and nature but that does not take into account an issue of fundamental importance to feminists: abortion.[1]

Given that the ecofeminist critique of contemporary Western society centers on the connection between the domination of nature and the domination of women, Mills feels that the general silence of ecofeminists on the issue of abortion and their failure to outline a politics of reproduction represents a serious omission indeed. Mills actually goes as far as claiming that this gap in ecofeminist theory "paves the way for the erosion of women's reproductive freedom." She is concerned that the growing popularity of ecofeminism within the mainstream women's movement may have a depoliticizing effect on feminism, reducing it to "merely a handmaid of the ecology movement."[2]

Since Mills's article was written, in 1991, there have been some limited attempts by ecofeminists to elaborate a politics of reproductive freedom. For example, in a chapter entitled "Abortion Rights and Animal Rights," which is included in her 1994 text *Neither Man nor Beast,* ecofeminist Carol Adams develops a defense of abortion that is informed by cultural feminist ethics as well as by a liberal extentionist animal rights theory. Yet, as even Adams herself acknowledges, this is a narrow approach that represents merely a first step in the formulation of a more extensive theory of reproductive freedom.[3] I have argued elsewhere, moreover, that, with its celebration of motherhood and its tendency to dualistically align women with a feminized *nature* in opposition to a male *culture,* the countercultural standpoint of cultural (as distinct from radical) feminism that currently enjoys hegemony in the women's and ecofeminist movements constitutes a questionable position in which to root ecofeminism,[4] let alone an ecofeminist politics of abortion. Irene Diamond's *Fertile Ground,* also published in 1994, is a much more comprehensive effort to outline what she refers to as a "politics of fertility."[5] Diamond's work, however, is informed by the poststructuralism of Michel Foucault—an anti-Enlightenment and antidialectical stance that I have argued is also ultimately problematic for ecofeminists.[6]

According to Mills, part of the reason why abortion is an issue in contemporary society is because of the nature/culture dualism at the root of Western thought. In order to move beyond this dualism, she stresses, what is needed is a dialectical understanding of feminist issues such as abortion:

> Without a more careful analysis of the dialectical relation between nature and history [or culture], the "natural" event of a pregnancy emerges as an end in itself, rather than part of the human historical enterprise. And this creates the possibility for interpreting abortion as a form of the domination of nature.[7]

Cognizant of the nature/culture dualism underlying both cultural and socialist feminisms, King calls for the formulation of a new dialectical feminism. In so doing, she looks to the social ecology of ecoanarchist theorist Murray Bookchin for political inspiration.[8] Bookchin's social ecology represents an attempt to transcend the dualistic separation of culture from nature typical of modern capitalist society, Marxism, and liberal environmentalism. It seeks as well to offer an alternative to the reductionism inherent in a "deep ecology" that espouses a "biocentrism" that ultimately collapses human society into nature.[9] "Social ecology," claims Bookchin, "show[s] how nature slowly *phases* into society without ignoring the differences between society and nature, on the one hand, as well as the extent to which they merge with each other on the other."[10] In order to distinguish this approach from the dialectical idealism of Hegel and the dialectical materialism of Marx, Bookchin calls his philosophy "dialectical naturalism."[11]

King's failure to clarify the status of abortion in relation to her critique of the domination of nature clearly indicates the limits of her attempt to develop a dialectical feminist theory. Given the dualism inherent in most contemporary feminist thought, the formulation of such a theory is, however, a project that holds considerable promise. This essay contributes to the development of such a dialectical feminism by attempting to elaborate, and then subject to a combined radical and socialist feminist critique, a politics of abortion and, more generally, reproductive freedom that is rooted in the dialectical naturalism of social ecology.

For a number of reasons, the ecoanarchist perspective of Bookchin represents an interesting starting point from which to investigate a dialectical approach to abortion and reproductive freedom. Anarchists have long favored sexual freedom—one of the earliest advocates of birth control and sexual liberation in North America, in fact, was the famous anarcho-feminist Emma Goldman. In addition, as Alison Jaggar and M. Hawkesworth point out, radical feminism and anarchism overlap to a certain extent with respect to the type of political strategies they choose to employ.[12] Moreover, Bookchin's differentiation of "freedom" from "justice" is helpful in clarifying the distinction radical and socialist feminists make between "reproductive freedom" and "abortion rights." Finally, as Jaggar states: "Because it cannot be achieved within the existing order, reproductive freedom is in fact a revolutionary demand."[13] Such an investigation, therefore, also serves as a means of evaluating the potential of social ecology to "remake" society in such a way that women as well as men are able to achieve real freedom.

This essay will be organized in the following manner. First, I will examine the position of social ecology on abortion. Since Bookchin himself says nothing on the issue, this position will be developed by expanding upon certain of his ideas, as well as by drawing upon the writings of social ecologist Janet Biehl who has broached the subject. Next, I will discuss the theme of reproductive freedom from a feminist perspective. Then I will subject the position of social ecology on reproductive freedom, elaborated in the second part of this chapter, to a feminist critique. I will conclude by offering a brief discussion of the implications of this feminist critique of social ecology for the project of developing a dialectical feminism.

Social Ecology and Abortion

In her book *Finding Our Way*, which is largely a critique of cultural ecofeminism, Janet Biehl raises (but does not develop to any significant extent) the subject of reproductive politics. Asserting that the "logic" of ecofeminism is "to deny women their reproductive freedom," Biehl advocates instead the approach of social ecology[14]:

> The ethics of social ecology has particularly profound implications for women. As human beings, women's lives are no more determined by biology than are men's. Unlike other female animals, the human female is capable of making decisive choices about when and under what circumstances she will reproduce. The distinction between the facile ethical proposals of ecofeminists and social ecology's ethics should be especially clear on the issue of abortion and reproductive freedom generally. An ethical prescription superficially drawn from first [external] nature that argues that all life is sacred would oblige us to oppose abortion on the grounds that it is destructive of "life" if its adherents are to be consistent. By contrast, in social ecology's ethics, in which first nature is a realm of increasing subjectivity out of which society [second nature] emerges, women would have a right to reproductive freedom that is grounded in the emergence of society and natural evolution. As human beings uniquely capable of making ethical choices that increase their freedom in the context of an ecological whole, women's reproductive freedom would be a given.[15]

The ethics of social ecology, in Biehl's view, grounds women's right to "choice" on the matter of abortion in the dialectical distinction of "first" or biological nature from "second" or social nature. As a biological but not yet a social entity, the fetus belongs to the realm of first nature. As both a social and a biological being, the pregnant woman is part of second nature—society—as well as first nature—biology. While the woman herself is an actuality, a person, the fetus she is carrying is, to use Bookchin's terminology, a "mere potentiality" that "has not 'come to itself.' "[16]

In elaborating his philosophy of dialectical naturalism, Bookchin notes "that what may be 'brought forth' is not necessarily 'developed.' " Using the example of an acorn, he points out that it "may become food for a squirrel or wither on a concrete sidewalk, rather than 'develop' into what it is potentially constituted to become—notably, an oak tree."[17] By comparing this example and the situation of a pregnant woman, it is possible to draw an analogy between the case of an acorn withering on the sidewalk and a (spontaneous) miscarriage. It is also possible to draw another analogy—but one that is ultimately even less satisfying—between the squirrel's biological need for food and the pregnant woman's need for both health on the biological level and happiness and fulfillment on the social level.

The squirrel's need for food can cause it to "abort" the potential of an acorn to develop into an oak tree by consuming it. By analogy, it is possible to argue that the pregnant woman's need for health, happiness, and fulfillment—all of which are interconnected—can cause her to abort the potential of the fetus she is carrying to develop into a human being by terminating her pregnancy.[18] Indeed, according to this line of argument, a woman who reproduces against her will is functioning as a biological rather than as a social being. Giving birth then becomes a truly human act only when it is volitional.

The argument and analogy above can be further strengthened by expanding it to include Bookchin's idea of a "natural tendency" or "nisus" in organic evolution toward "greater complexity and subjectivity."[19] Although the squirrel terminates the potential to develop into oak trees of those acorns it consumes, in the process of collecting acorns to eat, it moves some, which it subsequently loses, away from where they fell near the base of the tree. If these acorns develop into oaks, they will do so in locations where they will not compete with the parent tree for nutrients. Thus, in the process of collecting nuts to eat, a squirrel helps trees to achieve "greater complexity," while trees function in a like manner for it.

By analogy it can be argued that a woman who chooses to abort a fetus, in order to be better able to fulfill her own social and biological potentials, becomes capable of making a greater contribution to society. She can do this by having the time and energy to devote to developing her own unique talents which she can then apply to various social projects. She can also do this by raising the children she already has, or may have in the future, under conditions that are more amenable to the achievement of their individual potentials as well. Thus, while a woman who chooses to terminate a pregnancy is aborting the potentiality of a fetus, a woman who reproduces against her will may be aborting her own potentiality—the further potentiality of an existing actuality.

What distinguishes human beings, as part of second nature, from beings that exist in the realm of first nature only is, for Bookchin, their self-consciousness. The appearance of human life on earth marked a crucial change in the direction of evolution, he argues, a change from largely adaptive life forms to a life form that is at least potentially moral and creative.[20] Thus humans are able to "choose, alter, and reconstruct their [natural as well as social] environment and raise the moral issue of what *ought* to be, not merely live unquestioningly with the what *is*."[21] Applying this claim to the case of abortion, it is possible to argue, following the same line of thought as Bookchin, that the willful termination of a pregnancy as an act of "human intervention into the natural world is not a sick aberration of evolution." Rather, it is an example of "the extent to which humanity actualizes a deep-seated nisus in evolution toward self-consciousness and freedom."[22]

One last principle from Bookchin's elaboration of social ecology that is relevant to the practice of abortion is his notion of the "self-governing" individual. "The revolutionary project must take its point of departure from a fundamental libertarian precept," asserts Bookchin, "[that] every normal human being is competent to manage the affairs of the community in which he or she is a member."[23] As competent individuals, women should be able not only to participate in the management of community affairs, but also to manage their own affairs. If they choose to terminate their pregnancies, when (i.e., at what stage), where, why, and how they choose to do so should be left up to them and to them alone.

Feminism and Reproductive Freedom

Anthropology and history provide evidence, insists Rosalind Petchesky, that the majority of human societies have in some way tried to regulate their fertility.[24] "Conscious activity to control human fertility is as intrinsic to the social being of human groups," she asserts, "as the activity to control and organize the production of food."[25] Due to the fact that it is effective, technically simple, and, above all, does not require male cooperation, abortion has, she claims, been the most prevalent and persistent of all methods of fertility control.[26]

Petchesky considers the common assumption that birth control and abortion were invented by modern industrial societies to be an "intellectual error of technological determinism." It confuses the activities of birth control and abortion with the techniques used to carry them out.[27] Hence, she insists, the contemporary demand for abortion is the stimulus for, rather than the consequence of, its legalization and availability.[28] Women who choose to induce or undergo an abortion, as conscious agents of their own fertility control, play an active role in the process.[29] In the context of patriarchal culture and an ideology that defines women by their biology and/or the needs of others, exercising "choice" constitutes, therefore, a conscious act of resistance.[30]

However, a woman's right to abortion is, as Caroline Whitbeck points out, at best a negative one: the right to terminate a pregnancy without interference.[31] Petchesky expands upon this notion as follows:

> Abortion in itself does not create reproductive freedom. It only makes the
> burdensome and fatalistic aspects of women's responsibility for pregnancy
> less total. It does not socialize that responsibility, empower a woman in
> her relations with men or society, or assure her of a liberated sexuality.[32]

What is needed, then, to achieve true reproductive freedom is not only
free access to abortion but also sexual freedom and social autonomy.

Reproductive freedom involves being in control of one's own sexuality as well as one's own fertility. For Nikki Colodny, this controls entails "determining what kind of sex we have, and when. Redefining our
sexuality as autonomous women means developing our own criteria for
convenience."[33] Petchesky suggests that what is called for is an "alternative culture of sexuality embracing passion and play as well as love"
and a "new morality integrat[ing] a broad ecumenical acceptance of
multiple forms of pleasure with the principle of respect for another's
body and well-being."[34]

Calls for abortion rights and sexual freedom are, however, most
often couched only in terms of "women's control over their own bodies." Jagger considers this to be a dualistic formulation that tends to
portray women as essentially "vaginas and wombs on legs." She suggests that a more appropriate goal would be "women's control over
their lives."[35] Social autonomy for women, in the context of reproductive politics, necessitates that they have access to those material and social requirements that would allow for the practice of true "choice"
and sexual liberation in the first place.[36] Marlene Fried has succinctly
summarized such basic material and social provisions as follows: "shelter, food, day care, health care, education, and the possibility of meaningful work, relationships, and engagement in social and political
life."[37]

Among the necessary social conditions for a meaningful reproductive freedom, one that is particularly important for many feminists is the need for a radical change in the politics of family life.
The hegemonic view of motherhood in contemporary Western society,
for example, is one of absolute dedication to and total self-sacrifice
for one's (biological) children.[38] Pointing out that the institution of
motherhood as a "women's sphere" is an ideological construction,
Petchesky emphasizes that "having and raising children is a fundamental dimension of *human*—as opposed to gender-specific—fulfillment and social life."[39] Therefore, in order to achieve a meaningful
reproductive freedom, feminists call for changes in the social relations
of procreation. Especially important is the sharing of parenting in a
gender-neutral way.[40]

Given that it would entail radical changes in the organization of all social institutions—especially, the family—a truly comprehensive approach to reproductive freedom, contends Jaggar, would be "incompatible with the maintenance of any traditional version of public/private distinction."[41] This is so because a truly feminist approach to reproductive freedom must be capable not only of transcending the sexual organization of society—particularly the sexual division of labor in public and private realms—but also of recognizing the particular role played by women in reproduction.[42]

Arguing that the personal is political, the women's movement has struggled to bring reproductive issues into the public sphere, emphasizes Fried. Yet contemporary public discourse tends to center on the notion of a "woman's right to choose," while women's needs for sexual freedom and social autonomy—and for the social changes required to make these possible—are largely ignored. Given this reality, she calls for a transcendence of the current abortion rights movement to one focusing instead on reproductive freedom.[43]

Social Ecology and Reproductive Freedom: A Feminist Critique

In the ethics of social ecology "women's reproductive freedom . . . [is] a given," Biehl claims.[44] However, while in certain respects the dialectical naturalism of social ecology provides a strong ground from which to defend a politics of reproductive choice, in other respects it is highly problematic for feminists.

Drawing on Bookchin's understanding of dialectical theory, Biehl explains that the overcoming of a dialectical contradiction entails a process in which some previously existing state must be negated, absorbed, and finally transcended in order that its potentiality might ultimately be fulfilled. In social ecology it is thus the potentiality—the "should be"—that constitutes the only true standard by which existing reality can be judged.[45] Indeed, Bookchin ventures so far as to insist that a process that unfolds as it "should" to its "logical" end "is more properly 'real' than a given 'what-is' that is aborted or distorted and hence [is] . . . 'untrue' to its possibilities."[46]

"That which we prize as most integral to our humanity [is] our extraordinary capacity to think on complex conceptual levels," asserts Bookchin.[47] In light of this statement and the claims made above for the "should be" as the standard of wholeness, what are we to make of the woman who chooses *not* to abort a fetus that she knows will develop into a child who will be severely intellectually disabled? Is such a fetus, which is not as it "should be," to be less valued as a potential human being than one which is "normal"? Is a woman who makes such a decision, one that is most likely strongly influenced by emotion as well as by "think[ing] on complex conceptual levels," herself less than human?

With its central goal of striving to transcend the immanent "what-is" in order to actualize the potential "what-should-be," a reproductive politics rooted in social ecology would seem to be vulnerable to eugenic-like appeals to raise only "quality" children. To make this claim is not to accuse Bookchin and Biehl of advocating eugenics. Bookchin has, in fact, been a strong critic of ideologies that have been associated with eugenic "ideals": social Darwinism, neo-Malthusianism, and sociobiology. It is, however, to point out the potential consequences of an ethics that undervalues the immanent in its quest to achieve the transcendent and, moreover, to emphasize the danger to which social ecology leaves itself open in its failure to be explicit on such issues.

As Petchesky points out, the nineteenth- and early-twentieth-century birth control and feminist movements did incorporate certain elements of eugenic ideology into their thinking. Even Emma Goldman was not immune to this trend, as the following remark illustrates:

> Woman no longer wants to be party to the production of a race of sickly, feeble, decrepit, wretched human beings, who have neither the strength nor the moral courage to throw off the yoke of poverty and slavery. Instead she desires fewer and better children, begotten and reared through love and free choice; not by compulsion as marriage imposes.[48]

Petchesky emphasizes that the result of such ideological calls for "improvement of the race" and "enlightened motherhood" was that fertility control came to be regarded as "women's moral duty" rather than as something that increased women's freedom.[49] Thus it is possible that, while strongly defending a woman's right to abortion, a reproductive ethics grounded in dialectical naturalism might potentially reduce a woman's reproductive freedom by discouraging her from carrying to term, even if she freely chooses to do so, a fetus that is not as it "should be."

There is one last point concerning social ecology's defense of women's right to abortion that should not go unmentioned. In the buildup to her explanation of how dialectical naturalism grounds women's right to reproductive choice, it is ironic that Biehl should choose to cite with approval the following passage from Bookchin:

> A thing or phenomenon in dialectical causality remains unsettled, unstable, in tension—much as a fetus ripening toward birth "strains" to be born because of the way it is constituted—until it develops itself into

what it "should be" in all its wholeness or fullness. It cannot remain in endless tension or "contradiction" with what it is organized to become without becoming warped or undoing itself. It must ripen into the fullness of its being.[50]

While an acorn, given favorable environmental conditions, "develops itself into what it 'should be,' " the situation of a human fetus is considerably different. Whether the fetus develops and how well it develops depend on the conscious decision of the pregnant woman to nourish, neglect, or abort it. Although the fetus that a pregnant woman chooses to nourish organizes its own development, it is ultimately the woman who strains to give it birth. In effacing the pregnant woman and the dependence of the fetus on her, Bookchin (even if inadvertently) ends up portraying the woman as, to quote Petchesky, "the 'maternal environment,' the 'site' of the foetus, a passive spectator in her own pregnancy."[51]

Although a politics of reproductive choice grounded in social ecology's dialectical naturalism is problematic from a feminist perspective for the reasons outlined above, such a politics would at least defend a woman's right to abortion. However, for radical and socialist feminists, reproductive freedom includes not only access to abortion but also sexual freedom and social autonomy for women. Therefore, in order to properly investigate the validity of Biehl's claim that in social ecology women's reproductive freedom is a given, the position of social ecology on sexual freedom and social autonomy for women must also be examined.

Bookchin does (briefly) treat the subject of sexual freedom in his book *Remaking Society*. Emphasizing the importance of the sensual, he writes of the necessity "to reconcil[e] the dualities of mind [and] body" in "civilized" societies and the need for the "emancipation of the body in the form of a new sensuousness."[52] However, what exactly it is that would constitute such a "reconciliation" of the mind–body dualism for social ecology is made much clearer in the following passage from Biehl: "A dialectical naturalist approach overcomes mind–body dualism not by rejecting the distinction between the two but by articulating the continuum through which the human *mind* has evolved."[53]

Feminist conceptions of reproductive freedom also stress the importance of sensuality. However, most feminists usually include among the preconditions of a meaningful sexual freedom two additional requisites: love and (in Petchesky's words) a "new morality" based on "respect for another's body and well-being." Bookchin and Biehl, in contrast, confine their discussions of sexual freedom and sexual liberation to the liberation of bodies. It is only in their deliberations on mind that they consider emotion which they downgrade as particularistic relative to the "universalism" of reason. Given the emphasis both place on overcoming the mind/body dualism, this is rather ironic. Their approach to sexual liberation, in fact, seems to be closer to that advanced by the largely male-dominated sexual liberation movement of recent decades than that of radical and socialist feminists.

What is the position of social ecology on social autonomy for women? Claiming that she is "defend[ing] the best ideals of feminism," Biehl stresses the importance of "appreciati[ng] . . . women's historical role in childbearing and childrearing, while at the same time emancipating women from regressive definitions that place them exclusively in that role."[54] The implications here are twofold: first, that in the future "free" society envisioned by social ecology, women will continue, within the private sphere or domestic realm, to bear the greater part of the responsibility for childrearing; second, that women's emancipation will consist in the opportunity to participate, on an "equal" footing with men, in the activities of the public sphere or political realm.

Biehl's position on the politics of private and public life can be best understood by examining Bookchin's conception of the development of the public realm out of the private. In early human societies, according to Bookchin, sexual differences defined the type of work one performed in the home and in the community. Women assumed responsibility for food gathering and food preparation; men hunted and acted as protectors for the community at large. Hence, while women controlled the domestic world, "men, in turn, dealt with what we might call 'civil affairs'—the administration of the nascent, loosely developed 'political' affairs of the community."[55]

In the early stages of societal development, states Bookchin, sororal and fraternal cultures "complemented each other."[56] However, there eventually came a time when "male 'civil' affairs simply upstaged female 'domestic' affairs without fully supplanting them." This situation he attributes not only to social but also to biological "facts": "Males . . . produce significantly greater quantities of testosterone than females—an androgen that . . . fosters behavioral traits that we associate with a high degree of physical dynamism." Consequently, as it expanded in influence, the "male world" of civil society, contends Bookchin, gradually became more agonistic and assertive due to invasion, intercommunal strife, and then systemic warfare.[57]

For Bookchin, the "degradation of women" was but a by-product of this hierarchicalization of the "male world" of civil society:

> Even woman's world . . . was reshaped, to a lesser or greater degree, in
> order to support him [the "big" man] with young soldiers or able serfs,

clothing to adorn him, concubines to indulge his pleasure, and, with the growth of female aristocracies, heroes and heirs to bear his name into the future. All the servile plaudits to his great stature, that are commonly seen as signs of feminine weakness, emerged, throwing into sharp contrast and prominence a cultural ensemble based on masculine strength.[58]

"Hierarchy will not disappear until we change these roots of daily life radically," asserts Bookchin. To "validate" his contention that these roots lie in the "male" civil sphere, he even cites Biehl's assertion that "male domination over other males generally preceded the domination of women."[59] (Biehl, of course, drew such ideas from her partner Bookchin in the first place.) The implication being made here is that, as the original site of the development of institutional hierarchy, the civic sphere must also be the primary site for political change.

This "male" public sphere was purportedly not only the site of the emergence of institutionalized hierarchy as such; at the same time it was the ground, claims Bookchin, for its potential contradiction: a politics rooted in a universal human interest.[60] Thus, in its "assum[ption of] a protective role for the community as a whole," the male civil sphere planted the seeds for the "idea of a shared *humanitas*, that could bring people of ethnically, even tribally, diverse backgrounds together in the project of building a fully cooperative society for all to enjoy."[61]

The corollary to Bookchin's thesis of the universal human interest is the idea that interests that are specific to any given group are instances of, to cite but a few of the terms Bookchin likes to employ, "particularism," "parochialism," and "socially ghettoized" behavior.[62] Reiterating this theme, Biehl asserts that it is only by abolishing hierarchy as such, rather than specific hierarchies, that the basis for a free society can be established.[63] Referring to the *oikos* as the "world of entrapment," she calls upon women to break out of the literal *oikos* of the domestic realm as well as the figurative *oikos* of "women's values" by working to fully develop their "*human* capacities." Women, states Biehl, should "become full participants in society—*citizens*—not remain domesticated drones."[64] In line with this she mentions that she has decided herself not to work in a "particularistic vein." Instead, she identifies primarily with social ecology because it speaks for the "*general* interest of human beings *as a whole*."[65]

While most feminists would have some reservations concerning Bookchin's and Biehl's conceptions of abortion rights and sexual freedom, the greatest difference between radical forms of feminism and social ecology centers on the issue of social autonomy for women. Social ecology's politics of private and public life, which is rooted in

Bookchin's views of the evolution of the public sphere, is problematic for such feminists for a number of reasons.

In Bookchin's conception of the rise of the public sphere, the role played by women was essentially a passive one. Women, he claims, were the "degraded bystanders" of a male-centered civilization, a civilization around which their activities were merely "reshaped." Given his assertion that evolution is a "participatory" process and that even "a mere amoeba . . . is not simply passive in its relationship to its environment,"[66] it seems rather ironic that Bookchin should, in the very same text, describe women as being no more than "spectators of the intra-community changes" originating in the male sphere.[67]

Moreover, Bookchin's assumption that, in all but the "early societies," male domination of the public sphere and female "degradation" constitutes a universal pattern is not in keeping with current anthropological evidence. In her *Female Power and Male Dominance*, anthropologist Peggy Reeves Sanday, on the basis of information collected from more than 150 different social groups, divides societies into three main types with respect to the issue of male dominance: male dominant, mythically male dominant, and sexually equal.[68] Sanday classifies societies as belonging to one of these categories on the basis of two general types of behavior: exclusion of women from economic and political decision making and male aggression against women. Societies in which both of the above behaviors occur she defines as male dominant. Those in which the latter is present but not the former she considers to be examples of mythical male dominance. Societies exhibiting neither of the above traits she terms equal.[69]

Sanday links the development of male dominance to cultural disruption and social stress. She emphasizes, however, that "male oppression of women is neither an automatic nor an immediate response to stress." Instead, generally speaking, it "is based on a prior foundation formed by an *outer orientation* [seeking power in the world] and sexual segregation." Sanday finds that, rather than being rooted in some prototypical ethos of a "shared humanitas," male solidarities, when they do arise in sexually segregated societies, are "usually held together by fear of women."[70]

While Bookchin insists that it is a phenomenon characteristic only of the early stages of societal development, Sanday finds that "women hold political and economic power or authority [i.e., there is either sexual equality or only mythical male dominance] in 53% of the advanced agricultural societies" included in her study.[71] Advanced agricultural societies that practice sexual equality usually organize themselves according to one of two sex-role plans. In societies such as Bali, which are practically unisexual, sex distinctions are irrelevant in many everyday activities and thus the sexes are quite often interchangeable in relig-ious, economic, political, and social affairs.[72] In dual-sex systems such as those in West Africa, each sex administers its own affairs and the in-terests of women are represented at every level.[73] Ifi Amadiume's de-scription of the life of the Nnobi Igbo of eastern Nigeria is very much in keeping with Sanday's findings. This group not only had (until very recently) a dual-sex political system with parallel men's and women's councils but also a flexible gender system whereby some women even gained access to positions of male power.[74]

The anthropological evidence discussed above makes Bookchin's thesis of the universal division of societies, at some early stage of their development, into a male "civic" domain oriented around the interests of the community at large and a female "domestic" realm concerned only with the "most immediate means of life" questionable. In fact, it suggests that the female "domestic" sphere, if and when it did arise, was initially not only concerned with particular and private interests but with the interests of the larger community as well. Furthermore, this and other evidence seems to indicate that if and when male domi-nance did develop in any given society, it necessitated the exercise of male civic as well as physical force against any possible resistance by women to attempts to erode their power.[75]

To bring this discussion back to the question of social autonomy for women, What, then, are the implications of the above analysis for a feminist politics of public and private life? If the public sphere in con-temporary Western society evolved from a realm that arose to deal not only with the general interest but also with specifically male interests (which, in a male dominant society such as modern Western society, in-clude an interest in limiting women's power), then women do not enter this sphere on an equal footing with men. In order to counterbalance this hidden subtext of male interests, various forms of feminist politics that give voice to women's multiple concerns are therefore essential. Bookchin and Biehl to the contrary, feminism as such is *not* a particu-larism.

Probably one of the most androcentric (if not androcratic) norms governing the public sphere in Western society today is the assumption that public and private can be neatly separated. However, the reality experienced by female citizens disproves the validity of this premise. Barriers to women's participation in the public forum exist not only at the level of public life but also, and especially, at the level of private life. Obstructions at the public level are informal—for example, modes of dialogue that fit male but not female patterns of socialization[76]—as well as formal—for example, discriminatory rules. Even when experi-enced at the public level, however, such informal structures remain rooted in private life. At the private level, of course, the major problem is the disproportionate share of childcare and domestic duties assigned to women. Thus, for women, the politics of public and private life are intimately intertwined.

In directing women to break out of the domestic sphere and employ the complete range of their human capabilities by becoming active citizens, Biehl ignores the gender context of the citizen role in modern Western society and the necessary interconnections between public and private spheres. Feminists, aware of this gender context and these interconnections, call also upon men to expand the full range of their human capacities by sharing equally with women the responsibilities of private life. Thus, contrary to Biehl, the problem for feminists with respect to childcare is not that women are "placed . . . exclusively in that role" but that it is exclusively women who are placed in that role. Hence, the politics of public life cannot be changed to allow both men and women real social autonomy without a corresponding change in the politics of private life. And without genuine social autonomy, women will never achieve authentic reproductive freedom.

Conclusion

Although social ecology provides a strong ground from which to defend the ethical permissibility of abortion, promote the liberation of the body, and encourage the opening of the public sphere to all, a politics of reproductive freedom grounded in social ecology omits certain elements that many feminists consider to be essential. These feminists tend to view freedom in terms of a reproductive choice that values *both* the immanent "what-is" and the transcendent "what-should-be," a sexual liberation that is concerned with the emotional *as well as* the sensual and the rational, and a politics that deals with the private and particular *in addition to* the public and the universal.

Bookchin's tendency to value transcendent over immanent, rational over emotional and sensual, public and universal over private and particular is not accidental. This tendency is, as early Frankfurt School critical theorist Theodor Adorno and certain feminists influenced by his work have argued, characteristic of "philosophies of identity" in which an "idealistically prejudiced" predominant moment "swallows . . . by [dialectical] subsumption" whatever is not identical to itself.[77] Taking both Hegel and Marx to task for their lack of concern with "matters of true philosophical interest at this point in history . . . [including] the nonconceptual, the individual, the particular,"[78] Adorno rejected Hegel's idealist dialectics along with Marx's materialist but still identitarian one.

Like the dialectical formulations of Hegel and Marx, Bookchin's dialectical naturalism is highly problematic in that it too "leaves no . . . room for evolving otherness."[79] Indeed, it can be argued that such conceptions of dialectical theory create inherent difficulties not just for radical forms of feminist politics but ultimately for radical approaches to ecological politics as well.[80] Hence, Ynestra King's project to found a new dialectical ecofeminism that is rooted in Bookchin's dialectical naturalism can be challenged not only for its neglect of substantive issues like reproductive freedom but also, at the metatheoretical level, for being grounded in a philosophy of identity.

In her text *The Man of Reason*—a work that is noticeably influenced by the philosophical approach of Adorno—feminist philosopher Genevieve Lloyd discusses some of the implications for women of theoretical frameworks rooted in a philosophy of identity. By presenting certain modes of consciousness as immature relative to others while still allowing the former modes to be acknowledged and preserved, explains Lloyd, an identitarian dialectic "lends itself to the accommodation, containment, and transcending of 'feminine' consciousness in relation to more mature 'male' consciousness."[81] While recognizing the importance of affirming women's rationality and women's right to participate in the public realm, identity-based political theories, which incorporate women only at this level, are not really capable of dealing adequately with the conceptual complexities surrounding gender difference. Such approaches "see[m] implicitly to accept the downgrading of the excluded character traits traditionally associated with femininity" while at the same time "endors[ing] the assumption that the only human excellence and virtues which deserve to be taken seriously are those exemplified in the range of activities and concerns that have been associated with maleness."[82]

The approach that Lloyd critiques and that Biehl advocates—a politics that incorporates women by arguing for their identity with men on the basis of rationality and thus for their equal right to participate with men in the public sphere—is ultimately the approach of liberal feminism.[83] Biehl, in fact, conflates liberal feminism, one form of feminism, with feminism per se throughout her text. Claiming that the feminist project since Mary Wollstonecraft has been to "refute sexist ideologies," that early radical feminists demanded "equality," and that (following Bookchin) contemporary society is "patricentric," Biehl implies that the problems women experience as women are due merely to sexism and "patricentrism," that is, discrimination and exclusion.[84] Denigrating women who remain in the home by labeling them "domesticated drones" while acclaiming "citizens" (who may or may not participate in childcare and tasks connected to maintaining daily life) as "full participants in society,"[85] Biehl falls into the very trap that Lloyd cautions against.

Starting from a critique that overlaps with Lloyd's, Patricia Jagentowicz Mills proceeds further. She suggests an alternative framework for feminist theory—one rooted in Adorno's conception of a dialectics of nonidentity or a negative dialectics. Following Adorno, she insists that philosophies of identity have always been implicated in projects of domination—of object by subject, of matter by mind, of particular by universal, and of nature by history. Hence, rather than seeking some final "reconciliation" in which an ideologically predetermined dominant moment subsumes its nonidentical other by recasting it as a lower stage of itself, Adorno's dialectic does not posit any final "positive" moment of subsumption and identity. It constitutes a reconciliation in which two distinct but ultimately interrelated moments remain in constant dialectical tension. As such, it represents a dialectic of freedom and not of domination.[86]

In *Remaking Society,* Bookchin formulates what is, in essence, a critique of liberalism by distinguishing between the ideals of justice and freedom. Justice, he states, involves the "inequality of equals"—treating people of different mental and physical conditions as if they were "juridically equal." Freedom, on the other hand, entails the "equality of unequals"—the "attempt to equalize unavoidable inequalities."[87] While recognizing that what women are demanding entails some form of equality of unequals, he roots his dialectical naturalism in an identity theory that downgrades activities and character traits traditionally associated with women. Contrary to Biehl, social ecology does not offer women reproductive freedom—this requires a politics of nonidentity—but merely reproductive justice. From a feminist perspective, social ecology thus turns out to be yet another form of liberalism.

Acknowledgments

An earlier version of this essay was presented at the Canadian Women's Studies Association (CWSA) Conference at the University of Quebec, Montreal, in June 1995. I would like to acknowledge Andrew Light's very helpful comments and his editorial assistance in the preparation of the final manuscript. This work was supported by a doctoral fellowship from the Social Sciences and Humanities Research Council of Canada. I would also like to thank several anonymous reviewers for helpful comments on this essay.

Notes

1. Patricia Jagentowicz Mills, "Feminism and Ecology: On the Domination of Nature," *Hypatia,* Vol. 6, No. 1, Spring, 1991, p. 163.
2. Ibid., pp. 163, 175. In claiming that ecofeminism may lead to the erosion of women's reproductive freedom and that it may have a depoliticizing effect

on feminism, Mills seems to be collapsing ecofeminism per se with cultural eco-feminism, one particular form of ecofeminism—albeit the most prominent form at present—and/or with King's social ecofeminism. More critical forms of eco-feminism, I would argue, can offer additional support to the feminist project of reproductive freedom and, in this and other ways, critical forms of ecofeminism can actually have a politicizing effect on the larger feminist movement.

3. Carol J. Adams, *Neither Man nor Beast: Feminism and the Defense of Animals* (New York: Continuum, 1994), pp. 55–70, 212n1. To clarify the distinction between radical and cultural feminism, see Alice Echols, *Daring to Be Bad: Radical Feminism in America, 1967–1975* (Minneapolis: University of Minnesota Press, 1989).

4. This is discussed extensively in my paper "Conquering 'Female' Nature: Machiavelli, Ecofeminism, and the Dialectic of Enlightenment," which was pre-sented at a joint session of the CWSA (Canadian Women's Studies Association) and the ESAC (Environmental Studies Association of Canada) Conferences, held at Memorial University, June 1997, St. John's, Newfoundland. It will be included in Maria J. Falco (Ed.), *Feminist Interpretations of Machiavelli*, in press.

5. Irene Diamond, *Fertile Ground: Women, Earth, and the Limits of Control* (Boston: Beacon Press, 1994).

6. This is discussed briefly in my paper "Ecofeminism and Cyborg Poli-tics: From Utopia/Dystopia to a Negative Dialectics of Hope," which was pre-sented at a joint session of the CWSA and the Canadian Political Science Asso-ciation Conferences, held at the University of Ottawa, Ottawa, Ontario, May 1998. This paper and the one cited in note 4 above form part of my PhD thesis *Feminism, Ecology, and Negative Dialectics: Toward a Feminist Green Politi-cal Theory* (York University, Toronto, Ontario, 1998). This thesis argues that an ecofeminist appropriation of the early Frankfurt School critical theory of Theodor Adorno provides the theoretical resources needed in order to tran-scend the debate between, on the one hand, ecocentric deep and postmodern ecologists, and, on the other hand, anthropocentric social and socialist ecolo-gists while, at the same time, resonating with and facilitating a more adequate elaboration of the critical ecofeminism that is being outlined by ecofeminists such as philosopher Val Plumwood.

7. Mills, "Feminism and Ecology," p. 167.

8. Ynestra King, "Feminism and the Revolt of Nature," *Heresies*, Vol. 13, No. 4, 1982, p. 14.

9. See, e.g., Bill Devall and George Sessions, *Deep Ecology: Living as if Nature Mattered* (Layton, UT: Gibbs Smith, 1985).

10. Murray Bookchin, *Remaking Society* (Montreal: Black Rose Books, 1989), p. 30; emphasis in original.

11. Murray Bookchin, *The Philosophy of Social Ecology: Essays on Dia-lectical Naturalism* (Montreal: Black Rose Books, 1990), p. 16.

12. Alison Jaggar, *Feminist Politics and Human Nature* (Totowa, NJ: Rowman & Allenheld, 1983), pp. 280–286; M. E. Hawkesworth, *Beyond Op-pression: Feminist Theory and Political Strategy* (New York: Continuum, 1990), pp. 151–153.

13. Jaggar, *Feminist Politics*, p. 169.

14. Janet Biehl, *Finding Our Way: Rethinking Ecofeminist Politics* (Montreal: Black Rose Books, 1991), p. 99. This book was also published as *Rethinking Ecofeminist Politics* (Boston: South End Press, 1991).

15. Biehl, *Finding Our Way*, p. 128.

16. Bookchin, *Philosophy of Social Ecology*, pp. 27–28.

17. Ibid., p. 28.

18. In drawing an analogy between a fetus and an acorn, I do *not* wish to suggest that they have the same ontological status. I am merely extending an example that Bookchin uses repeatedly to illustrate his conception of dialectical naturalism.

19. Bookchin, *Philosophy of Social Ecology*, pp. 43–44.

20. Bookchin, *Remaking Society*, p. 72.

21. Ibid., pp. 38, 41.

22. Ibid., p. 203.

23. Ibid., p. 174.

24. Rosalind P. Petchesky, *Abortion and Woman's Choice: The State, Sexuality, and Reproductive Freedom* (London: Verso, 1986), p. 27.

25. Ibid., p. 25.

26. Ibid., pp. 28–29.

27. Ibid., p. 27.

28. Ibid., p. 112.

29. Ibid., p. 27.

30. Ibid., p. 373.

31. Caroline Whitbeck, "The Moral Implications of Regarding Women as People: New Perspectives on Pregnancy and Personhood," in William B. Bondeson et al., eds., *Abortion and the Status of the Fetus* (Dordrecht, The Netherlands: D. Reidel, 1983), p. 253.

32. Petchesky, *Abortion and Women's Choice*, p. 385.

33. Nikki Colodny, "The Politics of Birth Control in a Reproductive Rights Context," in Christine Overall, ed., *The Future of Human Reproduction* (Toronto: Women's Press, 1989), p. 36.

34. Petchesky, *Abortion and Women's Choice*, pp. 391–392.

35. Jaggar, *Feminist Politics*, p. 293.

36. Marlene Fried, "From Privacy to Autonomy: The Conditions for Sexual and Reproductive Freedom," in Marlene Fried, ed., *From Abortion to Reproductive Freedom: Transforming a Movement* (Boston: South End Press, 1990), p. 28.

37. Ibid., p. 39.

38. Ibid., pp. 328, 340–341.

39. Petchesky, *Abortion and Women's Choice*, pp. 377, 388.

40. Ibid., pp. 390–391.

41. Jaggar, *Feminist Politics*, p. 306.

42. Petchesky, *Abortion and Women's Choice*, p. 378.

43. Marlene Fried, "Transforming the Reproductive Rights Movement: The Post-Webster Agenda," in Fried, ed., *From Abortion to Reproductive Freedom*, p. 6.

44. Biehl, *Finding Our Way*, p. 128.

45. Ibid., pp. 123, 127.

46. Bookchin, *Philosophy of Social Ecology*, pp. 31–32.

47. Bookchin, *Remaking Society*, p. 31.

48. Emma Goldman as cited in Petchesky, *Abortion and Women's Choice*, p. 42.

49. Petschesky, *Abortion and Women's Choice*, p. 44.

50. Bookchin as cited in Biehl, *Finding Our Way*, p. 120.

51. Rosalind Petchesky, "Foetal Images: The Power of Visual Culture in the Politics of Reproduction," in Michelle Stanworth, ed., *Reproductive Technologies: Gender, Motherhood, and Medicine* (Minneapolis: University of Minnesota Press, 1987), p. 70.

52. Bookchin, *Remaking Society*, pp. 121, 125.

53. Biehl, *Finding Our Way*, p. 121; emphasis added.

54. Ibid., p. 1.

55. Bookchin, *Remaking Society*, p. 52.

56. Ibid., p. 53.

57. Ibid., pp. 55–56.

58. Ibid., p. 56.

59. Ibid., p. 65. According to Bookchin, gerontocracy involving both men and women was the earliest form of hierarchy (pp. 53–55). If gerontocracy is something to which *everyone* can aspire and in which *all* who reach the requisite age can participate, it cannot, therefore, be a true form of institutional hierarchy.

60. Ibid., p. 156.

61. Ibid., p. 80.

62. Ibid., pp. 166, 194.

63. Biehl, *Finding Our Way*, p. 54.

64. Ibid., pp. 142, 155.

65. Ibid., p. 5.

66. Bookchin, *Remaking Society*, pp. 200–201.

67. Ibid., p. 65.

68. Peggy Reeves Sanday, *Female Power and Male Dominance: On the Origins of Sexual Inequality* (New York: Cambridge University Press, 1981), p. 8.

69. Ibid., pp. 164–165.

70. Ibid., pp. 5, 9. An inner orientation, according to Sanday, exists "where the forces of nature are sacralized . . . and there is a reciprocal flow between the power of nature and the power inherent in women."

71. Ibid., p. 131.

72. Ibid., pp. 16–18.

73. Ibid., p. 88.

74. Ifi Amadiume, *Male Daughters and Female Husbands: Gender and Sex in an African Society* (London: Zed Books, 1987), pp. 15, 42, 65–67, 89. Amadiume points out that subsequent penetration of Igbo society by Western attitudes and practices regarding women ultimately led to a situation where "local men now manipulate a rigid gender ideology in contemporary politics and thereby succeed in marginalizing women's political position, or excluding them from power altogether" (p. 194).

75. An interesting example of women's resistance to male attempts to usurp their power was the "women's war" in Igboland in 1929. When British colonial administrators imposed a taxation system, a large number of Igbo women, threatened by this because they controlled the markets, marched on Aka, looting and attacking colonial stores and the bank. Such actions were in line with their traditional right to "sit upon" (punish) men who broke their laws and thus endangered the community at large. See Sanday, *Female Power and Male Dominance*, pp. 136–140.

76. Nancy Fraser, *Unruly Practices: Power, Discourse, and Gender in Contemporary Social Theory* (Minneapolis: University of Minnesota Press, 1989), pp. 120, 126.

77. Theodor Adorno, *Negative Dialectics*, trans. E. B. Ashton (New York: Seabury Press, 1973), pp. 120, 135.

78. Adorno as cited in Mills, "Feminism and Ecology," p. 166.

79. Adorno, *Negative Dialectics*, p. 337.

80. This argument is developed in the paper "Social Ecology and Teleology: Critiquing Bookchin's Critique of Kant" that I presented at the Environmental Studies Association of Canada (ESAC) Conference, held at Brock University in June 1996. Until recently, Bookchin espoused some allegiance to Adorno. Indeed, in *The Philosophy of Social Ecology*, he ended his essay "Toward a Philosophy of Nature" by stating: "Here is the faith that social ecology keeps with the promise opened by Adorno" (p. 89). Yet in the new 1995 edition of this text, all favorable references to Adorno are excised and his texts are portrayed as "mixed farragoes of convoluted neo-Nietzschean verbiage, often brilliant . . . but often confused, rather dehumanizing and, to speak bluntly, irrational" (p. 175). In conflating Adorno's negative dialectics with postmodernism instead of realizing that the former can actually offer a strong critique of the latter, as Fredric Jameson and others have pointed out, it seems that it is really Bookchin himself who is confused. See, for example, Fredric Jameson, *Late Marxism: Adorno, or, The Persistence of the Dialectic* (New York: Verso, 1990), and Jameson, *Postmodernism: or, The Cultural Logic of Late Capitalism* (Durham: University of North Carolina Press, 1992).

81. Genevieve Lloyd, *Man of Reason: "Male" and "Female" in Western Philosophy* (Minneapolis: University of Minnesota Press, 1984), pp. 72–73.

82. Ibid., p. 104.

83. Jaggar, *Feminist Politics*, p. 35–39.

84. Biehl, *Finding Our Way*, pp. 3, 6, 9, 23, 111, 155.

85. Ibid., p. 142.

86. Mills, "Feminism and Ecology," p. 166. A feminist theory rooted in negative dialectics would necessarily advance a feminist critique of Adorno. In line with this, see Patricia Jagentowicz Mills, *Woman, Nature, and Psyche* (New Haven, CT: Yale University Press, 1987). It would also be *highly* critical of many aspects of contemporary ecofeminism. This is the subject of my paper "Conquering 'Female' Nature."

87. Bookchin, *Remaking Society*, pp. 98–99.

Part II

Politics and Material Culture

Chapter 5

Municipal Dreams
A Social Ecological Critique
of Bookchin's Politics

JOHN CLARK

Introduction

In the following discussion, I will analyze Murray Bookchin's libertarian municipalist politics from the perspective of social ecology. This analysis forms part of a much larger critique in which I attempt to distinguish between social ecology as an evolving, dialectical, holistic philosophy and the increasingly rigid, nondialectical, dogmatic version of that philosophy promulgated by Bookchin. An authentic social ecology is inspired by a vision of human communities achieving their fulfillment as an integral part of the larger, self-realizing earth community. Ecocommunitarian politics, which I would counterpose to Bookchin's libertarian municipalism, is the project of realizing such a vision in social practice. If social ecology is an attempt to understand the dialectical movement of society within the context of the larger dialectic of society and nature, ecocommunitarianism is the project of creating a way of life consonant with that understanding. Setting out from this philosophical and practical perspective, I argue that Bookchin's politics is not only riddled with theoretical inconsistencies, but also lacks the historical grounding that would make it a reliable guide for an ecological and communitarian practice.[1]

One of my main contentions in this critique is that because of its ideological and dogmatic aspects, Bookchin's politics remains, to use Hegelian terms, in the sphere of morality rather than reaching the level of the ethical. That its moralism can be compelling I would be the last to deny, since I was strongly influenced by it for a number of years.

Nevertheless, it is a form of abstract idealism, and tends to divert the energies of its adherents into an ideological sectarianism, and away from an active and intelligent engagement with the complex, irreducible dimensions of history, culture, and psyche. The strongly voluntarist dimension of Bookchin's political thought should not be surprising. When a politics lacks historical and cultural grounding, and the real stubbornly resists the demands of ideological dogma, the will becomes the final resort. In this respect, Bookchin's politics is firmly in the tradition of Bakuninist anarchism.

Democracy, Ecology, and Community

The idea of replacing the state with a system of local political institutions has a long history in anarchist thought. As early as the 1790s, William Godwin proposed that government should be reduced essentially to a system of local juries and assemblies that would perform all the functions that could not be carried out voluntarily or enforced informally through public opinion and social pressure.[2] A century later, Elisée Reclus presented an extensive history of the forms of popular direct democracy, from the era of the Athenian *polis* to modern times, and proposed that their principles be embodied in a revolutionary system of communal self-rule.[3] Today, the most uncompromising advocate of this tradition of radical democracy is Murray Bookchin, who has launched an extensive and often inspiring defense of local direct democracy in his theory of libertarian municipalism.[4] Bookchin's ideas have contributed significantly to the growing revival of interest in communitarian democracy. For many years, he was one of the few thinkers to carry on the tradition of serious theoretical exploration of the possibilities for decentralized, participatory democracy. Perhaps the only comparable recent work has been political theorist Benjamin Barber's defense of "strong democracy." But although Barber offers a highly detailed presentation of his position and often argues for it persuasively, he undercuts the radicality of his proposals by accepting much of the apparatus of the nation-state.[5] Thus, no one in contemporary political theory has presented a more sustained and uncompromising case for the desirability of radical "grassroots" democracy than has Bookchin. Furthermore, he has been one of the two contemporary theorists of his generation (the other is Cornelius Castoriadis) to raise the most important philosophical issues concerning radical democracy.[6] This critique recognizes the importance of Bookchin's contribution to ecological, communitarian, democratic theory and investigates the issues that must be resolved if the liberatory potential of certain aspects of his thought is to be freed from the constraints of sectarian dogma.

One of the strongest points in Bookchin's politics is his attempt to ground it in ethics and a philosophy of nature. In viewing politics fundamentally as a sphere of ethics his political theory carries on the Aristotelian tradition. Aristotle saw the pursuit of the good of the *polis*, the political community, as a branch of ethics, the pursuit of the human good as a whole. He called this ultimate goal for human beings *eudaimonia*, which is often translated as "the good life." Bookchin expands this concept of the larger good even further to encompass the natural world. Beginning with his early work, he has argued that the development of a political ethics implies "a moral community, not simply an 'efficient' one," "an ecological community, not simply a contractual one," "a social praxis that enhances diversity," and "a political culture that invites the widest possible participation."[7]

For Bookchin, politics is an integral part of the process of evolutionary unfolding and self-realization spanning the natural and social history of this planet. Social ecology looks at this history as a developmental process aiming at greater richness, diversity, complexity, and rationality. The political, Bookchin says, must be understood in the context of humanity's place as "nature rendered self-conscious."[8] From this perspective, the goal of politics is the creation of a free, ecological society in which human beings achieve self-realization through their participation in a creative, nondominating human community, and in which planetary self-realization is furthered through humanity's achievement of a balanced, harmonious place within the larger ecological community of the earth. A fundamental political task is thus the elimination of those forms of domination that hinder the attainment of greater freedom and self-realization, and the creation of new social forms that are most conducive to these ends.

This describes "politics" in the larger, classical sense of a political ethics, but leaves open the question of which "politics" in the narrower sense of determinate social practice best serves such a political vision. While Bookchin has always emphasized the importance of such political precedents as the *polis* and the Parisian sections of the French Revolution, it has not always been clear what specific politics was supposed to follow from this inspiration. He has expressed considerable enthusiasm at different times for a variety of approaches to political, economic, and cultural change. In "The Forms of Freedom" (1968) he envisions a radically transformative communalism rapidly creating an alternative to centralized, hierarchical, urbanized industrial society. Employing terms reminiscent of the great utopian Gustav Landauer, he suggests that "we can envision young people renewing social life just as they renew the human species. Leaving the city, they begin to found the nuclear ecological communities to which older people repair in increasing numbers," as "the modern city begins to shrivel, to contract and to disappear."[9] The almost apocalyptic and millenarian aspects of Bookchin's views in this period reflect both the spirit of the time and his strong identification with the utopian tradition.

Several years later, in "Spontaneity and Organization," he sees the "development of a revolutionary movement" as depending on "the seeding of America" with affinity groups, communes, and collectives. His ideas are still heavily influenced by the 1960s counterculture (which his own early works in turn theoretically influenced), and he lists as the salient points of such entities that they be "highly experimental, innovative, and oriented toward changes in life-style as well as consciousness."[10] They were also to be capable of "dissolving into the revolutionary institutions" that were to be created in the social revolution that he believed at the time to be a real historical possibility.[11] Indeed, he could write in 1971 that "this is a revolutionary epoch" in which "a year or even a few months can yield changes in popular consciousness and mood that would normally take decades to achieve."[12]

Revolution in America (1969–1998)

Statements like this one express Bookchin's deep faith in revolutionary politics, a faith which, while far from being spiritual, is certainly "religious" in the conventional sense of the term. Like religious faith, it shows great resiliance in the face of embarassing evidence from the merely temporal realm. One of the most enduring aspects of Bookchin's thought is his hope for apocalyptic revolutionary transformation; his quest is to create a body of ideas that will inspire a vast revolutionary movement and lead "the People" into their great revolutionary future. His exaggerated assessment of the revolutionary potential of U.S. society a quarter century ago is not an isolated abberation in his thought. It prefigures many later analyses, including his recent discovery of supposedly powerful tendencies in the direction of his libertarian municipalism.

Bookchin himself points to his article "Revolution in America" for evidence of his astuteness concerning historical trends in the earlier period.[13] A careful examination of that text indicates instead a disturbing ideological tendency in his thought. In that article, published in February 1969 under the pseudonym "Robert Keller," Bookchin wisely denies that there was at that time a "revolutionary situation" in the United States, in the sense of an "immediate prospect of a revolutionary challenge to the established order."[14] However, he contends that we *have* entered into a "revolutionary epoch." His depiction of this epoch betrays the unfortunate theoretical superficiality that was endemic to the 1960s counterculture and shows a complete blindness to the ways in which the trends that he embraced so uncritically were products of late capitalist society itself. Furthermore, it harkens back in the anarchist tradition to Bakuninism, with its idealization of the marginalized strata, its voluntarist overemphasis on the power of revolutionary will, and its Manichaean view of the future.

According to Bookchin, "The period in which we live closely resembles the revolutionary Enlightenment that swept through France in the eighteenth century—a period that completely reworked French consciousness and prepared the conditions for the Great Revolution of 1789."[15] Interestingly, what he sees as spreading through U.S. society in a seemingly inexorable manner was a questioning of "the very existence of hierarchal power as such," a "rejection of the commodity system," and a "rejection of the American city and modern urbanism."[16] He finds symptoms of these trends in the fact that "the society, in effect, becomes disorderly, undisciplined, Dionysian" and that "a vast critique of the system" is expressed for example in "an angry gesture, a 'riot' or a conscious change in life patterns," all of which he interprets as "defiant propaganda of the deed."[17] He praises various social groups, including, "most recently, hippies"[18] for their contribution to the "new Enlightenment."

However, what is most interesting for those interested in Bookchin's anarchism are his Bakuninesque statements concerning the transformative virtues of spontaneous violence. He claims that "the 'rioter' and the 'Provo' have begun to break, however partially and intuitively, with those deep-seated norms of behavior which traditionally weld the masses to the established order," and that "the truth is that 'riots' and crowd actions represent the first gropings of the mass toward individuation."[19] Elsewhere, he praises the "superb mobile tactics" in a demonstration in New York, calls for "the successful intensification of these street tactics," and stresses the need for these tactics to "migrate" to other major cities.[20] Overall, he takes a rather mechanistic view of the "revolutionary" movement that he sees developing. According to his diagnosis, the problem is that "an increasing number of molecules" (as the result of what he calls the "seeping down" of the "vast critique" mentioned earlier) "have been greatly accelerated beyond the movement of the vast majority."[21] Switching rapidly from physical to biological imagery, he concludes that the challenge is for radicalized groups to "extend their own rate of social metabolism to the country at large."[22]

Certain tendencies that have always impeded Bookchin's development of a truly communitarian outlook are already evident in his conclusions on the place of "consciousness" in this process. "What consciousness must furnish *above all things* is an extraordinary flexibility of tactics, a mobilization of methods and demands that make exacting use of the opportunities at hand."[23] In this analysis, Bookchin expresses a Bakuninism (or anarcho-Leninism) that has been a continuing undercurrent in his thought and which has recently come to the surface in his programatic municipalism. His conception of consciousness at the service of ideology stands at the opposite pole from an authentically communitarian view of social transformation, which sees more elaborated, richly developed conceptions of social and ecological interrelatedness (not as mere abstract "Oneness," but rather as concrete unity-in-diversity) as the primary challenge for consciousness as reflection on social practice.

"Revolution in America" illustrates very well Bookchin's enduring tendency to interpret phenomena too much in relation to his own political hopes and too little in relation to specific cultural and historical developments. In this case, he fails to consider the possibility that the erosion of traditional character structures and the delegitimation of traditional institutions could be "in the last instance" the result of the transition from productionist ("early," "classical") capitalism to consumptionist ("late," "postmodern") capitalism. For Bookchin, "What underpins every social conflict in the United States, today, is the demand for the self-realization of all human potentialities in a fully rounded, balanced, totalitistic way of life."[24] He asserts that "we are witnessing" nothing less than "a pulverization of all bourgeois institutions," and contends that the "present bourgeois order" has nothing to substitute for these institutions but "bureaucratic manipulation and state capitalism."[25] Amazingly, he makes no mention of the vast potential for manipulation through mass media and commodity consumption—presumably because the increasingly enlightened populace was in the process of rejecting both.

Bookchin concludes with the Manichaean pronouncement that the only alternatives at this momentous point in history are the realization of "the boldest concepts of utopia" through revolution or "a disastrous [*sic*] form of fascism."[26] This theme of "utopia or oblivion" continued into the 1970s and beyond with his slogan "anarchism or annihilation" and the enduring message that ecoanarchism is the only alternative to ecological catastrophe. The theme takes on a new incarnation in his recent "Theses on Municipalism," in which he ends with the threat that if humanity turns a deaf ear to his political analysis (social ecology's "task of preserving and extending the great tradition from which it has emerged"), then "history as the rational development of humanity's potentialities for freedom and consciousness will indeed reach its definitive end."[27] While Bookchin is certainly right in saying that we are at a crucial turning point in human and earth history, he has never presented a careful analysis of why some types of reformism (or *any* alternatives to his own politics) cannot possibly avoid ending in either fascism or global ecological catastrophe. His claims are reminiscent of those of Bakunin, who spent much of his career writing a long work whose major, yet quite unsubstantiated, thesis was that Europe's only options were military dictatorship or anarchist social revolution.[28]

Bookchin claims to be shocked (indeed, "astonished") by such criticism of the Bakuninist aspects of his work. What amazes him is that "a self-proclaimed anarchist would apparently deny a basic fact of historical revolutions, that both *during and after those revolutions* people undergo very rapid transformations in character."[29] However, while anarchism as a romanticist ideology of revolution might uncritically accept the inevitablity of such transformations, anarchism as a critique of domination will retain a healthy skepticism concerning claims of rapid changes in character structure among masses of people.

First, I would recommend a much more critical approach than Bookchin's toward accounts of the history of revolutions. To put the matter the way Bookchin likes to—that is, bluntly—revolutionaries have tended to idealize revolutions and explain away their defects, while reactionaries have tended to demonize revolutions and explain away their achievements. For example, anarchists have had a propensity to emphasize accounts of the Spanish Revolution written by anarchists and sympathizers and to ignore questions raised about extravagant claims of miraculous transformations. It is seldom mentioned, as Fraser's interviews in *Blood of Spain* reveal, that there were anarchists who believed that if the anarchists had won the war, they would have needed another revolution to depose the anarchist militants who were dominating the collectives.[30] Considering the problems of culture and character structure that existed, this second revolution might have really meant a long process of self-conscious personal and communal evolution. While ideological apologists always contend that revolutionary movements are betrayed by renegades, traitors, and scoundrels, a balanced critical analysis would also consider the limitations and, indeed, the contradictions inherent in a given form of revolutionary process itself.

Furthermore, it is necessary to point out that there is an important anarchist tradition that has stressed the fact that the process of "transformation in character" is one that can only progress slowly, and that what some, like Bakunin and Bookchin, would attribute to the alchemy of revolution is really the fruit of long and patient processes of social creativity. This is the import of Elisée Reclus's reflections on the relationship between "evolution and revolution," and even more directly, of Gustav Landauer's view that "the state is a relationship" that can only be undone through the creation of other kinds of nondominating relationships developed through shared communitarian practice. To overlook the continuity of development and to count on vast changes in human character during "the revolution" (or even through participation in institutions like municipal assemblies) leads to unrealistic expectations, underestimation of limitations, and ideological distortions and idealizations of revolutionary periods.

Finally, it should be noted that Bookchin misses the main point of the criticism of Bakunin's and his own revolutionism. Beyond their idealization of *revolutions* themselves, both exhibit a tendency to idealize *revolutionary movements* (and even potentially revolutionary movements and tendencies) so that they are seen as implicitly and unconsciously embodying the ideology of the anarchist theorist who interprets them (as exemplified by Bookchin's "Revolution in America," his more recent observations on an emerging "dual power,"[31] and by almost everything Bakunin wrote about contemporary popular movements in Europe). Not only revolutions, but these social movements are depicted as producing very rapid changes in consciousness and character that would seem possible only through more organic processes of growth. Furthermore, the movements are attributed an inner "directionality" leading them to exactly the position the revolutionary theorist happens to hold, whatever the actual state of the social being and consciousness of the participants may be. Thus, Bookchin's conclusion that my analysis "raises serious questions about [Clark's] own acceptance of the possibility of revolutionary change as such"[32] is correct. Indeed, I question his or any uncritical revolutionism that abstractly, idealistically, and voluntaristically conceives of "revolutionary changes" as existing "as such" (*an sich*) and overlooks the many historical, cultural, and psychological mediations that are necessary for them to exist as self-realized, consciously developed social practices (*für sich*).

Bookchin is much more convincing when he returns from his revolutionary fantasies and proposes a comprehensive, many-dimensional program of social creation. His vision of an organically developing libertarian ecological culture has inspired many and has made an important contribution to the movement for social and ecological regeneration. In "Toward a Vision of the Urban Future," for example, he looks hopefully to a variety of popular initiatives in contemporary urban society. He mentions block committees, tenant associations, "ad hoc committees," neighborhood councils, housing cooperatives, "sweat equity" programs, cooperative day care, educational projects, food co-ops, squatting and building occupations, and alternative technology experi-

ments as making contributions of varying importance to the achievement of "municipal liberty."[33]

While Bookchin has always combined such proposals with an emphasis on the importance of the "commune" or the municipality in the
process of social transformation, the programs now associated with his
program of libertarian municipalism have taken precedence, while
other approaches to change have received less and less attention. The
municipality is now more explicitly recognized as the central political
reality and municipal assembly government becomes the preeminent expression of democratic politics. The present analysis will focus on this
libertarian municipalism as Bookchin's effort to make a distinctive contribution to political theory.

Citizenship and Self-Identity

Bookchin contends that the "nuclear unit" of a new politics must be
the citizen, "a term that embodies the classical ideals of *philia*, autonomy, rationality, and above all, civic commitment."[34] He rightly argues
that the revival of such an ideal would certainly be a vast political advance in a society dominated by self-images based on consumption and
passive participation in mass society.[35] To think of oneself as a citizen
contradicts the dominant representations of the self as egoistic calcula-
tor, as profit maximizer, as competitor for scarce resources, or as nar-
cissistic consumer of products, images, experiences, and even other persons. It replaces narrow self-interest and egoism with a sense of ethical
responsibility toward one's neighbors, and an identification with a
larger whole: the political community. Furthermore, it reintroduces the
idea of moral agency on the political level, through the concept that
one can in cooperation with others create social embodiments of the
good. In short, Bookchin's concept challenges the ethics and moral psychology of economistic, capitalist society and presents an edifying image of a higher ideal of selfhood and community.

Yet this image has serious limitations. To begin with, it seems unwise
to define any single role as such a "nuclear unit" or to see any as the privileged form of self-identity, for there are other important self-images with
profound political implications. A notable example is that of personhood.
While civic virtue requires diverse obligations to one's fellow citizens, respect, love, and compassion are feelings appropriately directed at all persons. If (as Bookchin has himself at times agreed) we should accept the
principle that "the personal is political," we must explore the political dimension of personhood and its universal recognition.[36]

Furthermore, the political significance of our role as members of the earth community can hardly be overemphasized. We might also conceive of this role as an expression of a kind of citizenship—if we think of ourselves not only as citizens of a town, city, or neighborhood, but also as citizens of our ecosystem, of our bioregion, of our georegion, and of the earth itself. In doing so, we look upon ourselves as citizens in the quite reasonable sense of being responsible members of a community. Interestingly, Bookchin believes that acceptance of such a concept of citizenship implies that animals, including insects, and even inanimate objects, including rocks, must be recognized as citizens.[37] This exhibits his increasingly rigid, unimaginative, and quite nondialectical approach to the life of concepts. Just as we can act as moral agents in relation to other beings that are not agents, we can exercise duties of citizenship in relation to other beings who are not citizens.[38] Furthermore, Bookchin himself uses the term "ecocommunities" to refer to what others call "ecosystems." By his own standards of rationalist literalism, one might well ask him how human beings could achieve "communal" or "communitarian" relationships with birds and insects—or, more tellingly, how the bird or insect might be expected to relate "communally" to (e.g.) Murray Bookchin.

Bookchin's personal preferences concerning linguistic usage notwithstanding, in the real world the term "citizen" does not have the connotation that he absolutizes. The fact is that it indicates membership in a nation-state and subdivisions of nation-states, including states that are in no way authentically democratic or participatory. While Bookchin may invoke the linguistic authority of famous dead radicals,[39] the vast majority of actually living people (who are expected to be the participants in the libertarian municipalist system) conceive of citizenship primarily in relation to the state, not the municipality. The creation of a shared conception of citizenship in Bookchin's sense is a *project* that must be judged in relation to the actually existing fund of meanings and the possibilities for social creation in a given culture.[40] The creation of a conception of citizenship in the earth community is no less a project, and one that has a liberatory potential that can only be assessed through cultural creativity, historical practice, and critical reflection on the result.[41]

Bookchin seems never to have gleaned from his readings of Hegel the distinction between an abstract and a concrete universal. While superficially invoking Hegel, he overlooks the philosopher's dialectical insight that any concept that is not developed through conceptual and historical articulation remains "vacuous." Much of the present critique of Bookchin's libertarian municipalism is a conceptual and historical analysis that draws out the implications and contradictions in his position, contradictions that are typically disguised through use of rhetorical devices, avoidance of difficult issues, and recourse to bombastic but irrelevant replies to criticism.[42] In short, his concepts often lack articulation. But just as often he seems to lack the ability to distinguish between what is and what is not articulated. He does not realize that, in themselves, concepts like "citizen of a municipality" and "citizen of the earth" are both "vacuous" and "empty"—that is, they are mere abstractions. Their abstractness cannot be negated merely by appealing to historical usage or to one's hopes for an improved usage in the future. They can be given more *theoretical content* by exploring their place in the history of ideas and in social history, by engaging in a conceptual analysis, and by reflecting on their possible relationship to other emerging theoretical and social possibilities. Yet they will still remain abstractions, albeit now more fully articulated ones. They gain *concrete content*, on the other hand, through their embodiment in the practice of a community—in its institutions, its ethos, its symbols, and its images.

Bookchin apparently confuses this historical concreteness with relatedness to concrete historical phenomena of the past. When he finds certain political forms of the past to be inspiring, they take on a certain numinous quality for him. Various models of citizenship become historically relevant today not because of their relation to real historical possibilities (including real possibilities existing in the social imaginary realm), but because they present an image of what our epoch assuredly *ought* to be. It is for this reason that he thinks that certain historical usages of the term "citizen" can dictate proper usage of the term today.

Of course, Bookchin is at the same time aware that the citizenship that he advocates is not a living reality, but only a proposed ideal. Thus, he notes that "today, the concept of citizenship has already undergone serious erosion through the reduction of citizens to 'constituents' of statist jurisdictions or to 'taxpayers' who sustain statist institutions."[43] Since he thinks above all of U.S. society in formulating this generalization, one might ask when there was a Golden Age in U.S. history when the populace were considered "citizens" in Bookchin's strong sense of "a self-managing and competent agent in democratically shaping a polity."[44] What has been "eroded" is presumably not the unrealized goals of the Democratic–Republican societies of the 1790s and other similar phenomena outside the mainstream of U.S. political history. This remarkable form of "erosion" (a phenomenon possible only in the realm of ideological geology) has taken place between discontinuous historical models selected by Bookchin and the actually existing institutions of contemporary society.

In addition to defending his concept of citizenship as the "true" meaning of the term, he also contends that its realization in society is a prerequisite for the creation of a widespread concern for the general good. He argues that "we would expect that the special interests that divide people today into workers, professionals, managers, and the like would be melded into a general interest in which people see themselves as *citizens* guided strictly by the needs of their community and region rather than by personal proclivities and vocational concerns."[45] Yet this very formulation preserves the idea of particularistic interest, that is, that which fulfills the needs of ones "community and region," which could—and in the real world certainly would—conflict with the needs of other communities and regions. There will always no doubt be communities that have an abundance of certain natural goods, all of which might fulfill real needs of the community, but some of which would fulfill even greater needs of other communities entirely lacking these goods or having special conditions that render their needs more pressing.

Of course, one might say that in the best of all possible libertarian municipalisms, the citizens would see their highest or deepest need as contributing to the greatest good for all—"all" meaning humanity and the entire planet. Bookchin does in fact hold that such a larger commitment would exist in his ideal system. But its existence would then imply a broadened horizon of citizenship. Each person would see a fundamental dimension of his or her political being (or citizenship) as membership in the human community and, indeed, in the entire earth community. There is a strong tension in Bookchin's thought between his desire for universalism and his commitment to particularism. Such a tension is inherent in an ecological politics that is committed to unity-in-diversity and which seeks to theorize the complex dialectic between whole and part. But for Bookchin this creative tension rigidifies into contradiction as a result of his territorializing of the political dimension at the level of the particular municipal community. In an important sense, Bookchin's "citizenship" is a regression from the universality of membership in the working class, whatever serious limitations that concept may have had. While one's privileged being qua worker consisted in membership in a *universal* class, one's being qua citizen (for Bookchin) consists of being a member of a *particular* group: the class of citizens of a given municipality.

Bookchin will, however, hear none of this questioning of the boundaries of citizenship. From his perspective, the concept of citizen "becomes vacuous" and is "stripped of its rich historical content"[46] when the limits of the concept's privileged usage are transgressed. Yet he is floundering in the waters of abstract universalism, since he is not referring to any historically actualized content, but merely to his idealized view of what that content *ought* to be. Citizenship is not developed (richly or otherwise) through some concept of "citizen" that

Bookchin or any other theorist constructs. Nor can it be "developed" through a series of historical instances that have no continuity in *concrete, lived* cultural history. It becomes "richly developed" when concept and historical precedent are given meaning through their relationship to the life of a particular community—local, regional, or global. Bookchin, like anyone concerned with the transformation of society, is faced with a cultural repertoire of meanings that must be recognized as an interpretative background, from which all projects of cultural creativity must set out to re-create meaning. We cannot re-create that background, or any part of it (e.g., the social conception of "citizenship") in our own image, or in the images of our hopes and dreams. Yet our ability to realize some of our hopes and dreams will depend in large part on our sensitivity to that background and our capacity to find in it possibilities for extensions and transformations of meaning.

The "Agent of History"

Bookchin asks at one point the identity of the "historical 'agent' for sweeping social change."[47] In a sense, he has already answered this question in his discussion of the centrality of citizenship. However, his specific response focuses on the social whole constituted by the entire body of citizens: "the People." Bookchin has described this emerging "People" as a " 'counterculture' in the broadest sense" and has stipulated that it might include "alternative organizations, technologies, periodicals, food cooperatives, health and women's centers, schools, even barter-markets, not to speak of local and regional coalitions."[48] While this concept is obviously shaped and in some ways limited by the image of the U.S. counterculture of the 1960s, it reflects a broad conception of cultural creativity as the precondition for liberatory social change. This is its great strength. It points to a variety of community-oriented initiatives that develop the potential for social cooperation and grassroots organization.

But just as problems arise from privileging a particular self-image, so do they stem from the privileging of any unique "historical agent," given the impossibility of analytical or scientific knowledge of the processes of social creativity. It is likely that such agency will always be exercised in many spheres and at many overlapping levels of social being. It is conceivable that in some sense "the person" will be such a historical agent, while in another "the earth community" will be. In addition, as will be discussed further, alternatives deemphasized in his view of what contributes to forming such agency (such as democratic worker cooperatives) may have much greater liberatory potential than those stressed by Bookchin. From a dialectical holistic viewpoint, it is obvious that there will always be a relative unity of agency and also a relative diversity, so that agency can never have any simple location. While political rhetoric may require a reifying emphasis on one or the other moments of the whole, political thought must recognize and theorize the complexity of the phenomena. Bookchin's concept is a seriously flawed attempt to capture this social unity-in-diversity.

The idea of "the People" as the preeminent historical agent is central to Bookchin's critique of the traditional leftist choice of the working class (or certain other economic strata) for that role. Bookchin, along with other anarchists, was far ahead of most Marxists and other socialists in breaking with this economistic conception of social transformation. Indeed, postmodern Marxists and other au courant leftists now sound very much like the Bookchin of thirty years ago when they go through the litany of oppressed groups and victims of domination who are now looked upon as the preeminent agents of change. Bookchin can justly claim that his concept is superior to many of these current theories, in that his idea of "the People" maintains a degree of unity within the diversity, while leftist victimology has often degenerated into incoherent, divisive "identity politics."

But perhaps Bookchin and, ironically, even some contemporary socialists go too far in deemphasizing the role of economic class analysis. Bookchin notes that while "the People" was "an illusory concept" in the eighteenth century, it is now a reality in view of various "transclass issues like ecology, feminism, and a sense of civic responsibility to neighborhoods and communities."[49] He is of course right in stressing the general, transclass nature of such concerns. But it seems clear that these issues are *both* class and transclass issues, since they have a general character, but also a quite specific meaning in relation to economic class—not to mention gender, ethnicity, and other considerations. The growing concern for environmental justice and the critique of environmental racism have made this reality increasingly apparent. Without addressing the class (along with ethnic, gender, and cultural) dimensions of an issue, a radical movement will fail to understand the question in concrete detail and will lose its ability both to communicate effectively with those intimately involved in the issue and, more importantly, to learn from them. The fact is that Bookchin's social analysis has had almost nothing to say about the evolution of class in either U.S. or global society. Indeed, Bookchin seems to have naïvely equated the obsolence of the classical concept of the working class with the obsolence of class analysis.

While "the People" are identified by Bookchin as the emerging subject of history and the agent of social transformation, he also identifies a specific group within this large category that will be essential to its successful formation. Thus, in the strongest sense of agency, the " 'agent' of revolutionary change" will be a "radical intelligentsia" which, according to Bookchin, has always been necessary "to catalyze" such change.[50] The nature of such an intelligentsia is not entirely clear, except that it would include theoretically sophisticated activists who would lead a libertarian municipalist movement. Presumably, as has been historically the case, it would also include people in a variety of cultural and intellectual fields who would help spread revolutionary ideas.

Bookchin is certainly right in emphasizing the need within a movement for social transformation for a sizable segment of people with developed political commitments and theoretical grounding. However, most of the literature of libertarian municipalism, which heavily emphasizes social critique and political programs, has seemed thus far to be directed almost exclusively at just such a group. Furthermore, it has assumed that the major precondition for effective social action is knowledge of and commitment to Bookchin's theoretical position. This ideological focus, which reflects Bookchin's theoretical and organizational approach to social change, will inevitably hinder the development of a broadly based social ecology movement, to the extent that this development requires a diverse intellectual milieu linking it to a larger public. Particularly as Bookchin has become increasingly suspicious of the imagination, the psychological dimension, and any form of "spirituality," and as he has narrowed his conception of reason, he has created a version of social ecology that is likely to appeal to only a small number of highly politicized intellectuals. Despite the commitment of social ecology to unity-in-diversity, his approach to social change increasingly emphasizes ideological *unity* over diversity of forms of expression. If the "radical intelligentsia" within the movement for radical democracy is to include a significant number of poets and creative writers, artists, musicians, and thoughtful people working in various professional and technical fields, a more expansive vision of the socially transformative practice is necessary.

Furthermore, a heavy emphasis on the role of a radical intelligentsia—even in the larger sense just mentioned—threatens to overshadow the crucial importance of cultural creativity by nonintellectuals. This includes those who create small cultural institutions, cooperative social practices, and transformed relationships in personal and family life. The nonhierarchical principles of social ecology should lead one to pay careful attention to the subtle ways in which large numbers of people contribute to the shaping of social institutions, whether traditional or newly evolving ones. Bookchin himself recognizes the importance of such activity when he describes the emergence of a "counterculture" that consists of a variety of cooperative and communitarian groups and institutions, and thereby promotes the all-important "reemergence of 'the People.'"[51] Why the intelligentsia and not this entire developing culture is given the title of "historical agent" is not clearly explained. One must suspect, however, that the answer lies in the fact that the majority of participants in such a culture would be unlikely to have a firm grounding in the principles of Bookchin's philosophy. The true agents of history, from his point of view, will require precisely such an ideological foundation.

The Municipality as the Ground of Social Being

The goal of the entire process of historical transformation is, of course, the libertarian municipality. Bookchin often describes the municipality as the fundamental political, indeed, the fundamental social, reality. For example, he states that "conceived in more institutional terms, the municipality is the basis for a free society, the irreducible ground for individuality as well as society."[52] Even more strikingly, he says that the municipality is "the living cell which forms the basic unit of political life . . . from which everything else must emerge: confederation, interdependence, citizenship, and freedom."[53] This assertion of the centrality of the municipality is a response to the need for a liberatory political identity that can successfully replace the passive, disempowering identity of membership in the nation-state, and a moral identity that can successfully replace the amoral identity of consumer. For Bookchin, the municipality is the arena in which political ethics and the civic virtues that it requires can begin to germinate and ultimately achieve an abundant flowering in a rich political culture. This vision of free community is in some ways a very inspiring one.

It is far from clear, however, why the municipality should be considered the fundamental social reality. Bookchin attributes to the municipality alone a role in social life that is in fact shared by a variety of institutions and spheres of existence. It is not only the ideologies of modern societies, which presuppose a division between private and public life, that emphasize the realm of *personal life* as central to social existence. Many anarchists and utopians take the most intimate personal sphere, whether identified with the affinity group, the familial group, or the communal living group, as fundamental socially and politically.[54] And many critical social analyses, including the most radical ones (e.g., Reich's classic account of fascism and Kovel's recent analysis of capitalist society) show the importance of the dialectic between the personal dimension and a variety of institutional spheres in the shaping of the self and values, including political values.[55]

One might suspect that Bookchin is using descriptive language to express his own prescriptions about what *ought to be* most basic to our lives. However, he sometimes argues in ways that are clearly an attempt to base his political norms in existing social reality. In his argument for the priority of the municipality he claims that it is "the one domain outside of personal life that the individual must deal with on a very direct basis" and that the city is "*the most immediate environment* which we encounter and with which we are obliged to deal, *beyond the sphere of family and friends,* in order to satisfy our needs as social beings."[56]

First of all, these statements really seem to be an argument for the priority of the family and, perhaps, the affinity group in social life, for the city is recognized as only the *next most important* sphere of life. But beyond this rather large problem, the analysis of the "immediacy" of the city seems to be a remarkably superficial and nondialectical one. To begin with, it is not true that the individual deals in a somehow more "direct" way with the municipality than with other institutions (even excluding family and friends). Millions of individuals in modern society deal more directly with the mass media, by way of their television sets, radios, newspapers, and magazines, until they go to work and deal with bosses, coworkers, and technologies, after which they return to the domestic hearth and further bombardment by the mass media.[57] The municipality remains a vague background to this more direct experience. Of course, the municipality is one *context* in which the more direct experience takes place. But there is also a series of larger contexts: a variety of political subdivisions, various natural regions, the nation-state, the society, the earth.[58] There are few "needs as social beings" that are satisfied uniquely by "the municipality" in strong contradistinction to any other source of satisfaction.

Bookchin has eloquently made points similar to these in relation to the kind of "reification" of the "bourgeois city" that takes place in traditional city planning. "To treat the city as an autonomous entity, apart from the social conditions that produce it . . . [is] to isolate and objectify a habitat that is itself contingent and formed by other factors. Behind the physical structure of the city lies the social community—its workaday life, values, culture, familial ties, class relations, and personal bonds."[59] It is important to apply this same kind of dialectical analysis to libertarian municipalism, and thereby to develop it even further (even if certain of its aspects are negated in the process). The city or municipality is a social whole consisting of constituent social wholes, interrelated with other social wholes, and forming a part of even larger social wholes. Add to this the natural wholes that are inseparable from

the social ones, and then consider all the mutual determinations between all of these wholes and all of their various parts, and we begin to
see the complexity of a dialectical social ecological analysis. Such an
analysis allows us to give a coherent account of what it is that we encounter with various degrees of immediacy, and what it is with which
we deal with various degrees of directness, in order to satisfy our needs
to varying degrees. This dialectical complexity is precisely what
Bookchin's dogmatic social ecology seeks to explain away through its
rigid and simplistic categories.[60]

The Social and the Political

Bookchin is at his weakest when he attempts to appear the most philosophical. This is the case with one of his most ambitious theoretical un-
detakings: his articulation of the concept of "the political." Much as
Aristotle announced his momentous philosophical discovery of the
Four Causes, Bookchin announces his discovery of the Three Realms.
He points out that he has "made careful but crucial distinctions between the three societal realms: the social, the political, and the
state."[61] In his own eyes, this discovery has won him a place of distinction in the history of political theory, for the idea "that there could be a
political arena independent of the state and the social . . . was to elude
most radical thinkers."[62] For Bookchin, the social and statist realm
cover almost everything that exists in present-day society. The *statist
sphere* subsumes all the institutions and activities—the "statecraft," as
he likes to call it—through which the state operates. The *social sphere*
includes everything else in society, with the exception of "the political."
This final category encompasses activity in the *public sphere,* a realm
that he identifies "with politics in the Hellenic sense of the term."[63] By
this, he means the proposed institutions of his own libertarian municipalist system and, to varying degrees, its precursors—the diverse
"forms of freedom" that have emerged at certain points in history. For
those who have difficulty comprehending this "carefully distinguished"
sphere, Bookchin points out that "in creating a new politics based on
social ecology, we are concerned with what people do in this *public or
political sphere*, not with what people do in their bedrooms, living
rooms, or basements."[64]

There is considerable unintentional irony in this statement. While
Bookchin does not seem to grasp the implications of his argument, this
means that, whatever we may hope for in the future, for the present we
should not be concerned with what people do *anywhere*, since the political realm does not yet exist to any significant degree. Except insofar

as it subsists in the ethereal realm of political ideas whose time has not yet come, the "political" now resides for Bookchin in his own tiny libertarian municipalist movement—though strictly speaking, even it cannot *now* constitute a "public sphere" considering how distant it is from any actual exercise of public power. Thus, the inevitable dialectical movement of Bookchin's heroic *defense* of the political against all who would "denature it," "dissolve it" into something else, and so on, culminates in the effective *abolition* of the political as a meaningful category in existing society.

There is, however, another glaring contradiction in Bookchin's account of the "social" and the "political." He hopes to make much of the fact (which he declares "even a modicum of a historical perspective" to demonstrate) that "it is precisely the *municipality* that most individuals must deal with directly, once they leave the social realm and enter the public sphere."[65] But since what he calls "the public sphere" consists of his idealized "Hellenic politics," it will be, to say the least, rather difficult for "most individuals" to find it in any actually existing world in which they might become politically engaged. Instead, they find only the "social" and the "statist" realms, into which almost all of the actually existing municipality has already been dissolved, not by any mere theorist, as Bookchin seems to fear, but by the course of history itself. Thus, unless Bookchin is willing to find a "public sphere" in the existing statist institutions that dominate municipal politics, or somewhere in that vast realm of "the social," there is simply no "public sphere" for the vast majority of people to "enter."

While such implications already show the absurdity of his position, his theoretical predicament is in fact much worse than this. For in claiming that the municipality is what most people "deal with directly," he is condemned to defining the municipality in terms of the social—precisely what he wishes most to avoid. Indeed, in a moment of theoretical lucidity he actually begins to refute his own position: "Doubtless the municipality is usually the place where even a great deal of *social* life is existentially lived—school, work, entertainment, and simple pleasures like walking, bicycling, and disporting themselves."[66] Bookchin might expand this list considerably, for almost *anything* that he could possibly invoke on behalf of the centrality of "the municipality" will fall in his sphere of the "social." The actually existing municipality will thus be shown to lie overwhelmingly in his "social" sphere, and his argument thus becomes a demonstration of the centrality of that realm. Moreover, what doesn't fall into the "social" sphere must lie in the actually existing "statist" sphere rather than in the nonexistent "political" one. In fact, his form of (fallacious) argumentation could be used with equal brilliance to show that we indeed "deal most directly" with the *state*, since all the phenomena he lists as lying within a municipality are also located within some nation-state. Indeed, this anarchist's argument works even more effectively as a defense of statism, since even when one walks, bicycles, "disports oneself," or whatever, *outside* a municipality one almost inevitable finds oneself within a nation-state.[67] Bookchin shows some vague awareness that his premises do not lead in the direction of his conclusions. After he lists the various *social* dimensions of the municipality, and as the implications of his argument begin to dawn on him, he protests rather feebly that all this "does not efface its distinctiveness as a unique sphere of life."[68] But that, of course, was not the point in dispute. It is perfectly consistent to accept the innocuous propositions that the municipality is "distinctive" and "a unique sphere of life" while rejecting every one of Bookchin's substantive claims about its relationship to human experience, the public sphere, and the "political."

Bookchin's entire project of dividing society into rigidly defined "spheres" belies his professed commitment to dialectical thought. One of the most basic dialectical concepts is that a thing always is what it is not and is not what it is. However, this is the sort of dialectical tenet that Bookchin *never* invokes, for he prefers a highly conservative conception in which the dialectician somehow "educes" from a phenomenon precisely what is inherent in it as a potentiality.[69] Were he an authentically dialectical thinker rather than a dogmatic one, he would, as soon as he posits different spheres of society (or any reality), consider the ways in which each sphere might be conditioned by and dependent upon those from which it is distinguished. In this connection, even the poststructuralists whom he dismisses with such uncomprehending contempt are more dialectical than Bookchin is, since they at least take the term "differ" in an active sense that implies a kind of mutual determination. In this, they work from the insight of Saussurean linguistics that the meaning of any signifier is a function of the entire system of significations. Bookchin, on the other hand, adheres to a dogmatic, nondialectical view that things simply are what they are, that they are different from what they are not, and that anyone who questions his rigid distinctions must be either a dangerous relativist or a fool.

Gundersen, in *The Environmental Promise of Democratic Deliberation,* suggests how a more dialectical approach might be taken to questions dealt with dogmatically by Bookchin. Gundersen discusses in considerable detail the significance of deliberation as a fundamental aspect of Athenian democracy, the most important historical paradigm for Bookchin's libertarian municipalism. He notes that while the official institutions of democracy consisted of such explicitly "political" forms as the assembly, the courts, and the council, the "political" must also be seen to have existed *outside* these institutions, if the role of deliberation is properly understood. As Gundersen states it, "Much of the deliberation that fueled their highly participatory democracy took place not in the Assembly, Council, or law courts, but in the agora, the public square adjacent to those places."[70] The attempt to constrain the political within a narrow sphere through the magic of definition is doomed to failure, not only when one begins to think dialectically, but also as soon as one carefully examines real, historical phenomena with all their mutual determinations. In the same way that Bookchin's non-dialectical approach flaws his theoretical analysis, it dooms his politics to failure, since it systematically obscures the ways in which the possibilities for "political" transformation are dependent on the deeply political dimensions of spheres that he dismisses as merely "social."

Bookchin also demonstrates his nondialectical approach to the social and the political in his discussion of Aristotle's politics and Greek history. He notes that "the two worlds of the social and political emerge, the latter from the former. Aristotle's approach to the rise of the *polis* is emphatically developmental. . . . The *polis* is the culmination of a political whole from the growth of a social and biological part, a realm of the latent and the possible. Family and village do not disappear in Aristotle's treatment of the subject, but *they are encompassed* by the fuller and more complete domain of the *polis*."[71] But there are two moments in Aristotle's thought here, and Bookchin tellingly sides with the nondialectical one. To the extent that Aristotle maintains a sharp division between the social and the political, his thought reflects a hierarchical dualism rooted in the institutional structure of Athenian society. Since the household is founded on patriarchal authority and a slave economy, it cannot constitute a political realm, a sphere of free interaction between equals. This dualistic, hierarchical dimension of Aristotle is precisely what Bookchin invokes favorably.

There is, on the other hand, a more dialectical moment in Aristotle's thought, which, though still conditioned by hierarchical ideology (as expressed in the concept of "the ruling part") envisions the polis as the realization of the self, family and village. Aristotle says that the polis is "the completion of associations existing by nature" and is "prior in the order of nature to the family and the individual" because "the whole is necessarily prior [in nature] to the part."[72] Implicit in this concept is the inseparable nature of the social and the political. Later, more radically dialectical thought has developed this second moment. An authentically dialectical analysis recognizes that as the political dimension emerges within society, it does not separate itself off from the rest of the social world to embed itself in an ex-

clusive sphere. Rather, as the social whole develops, there is a transformation and politicization of many aspects of what Bookchin calls
"the social" (a process that may take a liberatory, an authoritarian,
or even a totalitarian, direction). In Hegel's interpretation of this process, for example, the state emerges as the full realization of society,
yet it is also the means by which each aspect of society is transformed and achieves its fulfillment.

In a conception of the political that is less ideological than Hegel's
but equally dialectical (if we take the political as the self-conscious
self-determination of the community with its own good as the end), the
emergence of the political in any sphere will be seen both to presuppose
and also to imply its emergence in other spheres. For Bookchin, on the
other hand, the political remains an autonomous realm, and other
spheres of society can only be politicized by being literally absorbed
into that realm (as in the municipalization of production). This nondialectical approach to the political is central to Bookchin's development
of an abstract, idealist, and dogmatic conception of social transformation.

Paideia and Civic Virtue

One of the more appealing aspects of Bookchin's politics is his emphasis on the possibilities for self-realization through participation in political activity. His views are inspired by the Athenian polis, which
"rested on the premise that its citizens could be entrusted with 'power'
because they possessed the personal capacity to use power in a trust-
worthy fashion. The education of citizens into rule was therefore an
education into personal competence, intelligence, moral probity, and
social commitment."[73] These are the kind of qualities, he believes, that
must be created today in order for municipalism to operate successfully.
We must therefore create a similar process of *paideia* in order to combine individual self-realization with the pursuit of the good of the community through the instilling of such civic virtues in each citizen.

But there are major difficulties for this conception of *paideia*. The
processes of socialization are not now in the hands of those who would
promote the programs of libertarian municipalism or anything vaguely
related to it. Rather, they are dominated by the state and, above all, by
economic power and the economistic culture, which aim at training
workers (employees and managers) to serve the existing system of production and a mass of consumers for the dominant system of consumption. Municipalism proposes that a populace that has been so pro-

foundly conditioned by these processes should become a "citizenry," both committed to the process of self-rule and also fully competent to carry it out.

This is certainly a very admirable goal for the future. However, Bookchin's formulations sometimes seem to presuppose that such a citizenry has already been formed and merely awaits the opportunity to take power. He states, for example, that "the municipalist conception of citizenship assumes" that "every citizen is regarded as competent to participate directly in the 'affairs of state,' indeed what is more important, encouraged to do so."[74] But the success of the institutions proposed by Bookchin would seem to require much more than either an assumption of competence or the encouragement of participation in civic affairs. What is necessary is that the existing populace should be transformed into something like Bookchin's "People" through a process of *paideia* that pervasively shapes all aspects of their lives—a formidable task that would itself constitute and also presuppose a considerable degree of social transformation.

To equate this *paideia* primarily with the institution of certain elements of libertarian municipalism hardly seems to be a very promising approach. Indeed, to the extent that aspects of its program are successfully implemented before the cultural and psychological preconditions have been developed, this may very well lead to failure and disillusionment. A program of libertarian municipalism that focuses primarily on the decentralization of power to the local level might indeed have reactionary consequences within the context of the existing political culture of the United States and some other countries. One might imagine a "power to the people's assemblies" that would result in harsh anti-immigrant regulations, extension of capital punishment, institution of corporal punishment, expanded restrictions on freedom of speech, imposition of religious practices, repressive enforcement of morality, and punitive measures against the poor, to cite some proposals that have widespread public support in perhaps a considerable majority of municipalities of the United States. It is no accident that localism has appealed much more to the right wing in the United States than to the Left or the general population, and that reactionary localism is becoming more extremist and more popular. The far right has worked diligently for decades at the grassroots level in many areas to create the cultural preconditions for local reactionary democracy.

Of course, Bookchin would quite reasonably prefer to see his popular assemblies established in more "progressive" locales, so that they could become a model for a new democratic, and, indeed, a libertarian and populist, politics. But far-reaching success for such developments depends on a significant evolution of the larger political culture. To the extent that activists accept Bookchin's standpoint of hostility toward, or at best, unenthusiastic acceptance of the limited value of alternative approaches to social change, this will restrict the scope of the necessary *paideia,* impede the pervasive transformation of society, and undercut the possibilities for effective local democracy.[75]

The Municipalist Program

Libertarian municipalism has increasingly been presented not only as a theoretical analysis of the nature of radical democracy, but also as a programmatic movement for change. Indeed, Bookchin has proposed the program of libertarian municipalism as a basis for organization for the green movement in North America. However, a serious problem in Bookchin's political analysis is that it slips from the theoretical dimension to the realm of practical programs with little thoughtful consideration of how realistic the latter may be. His discussions of a postscarcity anarchist society seemed to refer to an ultimate ideal in a qualitatively different future (even if the coming revolution was sometimes suggested as a possible shortcut to that ideal). While the confederated free municipalities of libertarian municipalism sometimes also seem like a utopian ideal, this perspective has increasingly been presented as a strategy that is capable of creating and mobilizing activist movements in present-day towns and cities. Yet one must ask what the real possibilities for organizing groups and movements under that banner might be, given the present state of political culture, given the actual public to which appeals must be addressed, and not least of all, given the system of communication and information that must be confronted in any attempt to persuade.[76]

The relationship between immediate proposals and long-terms goals in libertarian municipalism is not always very clear. While Bookchin sees changes such as the neighborhood planning assemblies in Burlington, Vermont, as an important advance, even though these assemblies do not have policy-making (or law-making) authority, he does not see certain rather far-reaching demands by the green movement as being legitimate. He recognizes as significant political advance structural changes (like planning assemblies or municipally run services) that move in the direction of municipal democracy or economic municipalization, electoral strategies for gaining political influence or control on behalf of the municipalist agenda, and, to some degree, alternative projects that are independent of the state. On the other hand, he seems to reject, either as irrelevant or as a dangerous form of cooptation, any political proposal for reform of the nation-state beyond the local (or sometimes, the state) level.

Bookchin harshly criticizes, as capitulation to the dominant system, all approaches that do not lead toward municipal direct democracy and municipal self-management. This critique of reformism questions the wisdom of active participation by municipalists, social ecologists, left greens, and anarchists in movements for social justice, peace, and other "progressive" causes when the specific goals of these movements are not linked to a comprehensive liberatory vision of social, economic, and political transformation (or, more accurately, to the precisely correct vision). Bookchin often disparages such "movement" activity and urges activists to focus on working exclusively on behalf of the program of libertarian municipalism.

For example, he and Janet Biehl attack the left greens for their demand to "cut the Pentagon budget by 95 percent" and their proposals for "a $10 per hour minimum wage," "a thirty-hour work week with no loss of income," and a "workers' superfund."[77] The supposed error in these proposals is that they do not eliminate the last 5% of the budget for so-called defense of the nation-state, and that they perpetuate economic control at the national level. Bookchin later dismisses the left greens' proposals as "commonplace economic demands."[78] Furthermore, he distinguishes between his own efforts "to enlarge the *directly democratic* possibilities that exist within the republican system" and the left greens' "typical trade unionist and social democratic demands that are designed to render capitalism and the state more palatable."[79] It is impossible, however, to deduce a priori the conclusion that every institution of procedures of direct democracy is a historically significant advance, while all efforts to influence national economic policy and to demilitarize the nation-state are inherently regressive; the empirical evidence on such matters is far from conclusive. It is at least conceivable, for example, that improvement of conditions for the least privileged segments of society might lead them to become more politically engaged, and perhaps even make them more open to participation in grassroots democracy. In his sarcastic attacks on the left greens, we hear in Bookchin's statements the voice of dogmatism and demagogy.[80]

There is, in fact, an inspiring history of struggles for limited goals that did not betray the more far-reaching visions, and indeed revolutionary impulses, of the participants. To take an example that should be meaningful to Bookchin, the anarchists who fought for the eight-hour work day did not give up their goal of the abolition of capitalism.[81] There is no reason why left greens today cannot fight for a thirty-hour work week without giving up their vision of economic democracy. Indeed, it seems important that those who have utopian visions should also *stand with ordinary people* in their fights for justice and democracy—even when many of these people have not yet developed such visions, and have not yet learned how to articulate their hopes in theoretical terms. Unless this occurs, the prevailing dualistic split between reflection and action will continue to be reproduced in movements for social transformation, and the kind of "People" that libertarian municipalism presupposes will never become a reality. To reject all reform proposals at the level of the nation-state a priori reflects a lack of sensitivity to the issues that are meaningful to people now. Bookchin correctly cautions us against succumbing to a mere "politics of the possible." However, a political purism that dogmatically rejects reforms that promise a meaningful improvement in the conditions of life for many people chooses to stand *above* the actual people in the name of "the People" (who despite their capitalization remain merely theoretical).[82]

Bookchin is no doubt correct in his view that groups like the left greens easily lose the utopian and transformative dimension of their outlook as they become focused on reform proposals that might immediately appeal to a wide public. It is true that a left green proposal to "democratize the United Nations" seems rather bizarre (to say the least) from the decentralist perspective of the green movement. Yet it is inconsistent for Bookchin to dismiss all proposals for reform merely because they "propose" something less than the immediate abolition of the nation-state. Libertarian municipalism itself advocates, *for the immediate present,* working for change within subdivisions of the nation-state, as municipalities (and states, including small ones like Vermont) most certainly are. Bookchin has himself encouraged municipalists to work actively in a campaign against the extension of Vermont's gubernatorial term from two to four years. While this is a valid issue of degree of democratic control, its implications in regard to the power of the nation-state can certainly not be compared to those of a 95% reduction in national military spending.

Social ecological politics requires a dialectical analysis of social phenomena, which implies a careful analysis of the political culture (in relation to its larger natural and social context) and an exploration of the possibilities inherent in it. The danger of programmatic tendencies, which are endemic to the traditional Left and to all the heretical sectarianisms it has spawned, is that they rigidify our view of society; they reinforce dogmatism, inflexibility, and attachment to one's ideas; they limit our social imagination; and they discourage the open, experimental spirit that is necessary for creative social change.

While libertarian municipalism is sometimes interpreted in a narrower, more sectarian way (as it appears especially in Bookchin's polemics against other points of view), it can also be taken as a more general orientation toward radical grassroots democracy. Looked at in this broader sense, Bookchin's libertarian municipalism can make a significant contribution to the development of our vision of a free, cooperative community. Bookchin has sometimes presented a far-reaching list of proposals for developing more ecologically responsible and democratic communities. These include the establishment of community credit unions, community-supported agriculture, associations for local self-reliance, and community gardens.[83] Elsewhere he includes in the "minimal steps" for creating "Left Green municipalist movements" such activities as electing council members who support "assemblies and other popular institutions"; establishing "civic banks to fund municipal enterprises and land purchases"; creating community-owned enterprises; and forming "grassroots networks" for various purposes.[84] In a discussion of how a municipalist movement might be initiated in the state of Vermont, he presents proposals that emphasize cooperatives and even small individually owned businesses.[85] He suggests that the process could begin with the public purchase of unprofitable enterprises (which would then be managed by the workers), the establishment of land trusts, and support for small-scale productive enterprises. This could be done, he notes, without infringing "on the proprietary rights of small retail outlets, service establishments, artisan shops, small farms, local manufacturing enterprises, and the like."[86] He concludes that in such a system "cooperatives, farms, and small retail outlets would be fostered with municipal funds and placed under growing public control."[87] He adds that a "People's Bank" to finance the economic projects could be established, buying groups to support local farming could be established, and public land could be used for "domestic gardening."[88]

These proposals present the outline of an admirable program for promoting a vibrant local economy based on cooperatives and small businesses. Yet it is exactly the "municipalist" element of his program that might be less than practical for quite some time. It seems likely that for the present the members of cooperatives and the owners of small enterprises would have little enthusiasm for coming under "increasing public control," if this means that the municipality (either through an assembly or local officials) increasingly takes over management decisions. Whatever might evolve eventually as a cooperative economy develops, a program for change in the real world must either have an appeal to an existing public or must have a workable strategy for creating such a public. There is certainly considerable potential for broad support for "public control" in areas like environmental protection, health and safety measures, and greater economic justice for workers. However, the concept of "public control" of economic enterprises through management by neighborhood or municipal assemblies is, to use Bookchin's terminology, a "nonsense demand," since the preconditions for making it meaningful do not exist and are not even addresssed in Bookchin's politics.[89]

The Fetishism of Assemblies

While Bookchin sees the municipality as the most important political realm, he identifies the municipal assembly as the privileged organ of democratic politics and puts enormous emphasis on its place in both the creation and functioning of free municipalities. "Popular assemblies," he says, "are the minds of a free society; the administrators of their policies are the hands."[90] But unless this is taken as an attempt at poetry, it is in some ways a naïve and undialectical view. The mind of society—its reason, passion, and imagination—is always widely dispersed throughout all social realms. And the more that this is the case, the better it is for the community. Not only is it not *necessary* that most creative thought take place in popular assemblies, it is *inconceivable* that most of it should occur there. In a community that encourages creative thinking and imagination, the "mind" of society would operate through the intelligent, engaged reflection of individuals, through a diverse, thriving network of small groups and local institutions in which these individuals would express and embody their hopes and ideals for the community, and through vibrant democratic media of communication in which citizens would exchange ideas and shape the values of their community. And although in an anarchist critique of existing bureaucracy administrators might be depicted rhetorically as mindless, it does not seem desirable that in a free society they should be dismissed as necessarily possessing this quality. All complex systems of social organization will require some kind of administration and will depend not only on the good will but also on the intelligence of those who carry out policies. It seems impossible to imagine any form of assembly government that could formulate such specific directives on complex matters that administrators would have no significant role in shaping policy. Bookchin tellingly lapses into edifying rhetoric and political sloganeering when he discusses the supremacy of the assembly in policy making. Were he to begin to explore the details of how such a system might operate, he would immediately save others the trouble of deconstructing his system.

The de facto policy-making power of administrators might even be greater in Bookchin's system than in others, in view of the fact that he does not propose any significant sphere for judicial institutions that might check administrative power. Unless we assume that society would become and thereafter remain quite simplified—an assumption that is inconsistent with Bookchin's beliefs about technological development, for example—then it would be unrealistic to assume that all significant policy decisions could be made in an assembly, or even supervised directly by an assembly. A possible alternative would be a popular judiciary; however, the judicial realm remains almost a complete void in Bookchin's political theory, despite references to popular courts in classical Athens and other historical cases. One democratic procedure that could perform judicial functions would be popular juries (as proposed by Godwin two centuries ago) or citizens' committees (as recently suggested by Burnheim)[91] that could oversee administrative decision making. However, Bookchin's almost exclusive emphasis on the assembly—what we might call his "ecclesiocentrism"—precludes such possibilities.

Bookchin responds to these suggestions concerning popular juries and citizens' committees with what he thinks to be the devastating allegation that what I am "really calling for here" are "courts and councils, or bluntly speaking, *systems of representation*."[92] While it is far from clear that a "council" is inherently undesirable under all historical circumstances, what I discuss in the passage he attacks is citizens' committees, not councils.[93] What I "call for" is not some specific political form, but rather a consideration of various promising political forms whose potential can only be determined through practice and experimentation. Moreover, Bookchin's comments show ignorance of the nature of the proposals of Godwin and Burnheim that are cited, and unwillingness to investigate them before beginning his attack. Neither proposes a system of "representation." One of the appealing aspects of the jury or committee proposals is that since membership on juries or committees is through random selection (not election of "representatives"), all citizens have an equal opportunity to exercise decision-making power. Some of the possible corrupting influences of large assemblies (encouragement of egoistic competition, undue influence by power-seeking personalities, etc.) are much less likely to appear in this context. Furthermore, such committees and juries offer a way of *avoiding* the need for representation, since they are a democratic means of performing necessary functions that cannot possibly be carried out at the assembly level. As will be discussed, Bookchin's municipalism does not successfully address the question of how "confederal" actions can be carried out without representation; proponents of decentralized democracy would therefore be wise to consider various means by which the necessity for representation might be minimized in a less than utopian world.

In discussing his conception of "participatory democracy," Bookchin notes the roots of the concept in the politics of the New Left and
the counterculture of the 1960s. One implication of democracy in this
context was that "people were expected to be *transparent* in all their
relationships and the ideas they held."[94] He laments the fact that these
democratic impulses were betrayed by a movement toward dogmatism,
centralization, and institutionalization. Yet the concept of *transparency*,
like that of "the unmediated," requires critical analysis. Bookchin
might have achieved a more critical approach to such concepts had he
applyed a dialectical analysis to them. Unfortunately, the naïve expecta-
tion that people merely "be" transparent may become a substitute for
the more difficult and time-consuming but ultimately rewarding processes of self-reflection and self-understanding on the personal and
group levels. Values like "transparency" and "immediacy" often inhibit
understanding of group processes and function as an ideology that dis-
guises implicit power relationships and subtle forms of manipulation,
which are often quite opaque, highly mediated, and resistant to superficial analysis.

It is important that such disguised power relations should not find
legitimacy through the ideology of an egalitarian, democratic assembly,
in which "the People" act in an "unmediated" fashion and in which
their will is "transparent." The fact is that in assemblies of hundreds,
thousands, or even potentially tens of thousands of members (if we are
to take the Athenian polis as a model), there is an enormous potential
for manipulation and power-seeking behavior. If it is true that power
corrupts, as anarchists more than anyone else have stressed, then anarchists cannot look with complacency on the power that comes from being the center of attention of a large assembly, from success in debate
before such an assembly, and from the quest for victory for one's cause.
To minimize these dangers, it is necessary to avoid idealizing assemblies, to analyze carefully their strengths and weaknesses, and to experi-
ment with processes that can bring them closer to the highest ideals
that inspire them. In addition, there is the option of rejecting
Bookchin's proposal that all political power should be concentrated in
the assembly and separating it instead among various participatory institutions.

And whatever the strengths and weaknesses of assemblies as an organizational form, we must also ask whether it is even possible for sovereign municipal assemblies to be viable as the fundamental form of
political decision making in the real world. Bookchin concedes that local assemblies might have to be less than "municipal" in scope. He recognizes that given the size of existing municipalities, there will be a
need for more decentralized decision-making bodies. He suggests that

"whether a municipality can be administered by all its citizens in a single assembly or has to be subdivided into several confederally related assemblies depends much on its size," and he proposes that the assembly might be constituted on a block, neighborhood, or town level.[95] Since contemporary municipalities in much of the world range in population up to tens of millions, and neighborhoods themselves up to hundreds of thousands, the aptness of the term "municipalism" for a form of direct democracy should perhaps be questioned.[96] It would seem that in highly urbanized societies it would be much more feasible to establish democratic assemblies at the level of the neighborhood or even smaller units than at the municipal level, as Bookchin himself concedes.

The problem of defining neighborhood communities often poses difficulties. Bookchin claims that New York City, for example, consists of neighborhoods that are "organic communities."[97] It is true that there exists a significant degree of identification with neighborhoods that can contribute to the creation of neighborhood democracy. Yet to describe the neighborhoods of New York City or other contemporary cities as "organic communities" is a vast overstatement; indeed, one wonders if Bookchin is referring more to his idealized view of the past than to present realities. Contemporary cities (including New York City) have been thoroughly transformed according to the exigencies of the modern bureaucratic, consumerist society, with all the atomization and privatization that this implies. Natives of metropolitan centers such as Paris complain that traditional neighborhoods have been completely destroyed by commercialization, land speculation, and displacement of the less affluent to the suburbs. In the United States, much of traditional urban neighborhood life has been undermined by social atomization; institutionalized, structural racism; and the migration of capital and economic support away from the center. Bookchin correctly uses my own community of New Orleans as an example of a city that has a strong tradition of culturally distinct neighborhoods that have endured with strong identities until recent times.[98] However, it is also a good example of the culturally corrosive effects of contemporary society, which progressively transforms local culture into a commodity for advertising, real estate speculation, and tourism, while it destroys it as a lived reality. Thus, the neighborhood "organic community" seems to be much more of an imaginary construct (that is often entangled with nostalgic feelings and that reflects class and ethnic antagonisms) than an existing state of affairs. It is essential to see these limitations in the concept, and then to develop its imaginary possibilities as part of a liberatory process of social regeneration.

The apparently large size of assemblies (even at the neighborhood level) proposed for urban areas raises questions about how democratic these bodies could be. In Barber's discussion of such assemblies, he suggests that their membership would range from five to twenty-five thousand.[99] Bookchin says that they might encompass units of a single block up to dozens of blocks in an urban area, and thus might sometimes reach a similar level of membership. It is difficult to imagine the city block of present-day society as the fundamental political unit (though visionary proposals for a radically transformed future have made a good case for re-creating it as a small ecocommunity). And, in fact, libertarian municipalism is almost always formulated in terms of *municipal* and *neighborhood* assemblies. Therefore, in practical terms, it is proposing rather large assemblies for the foreseeable future in highly populated, urbanized societies.

Bookchin's discussion is curiously (and rather suspiciously) vague on the topic of the scope of decision making by assemblies. He does make it clear that he believes that *all* important policy decisions can and should be made in the assembly, even in the case of emergencies. He confidently assures us that, "given modern logistical conditions, there can be no emergency so great that assemblies cannot be rapidly convened to make important policy decisions by a majority vote and the appropriate boards convened to execute these decisions—irrespective of a community's size or the complexity of its problems. Experts will always be available to offer their solutions, hopefully competing ones that will foster discussion, to the more specialized problems a community may face."[100] But this mere affirmation of faith is hardly convincing. In a densely populated, technologically complex, intricately interrelated world, every community will face problems that can hardly be dealt with on an ad hoc basis by large assemblies.

Amazingly, Bookchin never explores the basic theoretical question of whether any formal system of local law should exist, and how the policy decisions of assemblies should be interpreted and applied to particular cases. However, his position seems to collapse were he to give any answer to this question. If general rules and policy decisions (i.e., laws) are adopted by an assembly, then they must be applied to particular cases and articulated programmatically by judicial and administrative agencies. It is then inevitable that these agencies will have some share in political power. This alternative is inconsistent with his many affirmations of the supremacy of the assembly. On the other hand, if no general rules are adopted, then the assembly will have the impossibly complex task of applying rules to all disputed cases and formulating all important details of programs. We are left with a purgatorial vision of hapless citizens condemned to listening endlessly to "hopefully competing" experts on every imaginable area of municipal administration. Given these two unpromising alternatives, Bookchin seems to choose the impossible over the inconsistent.

Furthermore, there are certain well-known dangers of large assemblies that would presumably threaten neighborhood or municipal assemblies too. Among the problems that often emerge are competitiveness, egotism, theatrics, demagogy, charismatic leadership, factionalism, aggressiveness, obsession with procedural details, domination of discussion by manipulative minorities, and passivity of the majority. While growth of the democratic spirit might reduce some of these dangers, they might also be aggravated by the size of the assembly, which would be many times larger than most traditional legislative bodies. In addition, the gap in political sophistication between individuals in local assemblies will no doubt be much greater than in bodies composed of traditional political elites. Finally, the assembly would lose one important advantage of representation: elected representatives or delegates can be chastised for betraying the people when they seem to act contrary to the will or interest of the community. On the other hand, those who emerge as leaders of a democratic assembly, and those who take power by default if most do not participate actively in managing the affairs of society, can be accused of nothing, since they are acting as equal members of a popular democratic body.[101]

To say the least, an extensive process of self-education in democratic group processes would be necessary before large numbers of people would be able to work together cooperatively in large meetings. And even if some of the serious problems mentioned here are mitigated, it is difficult to imagine how they could be avoided entirely in assemblies with thousands of participants, as are sometimes proposed, at least until institutions other than assemblies have radically changed personality structures. Indeed, the term "face-to-face democracy" that Bookchin often uses in reference to these assemblies seems rather bizarre when applied to these thousands of faces (assuming that most of them face up to their civic responsibilities and attend).

An authentically democratic movement will recognize the considerable potential for elitism and power seeking within assemblies. It will deal with this threat not only through procedures within assemblies, but above all by creating a communitarian, democratic culture that will express itself in decision-making bodies and in all other institutions. For the assembly and other organs of direct democracy to contribute effectively to an ecological community, they must be purged of the competitive, agonistic, masculinist aspects that have often corrupted them. They can only fulfill their democratic promise if they are an integral expression of a cooperative community that embodies in its institutions the love of humanity and nature. Barber makes exactly this point when he states that strong democracy "attempts to balance adversary politics by nourishing the mutualistic art of listening," and goir.g beyond mere toleration, seeks "common rhetoric evocative of a common democratic discourse" that should "encompass the affective as well as the cognitive mode."[102] Such concerns echo recent contributions in feminist ethics, which have pointed out that the dominant moral and political discourse has exhibited a one-sided emphasis on ideas and principles and neglected the realm of feeling and sensibility. In this spirit, we must explore the ways in which the transition from formal to substantive democracy depends not only on the establishment of more radically democractic forms, but also on the establishment of cultural practices that foster a democratic sensibility.

Municipal Economics

One of the most compelling aspects of Bookchin's political thought is the centrality of his ethical critique of the dominant economistic society and his call for the creation of a "moral economy" as a precondition for a just ecological society. He asserts that such a "moral economy" implies the emergence of "a productive community" to replace the amoral "mere marketplace" that currently prevails. Further, it requires that producers "explicitly agree to exchange their products and services on terms that are not merely 'equitable' or 'fair' but supportive of each other."[103] He believes that if the prevailing system of economic exploitation and the dominant economistic culture based on it are to be eliminated, a sphere must be created in gentlemen. which people find new forms of exchange to replace the capitalist market, and this sphere must be capable of continued growth. Bookchin sees this realm as that of the municipalized economy. He states that "under libertarian municipalism, property becomes part of a larger whole that is controlled by the citizen body in assembly as citizens."[104] Elsewhere, he explains that "land, factories, and workshops would be controlled by popular assemblies of free communities, not by a nation-state or by worker-producers who might very well develop a proprietary interest in them."[105]

However, for the present at least, it is not clear why the municipalized economic sector should be looked upon as the primary realm, rather than as one area among many in which significant economic transformation might begin. It is possible to imagine a broad spectrum of self-managed enterprises, individual producers and small partnerships that would enter into a growing cooperative economic sector that would incorporate social ecological values. The extent to which the communitarian principle of distribution according to need could be achieved would be proportional to the degree to which cooperative and communitarian values had evolved—a condition that would depend on complex historical factors that cannot be predicted beforehand. Book-chin is certainly right in his view that participation in a moral economy would be "an ongoing education in forms of association, virtue, and decency"[106] through which the self would develop. And it is possible that ideally "price, resources, personal interests, and costs" might "play no role in a moral economy" and that there would be "no 'accounting' of what is given and taken."[107] However, we always begin with a his-torically determined selfhood in a historically determined cultural con-text. It is quite likely that communities (and self-managed enterprises) might find that in the task of creating liberatory institutions within the contraints of real history and culture, the common good is attained best by preserving some form of "accounting" of contributions from citizens and distribution of goods. To whatever degree Bookchin's anarcho-communist system of distribution is desirable as a long-term goal, the attempt to put it into practice in the short run, without developing its psychological and institutional preconditions, would be a certain recipe for disillusionment and economic failure.

Bookchin attributes to municipalization an almost miraculous power to abolish egoistic and particularistic interests. He and Biehl at-tack proposals of the left greens for worker self-management on the grounds that such a system does not, as in the case of municipalization, "eliminate the possibility that *particularistic* interests of *any* kind will develop in economic life."[108] While the italics reflect an admirable hope, it is not clear how municipalization, or any other political pro-gram, no matter how laudable it may be, can assure that such interests are entirely eliminated. Bookchin and Biehl contend that in "a democ-ratized polity" workers would develop "a general public interest"[109] rather than a particularistic one of any sort. But it is quite possible for a municipality to put its own interest above that of other communities, or that of the larger community of nature. The concept of "citizen of a municipality" does not in itself imply identification with "a general public interest." To the extent that concepts can perform such a func-tion, "citizen of the human community" would do so much more ex-plicitly, and "citizen of the earth community" would do so much more ecologically.

Under Bookchin's libertarian municipalism, there is a possible (and perhaps inevitable) conflict between the *particularistic* perspective of the worker in a productive enterprise and the *particularistic* perspective of the citizen of the municipality. Bookchin and Biehl propose that "workers in their area of the economy" be placed on advisory boards that are "merely technical agencies, with no power to make policy deci-sions."[110] This would do little if anything to solve the problem of con-flict of interest. Bookchin calls the "municipally managed enterprise" at one point "a worker-citizen controlled enterprise,"[111] but the control is effectively limited to members of the community acting as citizens, not as workers.[112] Shared policy making seems on the face of it more of a real-world possibility, however complex it might turn out to be. In either case (pure community democracy or a mixed system of community and workplace democracy), it seems obvious that there would be a continual potential for conflict between workers who are focused on their needs and responsibilities as producers and assemblies that are in theory focused on the needs and responsibilities of the local community, not to mention those of the entire earth community, of which their own community is but a part.

Putting aside the ultimate goals of libertarian municipalism, Bookchin suggests that in a transitional phase, its policies would "not infringe on the proprietary rights of small retail outlets, service establishments, artisan shops, small farms, local manufacturing enterprises, and the like."[113] The question arises, though, why this sector should not continue to exist in the long term, alongside more cooperative forms of production. There is no conclusive evidence that such small enterprises are necessarily exploitative or that they cannot be operated in an ecologically sound manner. Particularly if the larger enterprises in a regional economy are democratically operated, the persistence of such small enterprises does not seem incompatible with social ecological values. This is even more the case to the degree that the community democratically establishes just and effective parameters of social and ecological responsibility.

However, Bookchin dogmatically rejects this possibility. He claims that if any sort market continues to exist, then "competition will force even the smallest enterprise eventually either to grow or to die, to accumulate capital or to disappear, to devour rival enterprises or to be devoured."[114] Yet Bookchin has himself noted that historically the existence of a *market* has not been equivalent to the existence of a *market-dominated society*. He has not explained why such a distinction cannot hold in the future. He has himself been criticized by "purist" anarchists who attack his acceptance of government as a capitulation to "archism." Yet he rightly distinguishes between the mere existence of governmental institutions and *statism*, the system of political domination that results from the centralization of political power in the state. Similarly, one may distinguish between the mere existence of market exchanges and *capitalism*--the system of economic domination that results from the concentration of economic power in large corporate enterprises. Bookchin asserts that the existence of any market sector is incompatible with widespread decentralized democratic institutions and cooperative forms of production. While he treats this assertion as if it were an empirically verified or theoretically demonstrated proposition, it is, until he presents more evidence, merely an article of (his) ideological faith.[115]

But whatever the long-term future of the market may be, it is in fact the economic context in which present-day experiments take place. If municipally owned enterprises are established, they will necessarily operate within the market, if only because the materials they need for production will be produced within the market economy. It is also likely that they would choose to sell their products within the market, since the vast majority of potential consumers, including those most sympathetic to cooperative experiments, would still be operating within the market economy. Indeed, it is not certain that even if a great many such municipal enterprises were created that they would choose to limit their exchanges entirely to the network of similar enterprises rather than continuing to participate in the larger market. In view of the contingencies of history, to make any such prediction would reflect a kind of "scientific municipalism" that is at odds with the dialectical principles of social ecology. But whatever may be the case in the future, to the extent that municipalized enterprises are proposed as a real-world practical strategy, they will necessarily constitute (by Bookchin's own criteria) a "reform" within the existing economy. Thus, it is inconsistent for advocates of libertarian municipalism to attack proposals for self-management, such as those of the left greens, as mere reformism. These proposals, like Bookchin's, are incapable of abolishing the state and capitalism by fiat. But if they were adopted, they would represent a real advance in expanding the cooperative and democratic aspects of production, while at the same time improving the economic position of the less privileged members of society.

Bookchin has increasingly downplayed the idea that social ecology should emphasize the importance of developing a diverse, experimental, constantly growing cooperative sector within the economy, and now focuses almost exclusively on the importance of "municipalization of the economy."[116] But while he has been writing about municipalism for decades, he has produced nothing more than vague and seemingly self-contradictory generalizations about how such a system might operate. He does not present even vaguely realistic answers to many basic questions. How might a municipality of about fifty thousand people (e.g., metropolitan Burlington, Vermont), over one million people (e.g., metropolitan New Orleans) or over eight million people (e.g., the Parisian region) develop a coherent municipal economic plan in a "directly democratic" way? Would the neighborhood or municipal assembly have even vaguely the same meaning in these diverse contexts (not to mention what it might mean in third world megalopolises such as Mexico City, Lagos, or Calcutta; in the villages of Asia, Africa, and Latin America; or on the steppes of Mongolia)? Could delegates from hundreds or thousands of block or neighborhood assemblies come to an agreement with "rigorous instructions" from their assemblies? Bookchin's municipalism offers no answers to these questions, and as we will see, neither does his confederalism. He is certainly right when he says that "one of our chief goals must be to radically decentralize our industrialized urban areas into humanly-scaled cities and towns" that are ecologically sound.[117] But a social ecological politics must not only aim at such far-reaching, visionary goals but also offer effective political options for the increasing proportion of human beings who live in highly populated and quickly growing urban areas, and who face serious urban crises requiring practical responses as soon as possible.

Bookchin's most fundamental economic principle also poses questions that he has yet to answer. He contends that with the municipalization of the economy, the principle of "from each according to his abilities and to each according to his needs" becomes "institutionalized as part of the public sphere."[118] How, one wonders, might abilities and needs be determined according to Bookchinist economics? Should a certain amount of labor be required of each citizen, or should the amount be proportional to the nature of the labor? Should those who have more ability to contribute, or whose work fulfills more needs, be required to work more? Of course, these questions can only be answered by specific communities through actual experiments in democratic decision making and self-organization. However, debate over these issues has a long history within ethics and political theory; socialists, communists, anarchists, and utopians have all devoted much attention to them (not to mention liberals such as Rawls). If the theory of libertarian municipalism is to inspire the necessary experiments, municipalists must at least suggest *possible* answers that might convince members of their own and other communities that the theory offers a workable future, or at least they must suggest what it might mean to try to answer such questions.

Bookchin finds it quite disturbing that I could find his invocation of the famous slogan concerning abilities and needs "problematical." One can almost hear his annoyance as he explains that "the *whole point* behind this great revolutionary slogan is that in a communistic post-scarcity economy, abilities and needs are not, strictly speaking, 'determined'—that is, subject to bourgeois calculation," which is to be replaced with "a basic decency and humaneness."[119] Once more one is tempted to ask how Bookchin can present himself as a staunch opponent of mysticism and yet orient his thought toward a final good that is an inexpressible mystery, not to mention a logical contradiction. It is clear that many of the revolutionaries who adhered to Bookchin's beloved slogan actually believed that needs and abilities could, at least in some general way, be "determined." However, Bookchin himself believes that certain acts should be performed and certain things should be distributed "according to" that which *cannot* be "determined." This may be an edifying belief, but it is also an absurdity, pure idealism, and an abdication of the "rationality" that Bookchin claims to value so highly.

But even if this particular form of mysticism were the correct standpoint toward some ultimately utopian society, it would not give us much direction concerning how to get there. Can anyone really take seriously a "libertarian municipalism" that proposes a municipalization of all enterprises, after which conditions of work and distribution of products would be determined (or perhaps we should say "nondetermined") by "basic decency and humaneness"? Once again, the problem of Bookchin's lack of mediations between an idealized goal and actually existing society becomes apparent. And this is not to say that his utopian goal is itself coherent. For despite his self-proclaimed role as the defender of "Reason," he scrupulously avoids consideration of the role of rationality in utopian distribution, in this case falling back instead on mere *feeling*, dualistically divorced from rationality according to the demands of ideological consistency. This is, of course, his only option short of a fundamental rethinking of his position. For reason, unfortunately for Bookchin, expresses itself in *determinations*, as tentative and self-transforming as these determinations may be.

Bookchin presents two additional arguments for his position, both of which have appeared many times in the Bookchinian oeuvre. Both reduce essentially to an appeal to faith. First, he claims that if "'primal' peoples" could "rely on usufruct and the principle of the irreducible minimum," then his ideal society could certainly do without "contractual or arithmetical strictures."[120] But this is merely a variation on the famous "If we can put a man on the moon, then we can do X" argument. According to this lunar fallacy, some proposal, the feasibility of which in no way follows from a moon landing, is argued to be a viable option because the latter achievement proved possible. What is true of tribal societies is that they have usually followed distinct rules of distribution and, indeed, often quite strict and complex ones based on kinship and the circulation of gifts. Whatever the content of these rules (which have often been very humane, ecological, etc.), it certainly does not follow from the fact that previous societies have adhered to these rules that some future society can get along *without rules of distribution*, quantitative or otherwise.

In his second argument, Bookchin notes that neither he nor I will make decisions for any future "post-scarcity society guided by reason," but only those who will actually live in it. This statement is undeniably true (assuming neither of us ever lives in it). However, this fact lends absolutely no support to Bookchin's position, since it is quite possible that these rational utopians might look back on his analysis of such a society and find it to be unconvincing or even absurd. If he wishes merely to express his faith that in his final rational utopia people will achieve things that we can hardly conceive of in our present fallen state, it would be difficult to argue with his position. However, if he intends to argue that a specific form of organization is a reasonable goal for a movement for social change, then he must be willing to offer evidence for this view rather than the merely edifying conception that "in utopia all things are possible."

Bookchin's Confederacy

Anarchist political thought has usually proposed that social cooperation beyond the local level should take place through voluntary federations of relatively autonomous individuals, productive enterprises, or communities. While classical anarchist theorists such as Proudhon and Bakunin called such a system "federalism," Bookchin calls his variation on this theme "confederalism." He describes its structure as consisting of "above all a network of administrative councils whose members or delegates are elected from popular face-to-face democratic assemblies, in the various villages, towns, and even neighborhoods of large cities."[121] Under such a system, we are told, power remains entirely in the hands of the assemblies. "Policymaking is exclusively the right of popular community assemblies," while "administration and coordination are the responsibility of confederal councils."[122] Councils exist only to carry out the will of the assemblies. Toward this end, "the members of these confederal councils are strictly mandated, recallable, and responsible to the assemblies that chose them for the purpose of coordinating and administering the policies formulated by the assemblies themselves."[123] Thus, while majority rule of some sort is to prevail in the assemblies, which are the exclusive policy-making bodies, the administrative councils are strictly limited to carrying out what these bodies decide.

However, it is not clear how this absolute division between policy making and administration could possibly work in practice. How for example, is administration to occur when there are disagreements on policy between assemblies? Libertarian municipalism is steadfastly

against delegation by assemblies of policy-making authority, so all collective activity must presumably depend on the *consensus* of assemblies,
as expressed in the "administrative councils." If there is a majority vote
on policy issues, then this would mean that policy would indeed be
made at the confederal level. Bookchin is quick to attack "the tyranny
of consensus" as a decision-making procedure within assemblies in
which each member of the group is free to compromise for the sake of
the common good. Ironically, he seems obliged to depend on it for decision making in bodies whose members are rigidly mandated to vote according to previous directions from their assemblies.

Or at least he seems to be committed to such a position until he
considers what will occur when some communities do not abide by the
fundamental principles or policies adopted in common. Bookchin states
that "if particular communities or neighborhoods—or a minority
grouping of them—choose to go their own way to a point where human rights are violated or where ecological mayhem is permitted, *the
majority* in a local or regional confederation has every right to prevent
such malfeasance through its confederal council."[124] However, this proposal blatantly contradicts his requirement that policy be made only at
the assembly level. If sanctions are imposed by a majority vote of the
council, this would be an obvious case of a quite important policy being adopted above the assembly level. A very crucial, unanswered question is by what means the confederal council would exercise such a
"preventive" authority (presumably Bookchin has in mind various
forms of coercion). But whatever his answer might be, such action
would constitute *policy making* in an important area. There is clearly a
broad scope for interpretation of what does or does not infringe on human rights, or what does or does not constitute an unjustifiable ecological danger. If the majority of communities acting confederally
through a council acts coercively to deal with such basic issues, then
certain statelike functions would emerge at the confederal level.

It appears that the only way to avoid this result is to take a purist
anarchist approach and assume that action can only be taken at any
level above the assembly through fully voluntary agreements, with full
rights of secession on any issue (including "mayhem"). According to
such an approach, a community would have the right to withdraw
from common endeavors, even for purposes that others might think un-
just to humans or ecologically destructive. Of course, the other communities would still be able to take action against the allegedly offending
community because of its supposed misdeeds. They would have had
this ability in any case, even if the offending community had never entered into the "non-policy-making" confederal agreement. Should
Bookchin choose to adopt this position, he would have to give up the

concept of enforcement at the confederal level. He would then be proposing a form of confederal organization in which everything would be decided by consensus, and in which the majority of confederating communities would have no power of enforcement in any area. His position would then have the virtue of consistency, though very few would consider it a viable way of solving problems in a complex world.

There are other aspects of Bookchin's confederalism that raise questions about the practicality or even the possibility of such a system. He proposes that activities of the assemblies be coordinated through the confederal councils, whose members must be "rotatable, recallable, and, above all, rigorously instructed in written form to support or oppose any issue that appears on the agenda."[125] But could such instruction be a practical possibility in modern urban society (assuming, as Bookchin seems to, that the arrival of municipalism and confederalism are not to be delayed until after the dissolution of urban industrial society)? Perhaps Paris might be taken as an example, in honor of the Parisian "sections" of the French Revolution that he mentions so often as a model for municipal politics. Metropolitan Paris has roughly eight and one-half million people. If government were devolved into assemblies for each large neighborhood of twenty-five thousand people, there would be three hundred and forty assemblies in the metropolitan area. If it were decentralized into much more democratic assemblies for areas of a few blocks, with about a thousand citizens each, there would then be eight thousand five hundred Parisian assemblies. If the city thus had hundreds or even thousands of neighborhood assemblies, and each "several" assemblies (as Bookchin suggests) would send delegates to councils, which presumably would have to form even larger confederations for truly municipal issues, could the chain of responsibility hold up? And if so, how?

When confronted with such questions, Bookchin offers no reply other than that he doesn't believe in the existence of the kind of centralized, urbanized society in which these problems arise. However, his political proposals are apparently directed at people living in precisely such a world. If municipalism is not practicable in the kind of society in which real human beings happen to find themselves, then the question arises of what other political arrangements might be practicable and also move toward the goals that Bookchin embodies in municipalism. Yet his politics does not address this issue. We are left with the abstract pursuit of an ideal and an appeal to the will that it be realized. Bookchin's late work in particular expresses a defiant will that history should become what it ought to be and a poorly contained rage at the thought that it stubbornly seems not to be doing so. Objections that his social analysis and political proposals lack an adequate relation to actual history are usually met with ridicule and sarcasm, seldom with reasoned argument.

Municipalizing Nature?

As Bookchin has increasingly focused on the concept of *municipalist* politics, the theme of *ecological* politics has faded increasingly further into the background of his thought. In fact, the idea of a *bioregional* politics has never really been developed in his version of social ecology. Yet, there are two fundamental social ecological principles that essentially define a bioregional perspective. One is the recognition of the dialectic of nature and culture, in which the larger natural world is seen as an active coparticipant in the creative activities of human beings. The other is the principle of unity-in-diversity, in which the unique, determinate particularity of each part of the whole is seen as making an essential contribution to the unfolding of the developing whole. While Bookchin has done much to stress the importance of such general principles, what has been missing in his discussion of politics is a sensitivity to the details of the natural world and the quite particular ways in which it can and does shape human cultural endeavors, and a sense of inhabiting a natural whole, whether an ecosystem, a bioregion, or the entire biosphere.

If one searches Bookchin's writings carefully, one finds very little detailed discussion of ecological situatedness and bioregional particularity, despite a theoretical commitment to such values. Typically, he limits himself to statements such as that there should be a "sensitive balance between town and country"[126] and that a municipality should be "delicately attuned to the natural ecosystem in which it is located."[127] In *The Ecology of Freedom* he says that ecological communities should be "networked confederally through ecosystems, bioregions, and biomes," that they "must be artistically tailored to their natural surroundings," and that they "would aspire to live with, nourish, and feed upon the life-forms that indigenously belong to the ecosystems in which they are integrated."[128] These statements show concern for the relationship of a community to its ecological context, but the terms chosen to describe this relationship do not imply that bioregional realities are to be central to the culture. Furthermore, Bookchin's discussions of confederalism invariably base organization on political principles and geographical proximity. He does not devote serious attention to the possibility of finding a bioregional basis for confederations or networks of communities.

It is possible that an underlying concern that discourages Bookchin from focusing on bioregional culture (and quite strikingly, on communal traditions also) is his mistaken perception that these realities somehow threaten the freedom of the individual. A bioregional approach places very high value on human creative activity within the context of a sense of place, in the midst of a continuity of natural and cultural history. Bioregionalism is based on a kind of commitment that Bookchin steadfastly rejects; that is, a giving oneself over to the other, a choosing without "choosing to choose," a recognition of the claim of the other on the deepest levels of one's being. Bookchin descibes his ideal community as "the commune that unites individuals by what they choose to like in each other rather than what they are obliged by blood ties to like."[129] But when one affirms one's membership in a human or natural community, one is hardly concerned with "choosing what to like and not to like" in the community (though one may certainly judge one's own human community quite harshly out of love and compassion for it). The community becomes, indeed, an extension of one's very selfhood. Individualist concepts of choice, rights, justice, and interest lose their validity in this context. It seems that Bookchin does not want to take the risk of this kind of communitarian thinking, and is satisfied with the weak communitarianism of libertarian municipalism, assembly government, and civic virtue.

Sometimes Bookchin seems to touch on a bioregional perspective, but he does not carry his thinking in this area very far. He says that in an ecological society, "land would be used ecologically such that forests would grow in areas that are most suitable for arboreal flora and widely mixed food plants in areas that are most suitable for crops."[130] Culture and nature would seemingly both get their due through this simple division. Yet a major ecological problem results from the fact that, except in the case of tropical rain forests, most areas that are quite well suited for forests (or prairies, or even wetlands) can also be used in a highly productive manner for crop production. A bioregional approach would stress heavily the importance of biological diversity and ecological integrity, and have much less enthusiasm for the further development of certain areas on grounds that they are "suitable for crops."[131]

Bookchin comes closest to an authentically bioregional approach when he explains that "localism, taken seriously, implies a sensitivity to speciality, particularity, and the uniqueness of place, indeed a sense of place or *topos* that involves deep respect (indeed, 'loyalty,' if I may use a term that I would like to offset against 'patriotism') to the areas in which we live and that are given to us in great part by the natural world itself."[132] These admirable general principles need, however, to be developed into a comprehensive bioregional perspective that would give them a more concrete meaning. This perspective would address such issues as the ways in which bioregional particularity can be brought back into the town or city, how it can be discovered beneath the transformed surface, and how it can be expressed in the symbols, images, art, rituals, and other cultural expressions of the community. Bioregionalism gives content to the abstract concept that the creation of the ecological community is a dialectical, cooperative endeavor between human beings and the natural world. A bioregional politics expands our view of the political, by associating it more with the processes of ecologically grounded cultural creativity and with a mutualistic, cooperative process of self-expression by the human community and nature. Libertarian municipalism tends to focus on politics as collective economic management and political processes as policy making and personal development through debate in assemblies. Unlike bioregionalism, it constitutes at best a rather "thin" ecological politics.

Conclusion: Social Ecology or Bookchinism?

The questions raised here about libertarian municipalism cast no doublt on the the crucial importance of participatory, grassroots democracy. Rather, they affirm that importance and point toward the need for diverse, multidimensional experiments in democratic processes and to the fact that many of the preconditions for a free and democratic culture lie in areas beyond the scope of what is usually called "democracy." Communes, cooperatives, collectives, and various other forms of organization are sometimes dismissed by Bookchin as "marginal projects" that cannot challenge the dominant system.[133] And indeed, this has often been true (though the weakness of the economic collectives in the Spanish Revolution, to mention an important counterexample, was hardly that they were *marginal* or *nonchallenging*). However, it is questionable whether there is convincing evidence—or indeed any evidence at all—that such approaches have less potential for liberatory transformation than do municipal or neighborhood assemblies or other municipalist proposals. An ecocommunitarianism that claims the legacy of anarchism (as a critique of domination rather than as a dogmatic ideology) will eschew any narrowly defined programs, whether they make municipalism, self-management, cooperatives, communalism, or any other approach the privileged path to social transformation. On the other hand, it will see experiments in all of these areas as valuable steps toward discovering the way to a free, ecological society.

Proposals for fundamentally restructuring society through local assemblies (and also citizens' committees) have great merit, and should be a central part of a left green, social ecological, or ecocommunitarian politics. But we must consider that these reforms are unlikely to become the dominant political processes in the near future. Unfortunately, partial adoption of such proposals (in the form of virtually powerless neighborhood assemblies and "town meetings," or citizens' committees with little authority) may even serve to deflect energy or diffuse demands for more basic cultural and personal changes. On the other hand, major cultural advances can be immediately instituted through the establishment of affinity groups, small communities, internally democratic movements for change, and cooperative endeavors of many kinds. Advocates of radical democracy can do no greater service to their cause than to demonstrate the value of democratic processes by embodying them in their own forms of self-organization. Without imaginative and inspiring examples of the practice of ecological, communitarian democracy by the radical democrats themselves, calls for "municipalism," "demarchy," or any other form of participatory democracy will have a hollow ring.

Bookchin has made a notable contribution to this effort insofar as his work has helped inspire many participants in ecological, communitarian, and participatory democratic projects. However, to the extent that he has increasingly reduced ecological politics to his own narrow, sectarian program of libertarian municiplism, he has become a devisive, debilitating force in the ecology movement and an obstacle to the attainment of many of the ideals he has himself proclaimed.

Notes

1. In the course of this critique, I will sometimes refer to Bookchin's response to some of the points I make. His criticisms are contained in a lengthy document entitled "Comments on the International Social Ecology Network Gathering and the 'Deep Social Ecology' of John Clark," published in *Democracy and Nature*, Vol. 3, No. 3, 1997, pp. 154–197. Bookchin wrote this polemic in response to a rough draft of the present chapter, excerpts of which were presented at the International Social Ecology Conference held in Dunoon, Scotland. While revisions of the text were later made, I quote Bookchin's comments only on those parts that remain unchanged. The term "deep social ecology" comes from a comment by editor David Rothenberg on an article I wrote for *Trumpeter: Journal of Ecosophy*. Bookchin mistakenly read Rothenberg's depiction of my ideas as my own self-description.

2. See John P. Clark, *The Philosophical Anarchism of William Godwin* (Princeton, NJ: Princeton University Press, 1977), pp. 192–193, 243–247.

3. See John P. Clark and Camille Martin, *Liberty, Equality, Geography: The Social Thought of Elisée Reclus* (forthcoming).

4. See esp. Murray Bookchin, "From Here to There," in *Remaking Society* (Montreal: Black Rose Books, 1989), pp. 159–207, and "The New Municipal Agenda," in *The Rise of Urbanization and the Decline of Citizenship* (San Francisco: Sierra Club Books, 1987), pp. 225–288.

5. See Benjamin Barber, *Strong Democracy: Participatory Politics for a New Age* (Berkeley and Los Angeles: University of California Press, 1984).

6. For Castoriadis's politics, see esp. *Philosophy, Politics, Autonomy* (New York: Oxford University Press, 1991).

7. Murray Bookchin, *Post-Scarcity Anarchism* (Berkeley, CA: Ramparts Press, 1971), p. 124.

8. This idea, like many of Bookchin's concepts, was expressed almost a century earlier by the great French anarchist geographer Elisée Reclus. Reclus begins his 3,500-page magnum opus of social thought, *L'Homme et la Terre*, with the statement that "L'Homme est la Nature prenant conscience d'elle-même," or "Humanity is Nature becoming self-conscious." For extensive translation of Reclus's most important work and commentary on its significance, especially in relation to social ecology, see Clark and Martin, *Liberty, Equality, Geography*.

9. Bookchin, *Post-Scarcity Anarchism*, p. 169.

10. Murray Bookchin, *Toward an Ecological Society* (Montreal: Black Rose Books, 1980), p. 263.

11. Ibid.

12. Ibid., p. 273. Admittedly, he was careful to note that he would not argue that the United States was (in 1971) "in a 'revolutionary period' or even a 'pre-revolutionary period'" (p. 263). But then again, who would have argued this? Richard Nixon's landslide reelection the next year and subsequent U.S. history suggest that the mood of actual people living through the epoch was somewhat less than revolutionary. Furthermore, despite the wishful thinking of dogmatic anarchists, studies of electoral abstainers has shown their outlook to be strikingly similar to that of voters.

13. Murray Bookchin, "Revolution in America," *Anarchos*, No. 1, 1968. I am grateful to Bookchin himself for his suggestion that I give this article more attention. Specifically, he stated of my earlier draft of the present analysis that "had [Clark] represented my views with a modicum of respect, he might have consulted 'Revolution in America'" ("Comments," p. 172). I readily admit that I have never considered that early article to be of much significance. I now recognize it, though, as a revealing statement of Bookchin's Bakuninist tendencies.

14. Bookchin, "Revolution in America," p. 3.

15. Ibid., p. 4.

16. Ibid.

17. Ibid.

18. Ibid., p. 5. Unfortunately, Bookchin has never produced a full-scale theoretical analysis of the relation between the hippies and the Enlightenment. His naïve enthusiasm for the hippy movement and similar cultural phenomena is reminiscent of the musings of another middle-aged utopian of the time, Charles Reich, who, in *The Greening of America*, lapsed into a similarly breathless misassessment of the significance of the American youth culture.

19. Bookchin, "Revolution in America," p. 5.

20. Ibid., pp. 11–12.

21. Ibid., pp. 10, 4, 10.

22. Ibid., p. 10.

23. Ibid., p. 12; emphasis in original.

24. Ibid., p. 7. While "underpinning" is not a very sophisticated theoretical category, the implication is clearly that there is a strong connection between the phenomena thus related.

25. Ibid. This was long before the think tanks of the bourgeois order finally discovered, as Bookchin has recently revealed, that it could perpetuate itself through deep ecology and "lifestyle anarchism."

26. Ibid.

27. Bookchin, "Theses on Social Ecology," *Green Perspectives,* No. 33, October 1995, p. 4

28. Michel Bakounine, *L'empire Knouto-Germanique et la revolution sociale,* ed. Arthur Lehning (Paris: Editions Champ Libre, 1982).

29. Bookchin, "Comments," p. 173.

30. See the interview with Fernando Aragon in Ronald Fraser, *Blood of Spain: An Oral History of the Spanish Civil War* (New York: Pantheon Books, 1979), pp. 367–369.

31. Bookchin, *Rise of Urbanization,* p. 256.

32. Bookchin, "Comments," p. 173.

33. Bookchin, *Toward an Ecological Society,* pp. 183–186.

34. Bookchin, *Rise of Urbanization,* p. 55.

35. Bookchin objects strongly to the concept of "self-image" as a fundamental concept in social theory. For him, there is the real world in which we live, on the one hand, and the imagined world that we might create with expansive vision, concerted effort, and correct organization, on the other. This simplistic division is part of Bookchin's dualism, which succeeds in combining reductionist and idealist elements. A dialectical analysis recognizes the centrality of the imaginary to all social reality. In particular, the way we imagine the self is seen as central to all our practical and theoretical activity.

36. Bookchin contends in his "Comments" that the statement just made implies that I want to "reduce 'citizenship' to personhood." Yet I think that it is clear that to analyze the political implications of personhood is not the same as equating personhood with citizenship. Bookchin seems to lapse into confusion by falsely projecting into my discussion his own premise that citizenship is the only form of self-identity with political implications, and then concluding invalidly that since I attribute political implications to personhood, I must consider it to be a form of citizenship. He also seems confused when he claims that after citizens have been reduced to taxpayers, I want to "further reduce" them to persons ("Comments," p. 166). While I do not in fact propose such a definition of citizenship, conceiving of someone as a "person" rather than a "taxpayer" hardly seems a reduction. In fact, the very concept of "reducing" human beings to persons seems rather confused and bizarre.

37. Bookchin, "Comments," p. 165. This attempt at reductio ad absurdum is reminiscent of Luc Ferry's antiecological diatribe *The New Ecological*

Order (Chicago: University of Chicago Press, 1996). For a critique of Ferry's inept efforts to pin the charge of insectocentrism on the ecology movement, see John Clark, "Ecologie aujourd'hui?," *Terra Nova* 1 (1996), pp. 112–119.

38. Presumably Bookchin's municipal citizens would have responsibilities in regard to the buildings, streets, soil, air, and other aspects (perhaps even the insects) of the municipality. Yet this does not imply that the buildings, etc., should be considered citizens, unless the sovereign assembly declares them to be so.

39. Bookchin, "Comments," p. 166. In an apparent argumentum ad verecundiam, he claims that "revolutionaries of the last century—from Marx to Bakunin—referred to themselves as 'citizens' long before the appellation 'comrade' replaced it." In fact, in Bakunin's voluminous correspondence he typically referred to himself as a "friend," or used some other conventional phrasing. His preferred term with his closest political collaborators was "brother," though he sometimes used "comrade," and Citizen Bakunin signed himself "Matrena," in writing to Nechaev, whom he addressed as "Boy."

40. It is a question of the social imaginary, to use a valuable concept that Bookchin contemptuously dismisses.

41. It is possible that the liberatory potential in the entire concept of "citizenship" is seriously limited, and that more inspiring communitarian self-images will play a more important role in the future. This is, however, a historical and experimental question, not one to be answered through stipulation, speculation, or dogmatic pronouncements.

42. When one uses a reductio ad absurdum argument against Bookchin he replies (and perhaps thinks) that one believes in the absurd.

43. Bookchin, "Comments," p. 166.

44. Ibid. The closest approximation of this conception was found in the radical democracy movement of the 1790s, which unfortunately extended it to only a minority of the population, and had a very limited influence on the course of American social history. See John Clark, "The French Revolution and American Radical Democracy," in Y. Hudson and C. Peden, eds., *Revolution, Violence, and Equality* (Lewiston, NY: Edwin Mellen Press, 1990), pp. 79–118.

45. Murray Bookchin, "Libertarian Municipalism: An Overview," *Green Perspectives*, No. 24, 1991, p. 4; emphasis in original. Note that in this statement Bookchin slips into admitting the possibility of "citizenship" in a region.

46. Bookchin, "Comments," p. 167.

47. Murray Bookchin, *The Last Chance: An Appeal for Social and Ecological Sanity* (Burlington, VT: Comment, 1983), p. 48.

48. Ibid., p. 48.

49. Bookchin, *Remaking Society*, p. 173.

50. Murray Bookchin, *The Modern Crisis* (Philadelphia: New Society, 1986), pp. 150–151.

51. Ibid., p. 152.

52. Bookchin, *Rise of Urbanization*, p. 249.

53. Ibid., p. 282.

54. Bookchin comments on this statement that the *civitas* of libertarian

municipalism "is the *immediate* sphere of public life—not the most 'intimate,'
to use Clark's crassly subjectivized word" ("Comments," p. 193; emphasis in
original). What a "crassly subjectivized word" may be will probably remain
one of the mysteries of Bookchinian linguistic analysis. What is clear, however,
is that nowhere do I contend that the municipality is the "most intimate"
sphere, nor do I imply that Bookchin does so. But his misrepresentation of my
claims gives him another opportunity to affirm exactly what I am questioning
about his politics: that he is positing a "sphere of public life" that he idealistically and nondialectically presents as "immediate" by systematically overlooking its cultural and psychological mediations.

55. See Wilhelm Reich, *The Mass Psychology of Fascism* (New York: Simon and Schuster, 1970), and Joel Kovel, *The Age of Desire* (New York: Pantheon Books, 1981). Kovel's analysis is an unsurpassed account of the complex
dialectic between individual selfhood, the family, productionist and consumptionist economic insitutions, the state, and the technological system. It would
be a mistake to privilege any psychological or institutional realm, as Bookchin
habitually does, and as he misinterprets critics as doing, when he projects his
own dualistic categories on their ideas.

56. Bookchin, *Remaking Society*, p. 183; emphasis added.

57. Bookchin's response to this statement reveals his propensity to misread texts very badly in his haste to refute them, and, more significantly, it
once more illustrates his idealist approach. According to Bookchin, "This reduction of the historico-civilizational domain introduced by the city simply to
individuals 'most directly' dealing 'with their television sets, radios, newspapers, and magazines' is not without a certain splendor, putting as it does our
'relationships' with the mass media on an equal plane with the relationships
that free or increasingly free citizens could have in the civic sphere or political
domain" ("Comments," p. 5). The reader will note that in reference to that
with which real, existing human beings "deal directly" I refer to the *actual*
shaping of consciousness in contemporary society, a process with which those
seeking social transformation are obliged to deal. Bookchin replies by invoking
an *abstract* "historico-civilizational domain" that for all its inspirational qualities does not count for much politically unless it is embodied in actual social
practice and actual cultural values. Otherwise, it retains a quite specific "splendor": that of the vaporous moral ideal unrelated to the historically real. Second, Bookchin's idealism becomes more explicit when he accuses me of placing
relationships that people *actually have* in the real world "on an equal plane"
with those that they *might* have in Bookchin's ideal world. Of course, I do not.
Rather, I distinguish between actually existing cultural realities, possibilities
that might be realized in the future, and Bookchin's idealist projections of what
he imagines "could be" onto the reality that presently "is."

58. I will return later to the contradictions entailed in Bookchin's hypos-
tatizing of the municipality.

59. Bookchin, *Toward an Ecological Society*, p. 137.

60. It is largely because of the complexity required by such an analysis
that a less-objectifying, more holistic, and more process-oriented *regional* approach to being is more adequate than is a territorial view. See Max Cafard,

"The Surre(gion)alist Manifesto," *Exquisite Corpse,* No. 8, 1990, pp. 1, 22–23.

61. Bookchin, "Comments," p. 158. Bookchin's distinction is heavily influenced by Arendt's distinctions in *The Human Condition* (Chicago: University of Chicago Press, 1958). See esp., Part II, "The Public and the Private Realm," pp. 22–78.

62. Bookchin, *Rise of Urbanization,* p. 33

63. Bookchin, "Comments," p. 158.

64. Ibid.

65. Ibid.

66. Ibid; emphasis in original.

67. Though there would, of course, be rare exceptions, as when one "disports oneself" in extraterritorial waters.

68. Bookchin, "Comments," p. 158.

69. Bookchin often uses "eduction" as a pseudodialectical ploy for attacking his opponents. By means of "eduction," he uncovers various unsavory implications in their ideas that could never be deduced through rigorous argumentation. In his lectures, Bookchin typically pronounces the term "eduction" while gesturing as if coaxing something into reality out of thin air. This is a striking example of revelatory nonverbal communication.

70. Adolph G. Gundersen, *The Environmental Promise of Democratic Deliberation* (Madison: University of Wisconsin Press, 1995), p. 4. Gundersen cites Mogens Herman Hansen, *The Athenian Democracy in the Age of Demosthenes,* in support of his interpretation.

71. Bookchin, *Rise of Urbanization,* p. 39; emphasis added.

72. Ernest Barker, trans., *The Politics of Aristotle* (London: Oxford University Press, 1946), pp. 5–6.

73. Bookchin, *Toward an Ecological Society,* p. 119.

74. Bookchin, *Rise of Urbanization,* p. 259.

75. One of the most yawning gaps in Bookchin's politics is the absence of any account of how participation in assemblies can effect such far-reaching changes in the character of human beings. Instead, we find vague generalizations, for example, that the assembly is the "social gymnasium" in which the self is developed. Yet one will find little philosophical psychology, philosophy of culture, and philosophy of education in Bookchin. Indeed, these fields endanger his municipalist politics, for the very discussion of the issues they pose leads to a consideration of the larger context of social questions that Bookchin seeks to answer within the confines of his artificially bracketed "political" sphere.

76. Bookchin considers the kind of questions that I raise here "galling in the extreme" ("Comments," p. 188). But those who have good answers to questions seldom respond to them with such anguish. In this case, the questions remind him of the troubling fact that a social movement will not succeed (or even emerge as a significant historical force) merely because a small number of proponents espouse some ideal and will vehemently that it be realized. The question of what might lead large numbers of people to share that ideal and to desire its attainment seems like a good one.

77. Murray Bookchin and Janet Biehl, "A Critique of the Draft Program of the Left Green Network," *Green Perspectives,* No. 23, 1991, p. 2. My references to the "left greens" refer in particular to the Left Green Network, a small coalition of ecoanarchists and ecosocialists within the U.S. green movement. Bookchin became disillusioned with the left greens when they failed to adopt his libertarian municipalism as their official ideology.

78. Bookchin, "Comments," p. 174.

79. Ibid., p. 175; emphasis in original.

80. Hawkins, the primary object of this attack on the left greens, was for years an ally of Bookchin and the latter must be, at least on some level of conceptual thought, aware of the fact that Hawkins's goal is not to bolster the legitimacy of capitalism and the state. But Hawkins has committed the one unpardonable sin: that of embracing the faith and then falling away from it. Conceptual thought therefore cedes its place to irrational denunciations. In a response common to both leftist sectarianism and religious fundamentalism, the charge is defection to the most hated of enemies. Hawkins now does the work of the Devil, seeking "to render capitalism and the state more palatable."

81. Bookchin does not, however, accept this example. He replies that the eight-hour demand was made only because it was part of the pursuit of "the goal of insurrection" and "was designed to reinforce what was virtually an armed conflict" ("Comments," p. 175). Even if this were correct, it would not support his argument that reformist demands mean capitulation to the status quo. However, Bookchin's explanation is a simplistic, inaccurate reading of history in support of his attack on the left greens. The goals of the anarchists in the eight-hour-day movement were complex. One aim was indeed the radicalization of the working class. In addition, the achievement of its limited goal as a real advance for the workers was also considered important to many. Finally, an important motivation was a feeling of solidarity with the workers and their struggles, apart from any pragmatic long- or short-term gains. This identification transcended the kind of strategic thinking that Bookchin emphasizes. A notable exponent of the later two justifications was Emma Goldman, who originally followed Johann Most in rejecting the significance of such limited demands as working *against* the radicalization of workers. She attributes her change in outlook to the moving words of a elderly worker in the audience at one of her lectures. See *Living My Life* (New York: Dover Books, 1970), Vol. 1, pp. 51–53.

82. It is noteworthy that almost all of Bookchin's allies over the past several decades who have become heavily involved in grassroots ecological, peace, and social justice movements have discarded narrowly Bookchinist politics, and this aspiring anarchist Lenin has been left stranded at the Finland Station along with his ideological baggage.

83. Bookchin, *Rise of Urbanization,* p. 276.

84. Bookchin, "Libertarian Municipalism," p. 4.

85. It is not always clear why his own endorsement of small businesses is legitimate while others who support them as part of a decentralized, localist, and regionalist economy are condemned for selling out to capitalism. Presumably, the difference is that despite his statements in favor of small businesses, he

holds the doctrinaire position that all private businesses and indeed every aspect of the market must be eliminated, while those he attacks accept the possibility of experimenting with various combinations of community-owned enterprises, self-management, and small private enterprises in pursuit of a just and democratic economic order.

86. Bookchin, *Rise of Urbanization*, p. 275.

87. Ibid.

88. Ibid., p. 276.

89. Social ecological proposals for grassroots democracy would appeal more to potential activists (with the exception of some theoretically oriented, politicized leftists) if the rhetoric of "libertarian municipalism" were dropped entirely and replaced with more populist concepts such as "neighborhood power" (in addition to more ecological concepts that will be discussed further). While municipalism is a nonconcept for most North Americans and Western Europeans, identification with one's neighborhood is sometimes fairly strong, and is capable of being developed much further in a liberatory direction. Similar localist tendencies exist in Latin America and many other places in which the urban neighborhood or the village are strong sources of identity. In fact, the idea of the creation of the urban village, incorporated into a larger bioregional vision, would be a social ecological concept that would be both radical and traditionalist in many cultural contexts.

90. Bookchin, *Remaking Society*, p. 175.

91. See John Burnheim, *Is Democracy Possible? The Alternative to Electoral Politics* (Berkeley and Los Angeles: University of California Press, 1985).

92. Bookchin, "Comments," p. 183; emphasis in original.

93. The only references to "councils" in the text attacked by Bookchin are in quotations from him or references to these quotations. While I have never "called for" councils, as if they were another panacea competing with Bookchin's assemblies, I have supported the expansion of the city council in my own city from seven to at least twenty-five members as one element in a comprehensive process of expanding local democracy (along with neighborhood assemblies, municipalized utilities, and other similar ideas). As we will see later, despite his apparent dislike for the concept, Bookchin himself "calls for" a kind of council, though in a form that seems entirely unworkable.

94. Bookchin, *Remaking Society*, p. 143.

95. Ibid., p. 181.

96. It is not only the size of the modern urban sprawl that brings into question Bookchin's "municipalist" outlook, but the qualitative changes that have taken place. Mumford pointed out in *The City in History* that what has emerged "is not in fact a new sort of city, but an anti-city" that "annihilates the city whenever it collides with it" (*The City in History* [New York: Harcourt, Brace & World, 1961], p. 505). Bookchin recognizes this change on the level of moralism, as an evil to be denounced, but he does not take it seriously as an object of careful analysis and a challenge to ideas of practice formed in previous historical epochs. Luccarelli, in *Lewis Mumford and the Ecological Region* (New York: Guilford Press, 1995), points out that Mumford's idea of the "anti-city" prefigured recent analyses of a "technurbia" that has emerged out of social transformations in a

"post-Fordist" regime that is "driven by telecommunications and computer-assisted design," that produces "forces that tend to disperse and decentralize production," and that results in a "diffused city" (p. 191). Bookchin's municipalism has yet to come to terms with these transformations and their effects on either organizational possibilities or subjectivity.

97. Bookchin, *Rise of Urbanization*, p. 246

98. Ibid., p. 102.

99. Barber, *Strong Democracy*, p. 269.

100. Bookchin, *Remaking Society*, p. 175.

101. It is certainly conceivable that an assembly of some size could function democratically without succumbing to these threats. Whether or not it does so to a significant degree depends in part on whether it confronts them openly and effectively, but even more on the nature of the larger culture and the way in which the character of the participants is shaped by that culture. But once again, the assembly itself can hardly be called upon as the primary agent of a *paideia* that would make noncompetitive, nonmanipulative assemblies *possible*.

102. Bookchin, *Remaking Society*, p. 176.

103. Bookchin, *Modern Crisis*, p. 91.

104. Bookchin, *Rise of Urbanization*, p. 263.

105. Bookchin, *Remaking Society*, p. 194.

106. Bookchin, *Modern Crisis*, p. 93.

107. Ibid., p. 92.

108. Bookchin and Biehl, "Critique of the Draft Program," p. 3; emphasis in original.

109. Ibid., p. 4.

110. Ibid.

111. Bookchin, *Modern Crisis*, p. 160.

112. It is not clear whether under libertarian municipalism citizens could work in a nearby enterprise that happened to be outside the borders of their municipality. If not, they would then have no voice in decision making concerning their workplace except as advisors to the citizens.

113. Bookchin, *Rise of Urbanization*, p. 275.

114. Bookchin, "Comments," p. 186. Bookchin calls these dismal consequences of the market a "near certainty," and by the next paragraph he has convinced himself, if not the reader, that they will "assuredly" occur.

115. Although Bookchin usually attacks Marx harshly, in this case he invokes Marx's "brilliant insights" that "reveal" what will "prevail ultimately" ("Comments," p. 186). Yet despite Marx's insights into the tendencies of historical capitalism, his ideas cannot validly be used to prejudge the role a market might play in all possible future social formations. This is not the first time that Marx's incisive critique has been used on behalf of heavy-handed dogmatism.

116. Bookchin, *Rise of Urbanization*, p. 262. He hastens to cite his "calls" for diversity when he is attacked for narrowness, but he then goes on to harshly attack anyone who questions the centrality of municipalism and the sovereign assembly.

117. Murray Bookchin and Dave Foreman, *Defending the Earth: A Dialogue between Murray Bookchin and Dave Foreman,* ed. Steve Chase (Boston: South End Press, 1991), p. 79. Bookchin says that these communities must be "artfully tailored to the carrying capacities of the eco-communities in which they are located." Unfortunately, this not only introduces the awkward metaphor of "tailoring" something to a "capacity," but, more seriously, utilizes the theoretically questionable concept of "carrying-capacity."

118. Bookchin, *Rise of Urbanization*, p. 264.

119. Bookchin, "Comments," p. 185; emphasis in original.

120. Ibid.

121. Murray Bookchin, "The Meaning of Confederalism," *Green Perspectives,* No. 20, 1990, p. 4.

122. Ibid.

123. Ibid.

124. Bookchin, "Libertarian Municipalism," p. 3; emphasis added.

125. Bookchin, *Rise of Urbanization*, p. 246.

126. Bookchin, *Remaking Society,* p. 168.

127. Ibid., p. 195.

128. Bookchin, *Ecology of Freedom*, p. 344.

129. Ibid.

130. Bookchin, *Remaking Society,* p. 195.

131. One of the challenges of a social ecological and bioregional perspective is to overcome one-sided approaches that undialectically focus on either production for human need or limiting production for the sake of ecological sustainability. Bookchin's social ecology has tended toward the former, especially as exhibited in his dogmatic, unrealistic statements concerning population, while some versions of deep ecology have tended toward the latter, as manifested in equally uncritical, reductionist analysis of population and "carrying capacity." In the resulting "debate," population is either the root of all evil, or no problem at all.

132. Bookchin, *Rise of Urbanization*, p. 253.

133. Bookchin, *Remaking Society,* p. 183.

Chapter 6

Bookchin's Ecocommunity as Ecotopia
A Constructive Critique

ADOLF G. GUNDERSEN

> The city at its best is an ecocommunity.
> —Murray Bookchin, *Urbanization without Cities*

Introduction

No discussion of the politics of scale, perhaps the central institutional problematic in environmental political philosophy, would be complete without reference to the work of the founder and principal defender of social ecology, Murray Bookchin. Bookchin's work anchors one pole in the environmental argument over appropriate political scale. His is the most systematic, rigorous, and creative defense now available of the position that environmental solutions depend upon a thoroughgoing decentralization of contemporary society. Bookchin's views are controversial, but that is hardly a liability. In the end, no environmental political philosophy, least of all one that claims to resolve the issue of scale, can ignore his challenge. Bookchin's tremendous influence within both the European and American green movements provides an additional, practical impetus to subject his defense of radical decentralization to careful scrutiny. This chapter's central aim is to do just that, in a skeptical but reconstructive spirit. I will not be examining every facet of Bookchin's argument. Instead, I will limit myself to his case for the *ecological* benefits of decentralization.

Murray Bookchin's vision of the ecocommunity is the culmination, both chronologically and theoretically, of a lifetime of social theorizing. One short essay cannot, therefore, do it full justice. That may turn out to be a blessing in disguise, however. Social theorists all too often fall prey to the temptation to "discuss" or "interpret" what other theorists have said without adding very much to our understanding of the issue at hand. Bookchin has always been a clear and enlightening exception to this tendency. I aim here to follow his example. Although there will be moments of exegesis, analysis, and criticism in what follows, I want to maintain a constructive focus in order to add something positive to the ongoing argument over ecological politics and, particularly, the question of appropriate scale. Hence I will keep the preliminaries to a minimum. I begin by briefly noting the place of the "ecocommunity" within the larger corpus of Bookchin's work and then go on to summarize its main features and the three central theses on which it is founded. But the bulk of my discussion is devoted to reworking these theses in light of the internal and external critiques I advance against them. In the end, I argue that all three of the theoretical propositions that support Bookchin's ecotopia are not so much wrong as they are *limited* by the oppositional elements that sully his attempt to develop a philosophy of "emergence."

The Ecocommunity as Culmination and Beginning

Bookchin's vision of the ecocommunity is both the culmination of Bookchin's theoretical project and a vision of a new beginning in human evolution.

Bookchin's earliest environmental works introduced the twin themes of society as the cause of environmental degradation (*OSE*, pp. 222, 225)* and radical decentralization as environmental cure (*OSE*; *CIOC*, Chapter 10). Twenty years later, Bookchin's vision of an ecological society or "ecocommunity" was well developed, and its "centrality" heavily emphasized (*MC*, p. 40). Even if we dispute Bookchin's

---

*For ease of presentation, I have used acronyms when citing Bookchin's book-length works, as follows: *Our Synthetic Environment* (*OSE*), *Crisis in Our Cities* (*CIOC*), *Post-Scarcity Anarchism* (*PSA*), *The Spanish Anarchists* (*SA*), *Urbanization without Cities* (*UWC*), *The Limits of the City* (*LOC*), *The Power to Create! The Power to Destroy!* (*PCPD*), *The Modern Crisis* (*MC*), *The Ecology of Freedom* (*EOF*), *Remaking Society* (*RS*), *Toward an Ecological Society* (*TES*), *The Philosophy of Social Ecology* (*PSE*), and *Defending the Earth* (*DE*). For full citations, the reader is advised to consult the list of references at the end of the chapter.

retrospective characterization of his own work "as a meaningful and logical sequence" (*TES*, p. 28), the ecocommunity certainly does represent the logical completion of Bookchin's philosophy as a whole. That philosophy, which Bookchin dubbed "social ecology," has two components: an "anthropology of hierarchy and domination" and a vision of "ecological society" (*EOF*, passim; see also *EOF*, p. xiv; *RS*, pp. 31, 155). In terms of Bookchin's thought as a whole, then, the ecocommunity completes Bookchin's social and ecological analysis by providing it with a final cause, a destination or goal. But for Bookchin, this destination, because it represents a new realm of freedom, volition, and self-consciousness, is itself fundamentally open-ended, "the point of departure for a new beginning . . . a new eco-social history marked by a participatory evolution within society and between society and the natural world" (*EOF*, p. 36; 1990a, p. 6; see also *EOC*, pp. 272, 279).

The Components of the Ecocommunity

In a telling aside, Bookchin notifies us that "'coherence' is my favorite word; it resolutely guides everything I write and say" (*EOF*, p. 14). Perhaps this is an overstatement, but in the present instance we need only remember that for Bookchin the ecocommunity is distinctive not only in its dedication to "libertarian municipalism" or "politics in Hellenic sense of wide public participation in the management of the municipality" (*LOC*, p. 10), but also in its ethics, technics, and very mode of thought.

> We should never lose sight of the fact that the project of human liberation has now become an ecological project, just as, conversely, the project of defending the Earth has also become a social project. Social ecology as a form of eco-anarchism weaves these two projects together, first by means of an organic way of thinking that I call *dialectical naturalism*; second, by means of a mutualistic social and ecological ethics that I call the *ethics of complementarity*; third, by means of a new technics that I call *eco-technology*; and last, by means of new forms of human association that I call *eco-communities* . . . A coherent ecological philosophy must address all of these questions (*DE*, pp. 131–132).

Likewise, ecological practice must weave these various elements into a coherent whole if they are to produce the "rational ecological society":

> Decentralization, localism, self-sufficiency, and even confederation—each taken singly—do not constitute a guarantee that we will achieve a rational ecological society. In fact, all of them have at one time or another supported parochial communities, oligarchies, and even despotic regimes.

> To be sure, without the institutional structures that cluster around our use
> of these terms and without taking them in combination with each other,
> we cannot hope to achieve a free ecologically oriented society. Decentral-
> ism and self-sustainability must involve a much broader principle of social
> organization than mere localism. Together with decentralization, approxi-
> mations to self-sufficiency, humanly scaled communities, ecotechnologies,
> and the like, there is a compelling need for democratic and truly commu-
> nitarian forms of interdependence—in short, for libertarian forms of con-
> federalism. (1990a, p. 4; see also *UWC*, pp. 296, 267)

The ecocommunity, then, is relatively self-sufficient, scaled to both human and ecological dimensions, organic in its thinking, complementary
in its ethics, and thoroughly participatory and confederal in both its
politics and economics.

The Ecological Argument for Bookchin's Ecocommunities: Three Theses

Bookchin's case for the ecological value of decentralization rests on
three distinct but related theses. The first holds that municipal scale is
ecologically efficient, the second that municipal scale is ecologically
educative, and the third that utopianism is necessary to ecological rationality. Each of these theses rests, in turn, on two separate claims. According to the first thesis, "small is ecologically beautiful" because it simultaneously lessens human *impacts* on the environment and improves
our ability to *monitor* them. The second thesis, that "small is ecologically educative," rests on Bookchin's claim that interaction—both *between citizens* and *with the natural world*—promotes ecological respect. The third thesis, that nothing short of a radically utopian version
of decentralism is ecologically adequate, hinges on what Bookchin sees
as the *motivational* and *practical* need for an ecological utopia.

Thesis 1: Municipal Scale Is Ecologically Efficient

Long before he coined the term "ecocommunity," Bookchin was a de-
votée of decentralization. Already in his first book, Bookchin wrote
that "the human scale is also the natural scale" (*OSE*, p. 239), a notion
he fleshed out some years later by equating "human scale" with "the
smallest ecosystem capable of supporting a population of moderate
size" (*PSA*, p. 102). Bookchin supports this contention by arguing,
first, that decentralized communities put less of a strain on the environment and, second, that decentralization also enhances our capacity to
monitor the environment. As he puts it, "In the long run, the attempt

to approximate self-sufficiency would, I think, prove more efficient than the exaggerated national division of labor that prevails today" (*PSA*, p. 102). Relatively self-sufficient communities are far superior to mass urban conglomerations embedded in a global division of labor, Bookchin argues, at

- promoting biological diversity (*OSE*, p. 242);
- fitting productive activities to local resources (*OSE*, pp. 211, 242; *PSA*, p. 96; *TES*, p. 92; *RS*, p. 186);
- alleviating concentrations of pollutants (*TES*, p. 69; *RS*, p. 186);
- eliminating unproductive administrative and bureaucratic activity (*OSE*, p. 243; *DE*, p. 80); and
- cutting energy use (*PSA*, p. 97; *TES*, pp. 91–92).

In short,

> A massive national and international division of labor is extremely wasteful in the literal sense of that term. Not only does an excessive division of labor make for overorganization in the form of huge bureaucracies and tremendous expenditures of resources in transporting materials over great distances; it reduces the possibilities of effectively recycling wastes, avoiding pollution that may have its source in highly concentrated industrial and population centers, and making the sound use of local or regional raw materials. (1990a, p. 2; see also *CIOC*, p. 194; *PSA*, pp. 131–132, 158)

An impressive set of claims, to be sure. But Bookchin also goes on to argue that decentralization promotes ecological efficiency in another way, namely by enhancing our abilities to monitor our (presumably reduced) environmental impacts. Bookchin suggests that the alternate technologies that decentralized locales would employ would "restore humanity's contact with soil, plant and animal life, sun and wind [thus] fostering a new sensibility toward the biosphere" (*TES*, p. 27; see also *TES*, pp. 31–32; *EOF*, p. 277), and that "the familiarity of each group with its local environment and its ecological roots would make for a more intelligent and loving use of its environment" (*PSA*, p. 102; see also *PSA*, p. 94; *DE*, p. 80). The terms "sensibility" and "intelligent" are important here; Bookchin distinguishes them clearly from "respect" and "love," thereby suggesting that something more than a moral or ethical dispensation will be encouraged by the new ecocommunity. Which is not to say that social ecology discounts the ethical impact of decentralization. Very much to the contrary, we are about to see that while ecocitizens will gain a new "sensibility" for their natural surroundings, Bookchin places even greater hope in their capacity for ethical learning.

Thesis 2: Municipal Scale Is Ecologically Educative

If Bookchin expects decentralization to encourage ecological sensibility and intelligence, he is positively convinced that decentralization will teach us ecological respect. On Bookchin's view, this will happen in two ways: through intensified interactions with nature and through heightened participation in the common life of the municipality. Not only do both of these educational claims surface repeatedly in Bookchin's writing, they are underlined over and over again by the priority Bookchin attributes to education more generally, whether as a revolutionary project (*PSA*, p. 41; *RS*, p. 197) or as "indispensable" to participatory democracy (1990a, p. 5).

Bookchin's claim that the proximity to nature that decentralization allows is ecologically educative is sometimes stated in tentative terms, as when in the mid-1960s he argued that "the relatively self-sufficient community, visibly dependent on its environment for its means of life, would *likely* gain a new respect for the organic interrelationships that sustain it" (*PSA*, p. 102; emphasis added). Perhaps as often, however, Bookchin is rather more categorical, at least in his choice of language: "Nature, and the organic modes of thought it *always* fosters, will become an integral part of human culture" (*PSA*, p. 141; emphasis added). And, by the mid-1980s, we find him arguing that, in the decentralized ecocommunity,

> nature will become an integral part of all aspects of the human experience, from work to play. *Only in this way* can the needs of the natural world become integrated with those of the social [world] to yield an authentic ecological consciousness that transcends the instrumentalist 'environmental' mentality of the sanitary engineer. (*TES*, p. 168; emphasis added)

Throughout his writings, however, the reasoning behind the claim remains the same: ecotechnologies and close proximity to the land remind us of our place in nature and thereby engender respect for it. Hands-on or immediate interaction with nature, in other words, teaches us ecological respect (*OSE*, pp. 63, 240; *MC*, p. 96; *PSA*, pp. 102, 151; *RS*, p. 92; *UWC*, p. x; 1990a, p. 2).

Decentralization also teaches respect for nature in a second way. On Bookchin's view, participatory politics engenders a participatory natural ethic. The end of human hierarchy will spell the end of hierarchical ways of interacting with nature; as human domination of human gives way to participatory modes of human sociation, human domination of nature will give way to participatory modes of interacting with nature (*EOF*, Chapter 12; *MC*, p. 71; *PCPD*, p. 4).

Thesis 3: Utopianism Is an Ecological Necessity

The third thesis that contributes to Bookchin's case for radical decentralization is at once obvious and rather counterintuitive. Bookchin has never made any bones about being a utopian; he invites the label and wears it proudly. Far from apologizing for the utopian character of his social ecology in general or of the decentralized ecocommunity in particular, he advances a two-pronged argument to the effect that utopianism is not only respectable, but also positively *necessary* to achieve the rational ecological society.

Bookchin's argument for the ecological necessity of utopianism begins by asserting that only utopian thinking can free us from instrumental thinking and help us recover a vision of a more rational future. Utopianism frees us from "captivity to the contemporary" (*PSA*, p. 11) and is "crucial to stir the imagination into creating radically new alternatives to every aspect of daily life" (*EOF*, p. 334). Thus utopianism helps counteract disempowerment, for Bookchin "the gravest single illness of our time" (*MC*, p. 123; see also *MC*, pp. 4, 34; *RS*, p. 120; *OSE*, p. 219, 220).

Utopianism is also ecologically necessary, on Bookchin's view, because of the very urgency of the ecological problems we face: "For us there are the alternatives only of utopia or social extinction" (*PSA*, p. 24). Over and over again, Bookchin insists that nothing less than radical, indeed revolutionary, change can prevent ecocatastrophe:

> We are reaching a point of almost cosmic finality in our affairs on this planet; that the recovery of an authentic politics and citizenship is not only a precondition for a free society. It is also a precondition for our survival as a species. . . . Either we will turn to seemingly "utopian" solutions . . . or we face the very real subversion of the material and natural basis for human life on the planet. (*UWC*, p. 288; *RS*, p. 185; see also *UWC*, p. xx; *LOC*, p. 25; *PSA*, p. 91; 1991b, p. 4; *DE*, pp. 78–79)

Reform, on the other hand, whether liberal or socialist, is doomed to failure. Reform efforts simply "lull people into a fase sense of security" (*DE*, p. 77) and act as a "safety valve for the established order" (*DE*, p. 76; see also *RS*, p. 94), a system which "is stacked against you" (*DE*, p. 78; see also *OSE*, p. lxix; *CIOS*, Chapter 10; 1986a, p. 2).

The Three Theses Critiqued

In my view, all three of the above theses are flawed. They all suffer from important internal inconsistencies. They are thus susceptible to what I will be calling an "internal" critique. Nor is that all. Each of Bookchin's principal theses can be challenged more directly. As we are about to see, this "external" critique is perhaps even more telling. In the end, however, these two critiques fail to add up to a refutation. Instead, they suggest that Bookchin's vision of the ecocommunity is rather more limited than he would have us believe.

Decentralization and Efficiency

Bookchin's argument that decentralized communities are more ecologically efficient is doubly problematic. To start with, nowhere does Bookchin ground his concern for ecological efficiency; it tends to float free from his dialectical metaphysics and his ethic of "complementarity." Meanwhile, there are compelling reasons to doubt whether ecological efficiency can be promoted through decentralization alone.

My internal critique of Bookchin's efficiency thesis comes to this: efficiency, however defined (and however important we think it to be), cannot be readily integrated with Bookchin's principle ecological norm, which is to "foster evolution" (*PSE*, p. 44). I do not have the luxury here of demonstrating this claim conclusively. Doing so would require a lengthy digression on Bookchin's dialectical naturalism, the source of his ethical view. But the contradiction between that ethical vision of a "free nature" and Bookchin's repeated emphasis on reducing human demands on the ecosystem, or "ecological efficiency," is plain enough. True, that concern is consistent with the notion of ecological "limits," which Bookchin occasionally acknowledges in a passing way (1991a, pp. 4, 6). But Bookchin's call for efficiency, his interest in reducing human demands on the natural world, hardly follows from his repeated ethical injunction that our primary ecological responsibility is to advance the evolutionary process, "to render nature more fecund, varied, whole, and integrated" (*EOF*, p. 342; see also *EOF*, pp. xxxii, 277; *RS*, p. 203). While one can certainly imagine circumstances in which efficiency may help us promote natural complexity (such as opting for locally generated solar energy rather than drilling for oil in wilderness areas), Bookchin gives us no reason to believe that "efficiency" will always promote "free nature"—or what to do when the two aims conflict. In the end, then, Bookchin fails to reconcile his ethical commitments to efficiency and to free nature. Whether or not it is true that this failing is rooted in a deeper tenson in Bookchin's view of nature itself (limiting on the one hand, fecund on the other), a certain ethical tension remains at the heart of his vision of a decentralized ecocommunity.

Even if we ignore this underlying problem, Bookchin's case for the ecological value of decentralization is far from compelling. That is, even if we confine our view to the less lofty value of "efficiency," there are plenty of reasons to question the ecological efficacy of radical decentralization. I do not doubt that significant gains would follow from ecological decentralization in some realms, such as agriculture and energy (Commoner 1976; Lovins 1977, Paehlke 1989, p. 81; Baldwin 1985, p. 120; Daly & Cobb 1990). Still, neither efficient monitoring nor efficient use of the environment are likely to occur in a thoroughly decentralized world.

In arguing for radical decentralization, Bookchin seems to forget one of the primary tenets of his own ecological philosophy, namely, that nature consists of relationships, not localized points on a map. Already in the mid-1960s, Bookchin had written that "the earth's atmosphere does not respect existing political divisions, subdivisions, and legal niceties" (CIOC, p. 181). But he seems to have forgotten that, for those interested in monitoring the natural world, the essential implication of such a view is anything but decentralist (TES, p. 104). From an ecological standpoint, the local environment really extends beyond the confines of what "can be taken in at one view," to use Aristotle's phrase. In fact, the locality's natural surroundings extend into the region, continent, and globe (Botkin, 1990). It may be true that residents of the locality can monitor their own local environment best and most closely. But to monitor (much less influence) those aspects of the environment that are extralocal (e.g., climate, hydrological cycles, wildlife movements), they must be in some *human* or *political* contact with that same extralocal environment.

There are problems, too, with Bookchin's claim that decentralization would also promote the efficient *use* of nature. There is good reason to believe that Bookchin is right about the potentially positive ecological dyanamics of local governance. Local governance may well be our first line of defense against the "tragedy of the commons" (Baldwin 1985, p. 279; Alexander 1990, p. 164). The problem is that localities by themselves cannot prevent the recurrence of the tragedy *between* one another, as the NIMBY phenomenon so amply illustrates (Brion 1991; Piller 1991; see also Dahl 1989, pp. 23, 229–230, 302, 303, 318–319, 322; Sandbach 1980, p. 192; Paehlke 1989, pp. 245–250). To his great credit, Bookchin is not only aware of the problem of local "parochialism" and the related problem of maintaining the integrity of the locality (RS, pp. 64, 78, 79; SA, p. 41; LOC, p. 183; EOF, p. 252; UWC, pp. 47–72, 89, 113, 175–176, 185, 293–294; *DE*, p. 62; 1986a, p. 3; see also 1990a, p. 1), but has devoted much of his most recent writing to an attempt to formulate a coherent response. Yet his solution—confederalism—ultimately fails.

Confederalism, "a network of administrative councils whose members or delegates are elected from popular face-to-face democratic assemblies" (1990a, p. 4), is designed precisely to counteract racial, cultural, and traditional sources of parchocialism (*UWC*, p. xix; see also *UWC*, Chapter 6, Appendix; *EOF*, p. lv; 1990a, p. 5; 1991b, p. 3). Bookchin's optimism about confederalism's potential is partly based on his reading of history. On his view,

> The problem of dealing with foreign intervention; the incorporation of sizable towns and the expansion of older ones into cities; the disparities in status, wealth, and power that developed, not to speak of local parochialism at one extreme and cosmopolitan "modernity" at the other, were never fully resolved. But they were held in remarkable balance for most of the Free State's history. (*UWC*, p. 230)

At the same time, Bookchin readily admits that confederalism rests on his belief in our capacity for moral regeneration, a belief rooted ultimately in his faith that "a basic sense of decency, sympathy and mutual aid lies at the core of human behavior" (*PSA*, p. 160). "There are no guaranteed solutions," he writes, "apart from the guiding role of consciousness and ethics in human affairs" (1986a, p. 3). When push comes to shove, then, confederations will triumph over parochialism because, quite simply, "free men will not be greedy" (*PSA*, p. 161). If this sounds uncharacteristic of Bookchin's more mature writing, consider this straightforward admission, made in a piece devoted entirely to the subject: "In a society that was radically veering toward decentralistic, participatory democracy, *guided by communitarian and ecological principles*, it is only reasonable to suppose that people would not choose such an irresponsible social dispensation as would allow the waters of the Hudson to be so polluted [by upstream communities]" (1990a, p. 2; emphasis added). Or again: "Libertarian municipalism . . . *presupposes* a genuinely democratic desire by people to arrest the growing powers of the nation-state and reclaim them for their community and their region" (1991b, p. 2; emphasis added). Notice just what is being admitted here: libertarian municipalities, united in confederal networks, are *assumed* to have the very moral outlook that Bookchin believes can only arise out of participatory sociation. In other words, Bookchin here puts the ethical cart before the political horse. In the end, Bookchin himself seems to sense the problem:

No one who participates in a struggle for social restructuring emerges from that struggle with the prejudices, habits, and sensibilities with which he or she entered it. Hopefully, then, such prejudices—like parochialism—will increasingly be replaced by a generous sense of cooperation and a caring sense of interdependence. (1991b, p. 3)

By this point, Bookchin's argument has largely collapsed. Hope is all that remains.

What is more, Bookchin's hope for moral regeneration is exceedingly fragile. If the doomsayers are right—and Bookchin has always at least partly endorsed their views—ecological devastation will bring with it struggle and competition: fallow ground, indeed, to cultivate intercommunal generosity, cooperation, and care. Worse still, because Bookchin fails to convince us that the decentralized community will be educative to the degree he claims, the future of confederalism seems largely moot.

Decentralization and Education

Bookchin's case for the educative impact of decentralization encounters much the same difficulties that beset his argument for the ecological efficiency of decentralization. It suffers from a similar internal tension, on the one hand, and from a similar tendency to beg certain important questions, on the other.

If anything, the internal tension that afflicts Bookchin's case for the educative impact of decentralization is far more problematic than the parallel tension between an ethic of efficiency and an ethic of fostering evolution. Indeed, it goes to the very heart of social ecology's primary emphasis on the social determinants of ecological consciousness. Put quite simply, Bookchin's emphasis on the educative value of the proximity to biological nature afforded by decentralization stands in flat contradiction to the claim that humans' view of nature reflects their view of each other—a proposition that is perhaps the defining thesis of his entire outlook. According to Bookchin the anthropologist, humans learn how to treat nature by learning about how to treat each other. According to Bookchin the utopian visionary, however, a new ecological outlook will follow, as we saw above, from the renewed proximity to nature afforded by the ecocommunity and its ecotechnologies.

Although the formula has varied, Bookchin has consistently argued that society is the source of our ecological ethic. He has argued variously that our ecological ethic "derives from" (*POSE*, p. 112), "stems from" (*RS*, p. 44; *EOF*, p. 1), "rests within" (*RS*, p. 60), is "a projection of" (*DE*, p. 57), or "emerges directly from" (*PSA*, p. 85; *TES*, p. 95) social relations—and that social relations "gave rise to" (*RS*, p. 44), "extended into" (*RS*, p. 60), or "carried over conceptually into humanity's relationship with nature" (*TES*, p. 40; see also *DE*, p. 129). Lest there be any doubt on this score, Bookchin has repeatedly underscored the priority of society as a source of ethics, both human and ecological. To choose but one representative example, Bookchin has argued that "a human nature *does* exist, but it seems to consist of proclivities and potentialities that become increasingly defined by the instillation of social needs" (*EOF*, p. 114; see also *TES*, pp. 48, 66, 81–82; *UWC*, Chapter 4; *UWC*, p. 292). Whatever the merits of this view, it can hardly be made consistent with the view that changing our relationship with nature, through decentralization (or anything else, for that matter) will have an independent effect on our perception of nature. Either we learn to love and respect nature through interacting with nature, or we do so through social interaction. Bookchin the social visionary sees it at least partly the first way; Bookchin the anthropologist sees it almost exclusively the second way.

Bookchin may waver between his analytical commitment to the educative priority of society over nature and his prescriptive commitment to the educative value of nature, but the more significant flaws in his belief in the educative value of decentralization lie elsewhere. To begin with, he grossly overestimates the transformative power of participatory politics. Although in his historical analyses he is generally quite sensitive to the degree to which fundamental ethical and cultural agreement underwrote participatory politics in the past (*UWC*, Chapter 4), he seems oblivious to the way in which such understandings will limit future participatory experiments. Political participation in ancient Athens flourished because Athenians believed in it; political participation did not *create* that belief. The same can be said about ecological values: participation will produce ecologically beneficent outcomes only if those who participate enter into political life with ecologically beneficent values.

Bookchin would have us believe that "as citizens, [individuals] would function in [municipal] assemblies at their highest level—their *human* level—rather than as socially ghettoized beings. They would express their general human interests, not their particular status interests" (*RS*, p. 194). And, of course, such interests would include the broadest view of their ecological "interests." But Bookchin gives us precious little reason to believe that this would actually occur. What is it about participation in local politics that expands our view of the natural world? More pointedly, what is it about local participation that expands our view beyond the locality? Bookchin never tells us. More damaging still, there are good reasons to suspect that any answer would be less than convincing. Face-to-face democracy need not produce harmony and equality; in fact, it very often produces precisely the reverse (Mansbridge 1983; see also Dahl 1989, pp. 20–21; Ophuls 1977, p. 228). As a result, Bookchin's social theory of ecological education amounts to little more than a fervently held faith in participation.

The Problems with Utopia

Bookchin's utopianism is certainly attractive, even inspiring. Unfortunately, upon closer inspection, the very purity that lends it its initial allure begins to look more formal than theoretical.

I do not mean to challenge Bookchin's claim that radical change represents the only answer to the ecological crisis of our time, though that claim can certainly be challenged. (Just what, after all, constitutes a "crisis"?) What I do find implausible, even preposterous, is Bookchin's definition of "radical change." As we have already seen, on Bookchin's view only utopian ecocommunities will do—a *world* of ecocommunities coordinated by confederal bodies. Nor is that all. Bookchin admits and even insists that this can only happen through the spontaneous and simultaneous action of communities *everywhere:* "We cannot hope to realize this vision in only one neighborhood, town, or city. Ours needs to be a confederal society based on the coordination of all municipalities" (*DE*, p. 84). Bookchin has even gone so far as to suggest that "after the revolution the planet would be dealt with as a whole" (*PSA*, p. 261; see also *RS*, p. 184; *TES*, p. 256). That such a widespread revolution, with all the profound changes it would entail, is a faint hope hardly seems worth emphasizing, especially given the tremendous number and complexity of the new intercommunal relationships that it presupposes. Small wonder, then, that despite Bookchin's efforts to plumb history for evidence of communal uprisings, he can find no parallel for such a society-wide popular uprising.

Bookchin's claim that a utopian vision is necessary to raise people out of their stupor and motivate real change on behalf of the environment is likewise questionable.

True, his critique of instrumentalism is compelling. But utopianism is not the only alternative. The ethical gap left by society's fixation on instrumental reason requires not utopian thinking, but purposive reason. Ideals, by definition, are "utopian," in the sense that they picture a world not yet real. And Bookchin occasionally seems to define reason in just this way, as in these lines from *Post-Scarcity Anarchism*: "Speculative philosophy is by definition a claim by reason to extend itself beyond the given state of affairs" (p. 157). From the standpoint of moral psychology, utopianism has nothing over more garden-variety ideals or visions. On the contrary, ideals that are too lofty tend to lose their power to motivate and thus become divorced from the very practice they are meant to inform and guide.

A related problem with Bookchin's emphasis on utopian thinking is that it depends quite directly on a prior belief in the power of ethics, ideals, and consciousness. The problem is not that Bookchin does not present us with reasons to endorse such a belief in consciousness. Indeed, he takes great pains to distance himself from Marx on this score (*UWC*, pp. 187, 268; *TES*, p. 261), and insists that "consciousness—not pat formulas—[will] ultimately determine whether humanity will achieve a rich sense of collectiveity without sacrificing a rich sense of individuality" (*UWC*, p. 254; see also *EOF*, p. 8). Rather, the problem is that Bookchin, like Marx before him, remains wed to the view that society rather than consciousness or reason creates values. That is, after all, what is "social" about social ecology. Bookchin cannot have it both ways: either visions create society, or society creates visions.

The Rational Ecological Society: From Oppositional Utopia to Emergent Ideal

In this final section I will not argue for an alternative ecological vision. Instead, I want to suggest how the commited social ecologist might meet the critiques I have just advanced. In a reconstructive spirit, I argue that the various problems I have identified in the preceding section can be addressed *with least damage to Bookchin's overall ecological vision* by a more consistent and systematic elucidation and application of his metaphysic of *emergence*. Bookchin needs, in short, to take more seriously his own dictum that "All phenomena are emergent" (*POSE*, p. 109). I leave to the reader the question of the ultimate success of the reconstruction that follows.

Emergence and Ecological Efficiency

Recall that the primary problems with Bookchin's efficiency thesis were that it rested on shaky ethical ground, both theoretically and politically, and that it ignored both limitations on localities' abilities to monitor ecological change and their tendency to inflict ecological havoc on their neighbors. As we are about to see, answers to all of these problems follow quite readily once we adopt an emergent view of political development—a view that I believe to be truer to Bookchin's own dialectical naturalism than the one that often informs his more explicitly political writing.

To begin with, an emergent perspective allows us to view the ten-

sion I detect between "efficiency" and what Bookchin calls "free nature" as a *historical* phenomenon, not an ethical one. For all of his
stress on the openness of history, Bookchin in the end hypostatizes his
"ecocommunities": they end up appearing fixed, in their practices and
in their beliefs. If, on the other hand, history—including the history of
human reason and subjectivity—is indeed an emergent process, the way
is open to arguing that a participatory orientation to "free nature" will
grow out of a concern for efficiency. To use Bookchin's favorite phrase,
efficiency can "grade into" a more differentiated and holistic concern
with fostering evolution. In precisely the same way, a more consistently
emergent view of political development avoids the chicken-and-egg
problem that besets all visionary political theories. If institutions and
political beliefs evolve together in mutual interaction, as Bookchin at
least occasionally allows (*EOF*, p. 8), there is no need to assume that
citizens arrive fully formed in the ecocommunity of the future. Their
values, again, can be seen to emerge from the very process of development.

A dialectic of emergence, rather than of opposition, also helps address the difficulties of translocal monitoring and restraint. If localism
is seen to emerge historically and politically out of a top-down, hierarchical organization of society rather than springing full-formed in opposition to it, vestiges of hierarchy will logically remain. As heinous as
hierarchy may be, viewing its abolition as a graduated phasing into a
thoroughly egalitarian society suggests that dealing with intercommunal
problems of ecological abuse and monitoring will still be able to draw
on the tainted instruments of hierarchy—arguably long enough to allow their self-transcendence by the emergent politics of community and
its attendant changes in ethics, values, and consciousness.

Emergence and Ecological Education

I argued above that Bookchin's theory of learning, as it now stands, is
flawed in three ways: it fails to acknowledge the need for foundational
agreement among citizens, it fails to explain *how* citizens learn through
participation, and it fails to reconcile participatory political learning in
the assembly with participatory experiential learning in nature. Here,
too, emergence offers a way out or, rather, parallel ways out.

That the concept of emergence can reconcile experiential and political learning is perhaps easiest to see. Bookchin, after all, has done a
careful job of showing how humanity emerged from biological nature
into social or "second" nature. What is remarkable about his account,
most fully developed in *The Ecology of Freedom,* is his insistence on
humanity's simultaneous differentiation *and* continuity with the bio-

logical realm of ("first") nature. Assuming for the moment that this account is persuasive, it follows rather directly that human *learning* will be both biological *and* social. As beings emergent from nature, we learn in an emergent fashion. Seen in this way, Bookchin can not only claim that closeness to nature is an aid to learning and that society contributes importantly to learning, but that these two processes are inextricably linked: "experiential" learning phases into "political" learning.

In a similar way, the notion of emergence offers at least a general answer to the complaint that Bookchin lacks a processual theory of social learning. To say that ethics are formed by society implies that they can be *re*formed in a new society. Still, Bookchin leaves us wondering just *how* this is to occur. True, he repeatedly invokes "discourse," "participation," "dialogue," and the like (*UWC*, pp. 58, 250, 258; *LOC*, pp. 49, 121; *RS*, p. 144; *EOF*, pp. 131–132; *TES*, pp. 103–104, 190)—but never specifies just what these processes involve. Still, there are at least the beginnings of a theory of political discourse in Bookchin's writings. Again, they are to be found in his ontology of emergence. Consistent with that ontology, Bookchin quite clearly states that consciousness is endowed with "an *emergent* dialectic" (*TES*, p. 267; emphasis in original). At the very minimum, this suggests that political dialogue involves an organic process of differentiation and growth, a process that results in a more encompassing whole. If each of these terms were carefully articulated and their interrelationships explored, the result would be a discursive theory that would do much to ground Bookchin's ultimate faith in the transformative power of participatory politics. (Whether it could ultimately sustain such a communal politics, either practically or theoretically, is, of course, another matter.)

Emergence and Utopia

I argued above that the motivational power of Bookchin's utopianism was undermined by his insistence on the social origins of ecological ethics. I also tried to show that Bookchin's framing of the ecological crisis as an either/or choice between ecological catastrophe and a spontaneous, worldwide libertarian revolution was wildly overdrawn. For these and other reasons, I concluded that Bookchin's hyperbole ends up putting his vision beyond the reach of the very people it is meant to inspire. Can a more careful articulation of emergence overcome these problems? I think so—though I hasten to add that I am under no illusion that Bookchin himself would be willing to soften or blur his utopianism, his occasional aside to the contrary notwithstanding (*TES*, p. 281).

The more coherent and well-articulated theory of the emergent nature of human consciousness that I argued for above on other grounds would also render Bookchin's radical idealism more coherent. Consciousness, seen as a culmination of a history of increasing subjectivity, does not lose its roots in nature or society. Yet it is still free, free to point toward a future—and to call us toward it.

Utopianism might give way to a more emergent vision of the future in a variety of different ways. But the one that seems most consistent with Bookchin's own philosophy of history, and with his stress on the motivational importance of radical thought, would be to stress the emergent ideals embedded in our own history and how they might work themselves out in the future rather than casting the future as a titanic dialectic between good and evil, totalitarianism and freedom, catastrophe and redemption. People need ideals. On that, Bookchin and I are in firm agreement. But people also pursue ideals *from somewhere*. The ideal does not mean much if it cannot be connected clearly with the real. And since, for Bookchin, "the real" is literally history itself, the most consistent and compelling argument he can make about the shape of the future should be rooted very firmly there—should be "educted" from history rather than *aufgehoben* by a cataclysmic form of speculation.

Conclusion

I want to close not with a ringing conclusion but with a careful caveat. I believe Bookchin's utopian vision can be improved in various ways. But I am far from convinced that even a more coherent and well-articulated version of social ecology will be compelling—theoretically or politically. Indeed, I would argue on both theoretical and empirical grounds that a broadly Aristotelian approach is preferable to the idealism of both participatory democrats like Bookchin and devotees of free markets like Milton Friedman and Friedrich Hayek. Aristotle believed that ideals were rarely best served by idealistic prescriptions, that some measure of compromise with reality was usually necessary if we were to transcend it (Gundersen, 1995, Chapter 2). Whereas compromise of any sort is anathema to Bookchin, to Aristotle it was the tragic price the idealist must pay. I would also challenge Bookchin's participatory view of political discourse: in my view, a dyadic model of political deliberation is far more robust (Gundersen, forthcoming).

Still, all such quarrels aside, there is still a great deal we can learn from Bookchin; some of us may even be inspired by him. In any case, it is far too soon to close the book on social ecology, especially if we can get beyond the strident oppositional thinking that sometimes mars its bests insights and work at developing the ethical and political implications of "emergence." If we do, we may yet contribute to Bookchin's dream of "an unceasing but gentle transcendence" (*TES*, p. 274).

References

Alexander, Donald. 1990. "Bioregionalism: Science or Sensibility?" *Environmental Ethics,* Vol. 12, No. 2, pp. 161–173.

Baldwin, John H. 1985. *Environmental Planning and Management.* Boulder, CO: Westview Press.

Bookchin, Murray. 1972. *Our Synthetic Environment,* Rev. ed. New York: Harper and Row. (Originally published 1962)

Bookchin, Murray (Lewis Herber). 1965. *Crisis in Our Cities.* Englewood Cliffs, NJ: Prentice Hall.

Bookchin, Murray. 1978. *The Spanish Anarchists.* New York: Harper Colophon.

Bookchin, Murray. 1980. *Toward an Ecological Society.* Montreal: Black Rose Books.

Bookchin, Murray. 1986a. "Municipalization: Community Ownership of the Economy." *Green Perspectives,* No. 2, February 1986, pp. 1–3.

Bookchin, Murray. 1986b. *The Limits of the City.* Montreal: Black Rose Books. (Originally published 1973)

Bookchin, Murray. 1986c. *Post-Scarcity Anarchism.* Montreal: Black Rose Books. (Originally published 1971)

Bookchin, Murray. 1986d. *The Power to Create! The Power to Destroy!* Reprint, Burlington, VT: Green Program Project. (Originally published 1969)

Bookchin, Murray. 1987. *The Modern Crisis,* 2nd ed. New York: Black Rose Books.

Bookchin, Murray. 1990a. "The Meaning of Confederalism." *Green Perspectives,* No. 20, November 1990, pp. 1–7.

Bookchin, Murray. 1990b. *The Philosophy of Social Ecology.* Montreal: Black Rose Books.

Bookchin, Murray. 1990c. *Remaking Society.* Boston: South End Press.

Bookchin, Murray. 1991a. *The Ecology of Freedom.* Montreal: Black Rose Books.

Bookchin, Murray. 1991b. "Libertarian Municipalism: An Overview." *Green Perspectives,* No. 24, October 1991, pp. 1–6.

Bookchin, Murray, with Dave Foreman. 1991. *Defending the Earth: A Dialogue between Murray Bookchin and Dave Foreman,* ed. Steve Chase. Montreal: Black Rose Books.

Bookchin, Murray. 1992. *Urbanization without Cities*. Montreal: Black Rose Books.

Botkin, Daniel. 1990. *Discordant Harmonies*. New York: Oxford University Press.

Brion, Deni J. 1991. *Essential Industry and the NIMBY Phenomenon*. New York: Quorum Books.

Commoner, Barry. 1976. *The Poverty of Power: Energy and the Economic Crisis*. New York: Knopf.

Dahl, Robert A. 1989. *Democracy and Its Critics*. New Haven, CT: Yale University Press.

Daly, Herman E., and John B. Cobb, Jr. 1990. *For the Common Good: Redirecting the Economy toward Community, the Environment, and a Sustainable Future*. Boston: Beacon Press.

Gundersen, Adolf G. 1995. *The Environmental Promise of Democratic Deliberation*. Madison: University of Wisconsin Press.

Gundersen, Adolf G. Forthcoming. *A Socratic Theory of Democracy*.

Lovins, Amory. 1977. *Soft Energy Paths*. Cambridge, MA: Ballinger.

Mansbridge, Jane. 1983. *Beyond Adversary Democracy*. Chicago: University of Chicago Press.

Ophuls, William. 1977. *Ecology and the Politics of Scarcity*. San Francisco: Freeman.

Paehlke, Robert C. 1979. *Environmentalism and the Future of Progressive Politics*. New Haven, CT: Yale University Press.

Piller, Charles. 1991. *The Fail-Safe Society*. New York: Basic Books.

Sandbach, Francis. 1980. *Environment, Ideology, and Policy*. Montclair, NJ: Allanheld and Osmun.

Chapter 7

Social Ecology and the Problem of Technology

DAVID WATSON

Technology, Neutral or Otherwise

Murray Bookchin is certainly correct in stressing the need for "a clearer image of what is meant by 'technics' " (1982, p. 223).[1] Unfortunately, Bookchin's own confusion about technics is palpable. "The industrial machine seems to have taken off without the driver," he writes in *The Ecology of Freedom*, but "the driver is still there." Sixty pages later we read: "A look at technics alone reveals that the car is racing at an increasing pace, with nobody in the driver's seat" (1982, pp. 239, 302). The problem of human agency is indeed thorny. In distinct ways a "driver" can be said to be and not be present. But Bookchin only stays on the surface of such an inquiry; confusion and contradiction plague his work.

In another passage he writes, for example, "Marx was entirely correct to emphasize that the revolution required by our time must draw its poetry not from the past but from the future" (1982, p. 20). Yet elsewhere he argues that "the 'tradition of all the dead generations' which Marx, in his effluvium of nineteenth-century progressivism, hoped to exorcise with the 'poetry' of 'the future' has yet to be recovered and explored in the light of the dead-end that confronts us. The future as we know it today . . . has no poetry to inspire us" (1986, p. 114).

In fact, Bookchin is not certain which poetry attracts him more —which is why his work is so problematic, and ultimately far less than the kind of holistic thinking social ecology claims to fulfill. Bookchin remains trapped within the transition from a red to a green radicalism.

211

His contradictions and his failure to advance the radical kernel within his own sensibility leave social ecology at an impasse—an impasse notable in his writings on rationality, history, and other areas, but most pointedly in his writings on technology. If the social ecology perspective is to realize its own potential as a radical philosophical and practical break from the context of modern, capitalist, alienated consciousness and instrumental reason, it will have to abandon its founder's compulsions for a new order of thinking.

Objecting to the contemporary "grim fatalism" about technology (1982, pp. 220–223), Bookchin always insists on its promise. From the beginning, his utopianism has been deeply rooted in the faith that the new technics created by modern industrial capitalism have brought about certain preconditions, if not necessarily the actual conditions, for a rational, free society. To be sure, he has also written, and occasionally quite eloquently, about the pathological destructiveness of modern technological arrangements. But if he believes that some forms (e.g., nuclear power) are inherently evil, for the most part he stresses that "technology as such" is not the problem but rather more fundamental "economic factors" are (1988; 1995, p. 28).

Bookchin sometimes explicitly rejects the idea that technics can be neutral or that "their impact [is] contingent merely on individual and social intentions" (1982, p. 289). "Modern technology," for example, "is intrinsically authoritarian" (1986, p. 117). Yet even though he has described himself as "practically a luddite" (1991, p. 35), Bookchin takes a recognizably Marxist position. "Technology in itself," he writes—without explaining what "in itself" might mean—"does not produce the dislocations between an antiecological society and nature, although there are surely technologies that, in themselves, are dangerous to an ecosystem. . . . To speak of 'technological society,' or an 'industrial society' . . . is to throw cosmic stardust over the economic laws that guide capital expansion which Marx so brilliantly developed" (1988, p. 23). Social relationships play "a decisive role in the technologies and industries society develops and the use to which they are put" (1988, p. 23).

According to Bookchin, capitalism misuses modern technology by exploiting its "malignant power to destroy instead of its benign power to create" (1986, p. 111). Oil spills and nuclear meltdowns arise from the "abuse of technology by a grow-or-die economy" (1989b). Technology only "*magnifies* more fundamental economic factors" (1990, p. 93; emphasis in original). "Every warped society," he says, "follows the dialectic of its own pathology of domination, *irrespective of the scale of its technics*" (1982, p. 241; emphasis added).[2] Capitalist social and economic relations "blatantly determine *how* technology will be used" (1995, p. 29; emphasis in original).

To those who recognize the fallacy that technology is a neutral tool to be used or abused by the one who wields it, Bookchin offers a disclaimer: because technology is shaped by social forces, our concepts about it "are *never* socially neutral" (1982, p. 226). This statement is simply an evasion; the idea that technology is *not* neutral logically implies not only that our concepts shape and determine technology, but that the technological relations and requirements imposed by our technology also shape our concepts and social relations. Technological arrangements themselves generate social change and shape human action, bringing about imperatives unanticipated by their creators. Technological *means* come with their own repertoire of *ends*.

The ecological crisis is a dramatic example of this phenomenon. No one but a Marxist of the crudest variety could believe that technological dysfunction and disaster are the results only of corporate capitalist greed. As Bookchin himself has noted about oil spills, "even the sturdiest ships have a way of being buffeted by storms, drifting off course, foundering on reefs in treacherous waters, and sinking" (1989b, p. 20). Not only capitalist grow-or-die economic choices, but a complex petrochemical grid itself makes disasters inevitable.[3]

Bookchin thinks it enough to say that technics are part of a social matrix—a point no serious critic of technology would dispute.[4] Technics, he avers, are "immersed in a social world of human intentions, needs, wills, and interactions" (1980, p. 128). The technics emerging from "the immanent dialectic" within hierarchical society reinforced hierarchy and domination, he says. He argues elsewhere:

> If I read the historical record correctly, it is fair to say that before mass manufacture came into existence, there had already been widespread destruction of community life and the emergence of uprooted, displaced, atomized and propertyless "masses"—the precursors of the modern proletariat. This development was paralleled by science's evocation of a new image of the world—a lifeless physical world composed of matter and motion that preceded the technical feats of the Industrial Revolution. (1982, p. 223)

He notes in still another essay:

> A blissful ignorance clouds the fact that several centuries ago, much of England's forest land . . . was deforested by the crude axes of rural proletarians to produce charcoal for a technologically simple metallurgical economy and to clear land for profitable sheep runs. This occurred long before the Industrial Revolution. (1989b, p. 22)

That earlier societies may have also been socially and ecologically destructive is not exactly pertinent to a critical discussion of technology. But Bookchin here falsely contrasts nascent forms of technological society with what was to come later. Failing to treat technics as a social and historical process, he retreats to a formalistic notion of what constitutes industrialism. "Technics does not exist in a vacuum," he says, "nor does it have an autonomous life of its own" (1982, p. 223). Capitalism does not exist in a vacuum either; modern capitalist civilization is as much the product of early industrial manufacture, timekeeping and the expansion of scientific-technical knowledge as it is of expanding trade and private property. In fact, as Lewis Mumford so brilliantly argued, the voracious markets and hierarchic work machines of industrial capitalism had their early forerunners in the ancient empires.[5]

It's simply confused to speak of a liberatory society as the unintended result of modern capitalist technics, as Bookchin has done from the beginning—to see the "means of production" outgrowing society (1980, p. 270)—and then to paint technics as little more than the passive recipient or result of human intentions and interactions. In fact, Bookchin's entire argument in his early essays was based on the idea that technological development brings about enormous, unforeseen social changes and new, problematic contexts. "Qualitatively new problems have arisen which never existed in [Marx's] day," he says; through "cybernation and other technological advances," modern technics have taken on an entirely new character, reconstituting the terms of revolutionary history. New developments, for example, have eroded the "strategic economic position" of the working class and its role as the agent of revolutionary change (1971, pp. 179, 183). Bookchin's flaw from *Post-Scarcity Anarchism* onward was to celebrate this technological transformation as the necessary precondition for a liberatory society, rather than the emergence of a qualitatively new stage of domination—what Jacques Camatte has described as "the runaway of capital" and the transformation of "formal domination" into "real domination."[6]

Capitalism and Technology

"We cannot avoid the use of conventional reason, present-day modes of science, and modern technology," Bookchin asserts (though he doesn't explain *why* we must put up with "present-day modes of science" and technics). "But we can establish new *contexts* in which these modes . . . have their proper place" (1982, p. 240; emphasis in original). Present-day modes of science and technology apparently never establish contexts: "The ecological impact of human reason, science, and technology depends enormously on the type of society in which these forces are shaped and employed" (1991, p. 32).

Not even the scale or the form of technics seem to matter. "The historic problem of technics," he declares, "lies not in its size or scale, its 'softness' or 'hardness,' much less the productivity or efficiency that earned it the naive reverence of earlier generations; the problem lies in how we can *contain* (that is absorb) technics within an emancipatory society" (1982, pp. 240, 260–261; emphasis in original). Our dialectician does not notice that at a certain level of size, scale, or "hardness" technology and its accompanying operational demands might "absorb" *us*. Yet here, too, Bookchin is ambiguous. It "is not mere Luddism [Bookchin always considers it "mere" luddism] to say that we would be safer as a species if we could restore a Paleolithic world of flints than if we were to 'advance' to a 'post-industrial' world of 'intelligent robots.' Not that the former is a desideratum in itself, but merely that it is less menacing and demonic in a society ruled by moral cretins and emotional brutes" (1986, pp. 111–112). Notwithstanding the fact that we have no reason to assume paleolithic society was ruled by "moral cretins," Bookchin seems to be saying that scale, indeed, is a critical factor.

Bookchin doesn't consider the possibility that a mass technological society might itself come to constitute a "type," or that technological development could shift life, qualitatively transform old pathologies, and generate wholly unprecedented problems, not just magnify old ones. Furthermore, his dichotomy of blaming technics *instead of* corporate and state institutions, or speaking of technological society *instead of* capitalism, is a false one; the matrix of social relations is more complex than he suggests. It also apparently escapes him that to speak of modern technological society is in fact to refer to *the technics generated within capitalism,* which have engendered new forms of capital in turn. To speak of a distinct realm of social relations that determines technology is not only ahistorical, it is neither dialectical nor holistic. It is to fall victim to a kind of simplistic base/superstructure schema.[7]

Instead of clarity Bookchin offers vague rhetoric and bluster. Thus even "appropriate technology" is inadequate for authentic ecological change and only a "liberatory technology" will do.[8] Yet the idea of "ecotechnologies" is highly ambiguous if the differences between high and low tech, "hard" and "soft" tech, large and small tech are secondary (as Bookchin has argued), and if advanced industrial technology, complex mass communications and energy systems, and even genetic engineering are all at least possibly allowable within the matrix of an ecological society.

Bookchin responds to such doubts with naïve rationalism and moral pieties. "The current social setup," he says, "means that the sci-

entific establishment is not *morally* capable of dealing with biotechnology. . . . Our society is so immoral that it can't be entrusted to
invent anything until we are able to sit down and decide, as a socially
responsible, ecologically sensitive community, how we're going to design and use our technology" (1991, pp. 34–35). A "moral" society, we
are to presume, could "sit down and decide" how to "use" bioengi-
neering without catastrophic results—despite the immense complexity;
the inherent social, technical, and ecological compartmentalization and
opacity of the processes; and the repressive epistemology (specialization
hierarchies, manipulation of nature, Cartesian–Baconian pseudomas-
tery) that such practices require and engender.

Society's technological ensemble—from concepts of technical practice (be they highly systematic and instrumental, or connotative, unconscious, even magical), to the vast urban-industrial environment and apparatus itself—is part of a historically evolved context, formed of
archaic social hierarchies, economic configurations, the state, patterns
of social organization and association, scientific practices and ideologies, and more. *But technology also forms a matrix,* through its syner-
gistic tendency to reshape the patterns within which it emerged. The
"social world of human intentions, needs, wills and interactions" in
which technical relations are immersed has itself become immersed in
technical relations.[9] Because he doesn't consider how technology might
become, socially and epistemologically, a centerpiece of the "social
setup," Bookchin retreats to an alternative version of the mastery of
nature—one more ecologically "sensitive" and "moral," perhaps, but
composed more or less of the same content of the society he claims to
oppose.

"It may well be," notes Bookchin in a moment of uncharacteristic
humility, "that we still do not understand what capitalism really is"
(1990, p. 128). Indeed, we need a larger definition of capitalism that includes not only market relations and the power of bourgeois and bureaucratic elites but the very structure and content of mass technics, reductive
rationality and the universe they establish: the social imaginaries of progress, growth, and efficiency; the growing power of the state; and the mate-
rialization, objectification, and quantification of nature, culture, and human personality. Only then can we see that commodification and the
objectification of nature and human beings are moments in the same social process. Market capitalism has been everywhere the vehicle for a
mass megatechnic civilization—the nuclear-cybernetic-petrochemical-
communications-commodity grid being developed globally. But technicization is integral to the economic-instrumental culture of capital now ex-
tinguishing vast skeins in the fabric of life, and transforming the planet

into an enormous megalopolis, with its glittering high-tech havens and wasted, contaminated sacrifice zones.

Bookchin acknowledges that expanding market relations combined with technical innovations to explode into qualitatively new social developments (1987a, p. 204; 1990, p. 132–134), and he reviews how states and markets, in combination with new technics and social arrangements *as* technologies (e.g., the factory and its "technics of supervision"), have brought about a unified if multifaceted process (1982, pp. 139, 250).[10] As Ellul observes, "The multiplicity of these techniques has caused them literally to change their character . . . they no longer represent the same phenomenon."[11]

While capitalist accumulation is at the center of this complex, it makes no sense to layer the various components in a mechanistic hierarchy of first cause and secondary effects. There is no simple or single etiology to this plague, but instead a synergy of vectors.[12] Humanity, Mumford noted, "is now in process of changing its quarters only to be moving to a modern wing in the same archaic prison whose foundations were laid in the Pyramid Age: better ventilated and more sanitary, with a pleasanter outlook [though only for a relative handful, we should add]—but still a prison, and even more difficult to escape from than ever before because it now threatens to incarcerate a much larger part of the human race."[13]

Intriguingly, just when Mumford was reaching his gloomiest conclusions about modern technology, Bookchin appeared as a febrile enthusiast.[14] While more recently Bookchin has tempered his enthusiasm for technological development, a celebration and a defense of technological progress continue to permeate his work. "For the first time in the long succession of centuries," he enthuses, "this century—and this one alone—has elevated mankind to an entirely new level of technological achievement and to an entirely new level of the human experience" (1971, p. 10). "Utopia . . . once a mere dream in the preindustrial world, increasingly became a possibility with the development of modern technology," "a development that opens the possibility of the transcendence of the domain of necessity" (1980, pp. 28, 270). Only the "technical limits of past eras" prevented utopia (1990, p. 121). Abundance, "indeed luxury, will be available to all to enjoy because technological development will have removed the economic basis for scarcity and coercion" (1982, pp. 330–331).

Bookchin's idea of progress proves almost indistinguishable from a Kruschevite threat to out-do capitalism. "Bourgeois society," he insists, "if it achieved nothing else, revolutionized the means of production on a scale unprecedented in history. This technological revolution, culminating in cybernation, has created the objective quantitative basis for a world without class rule, exploitation, toil or material want." Only "bourgeois control of technology" prevents its liberatory potential from being realized. With the new technology, "the means now exist for the development of the rounded man, the total man" (1971, pp. 33–34, 17).[15]

And what are these means? "The potential for technological development, for providing machines as substitutes for labor is virtually unlimited" (1971, p. 95). Technology can now "produce a surfeit of goods with a minimum of toil," he says; it is no longer a servant but about to become a "partner" in human creativity. The new technology "could largely replace the realm of necessity by the realm of freedom." He observes approvingly that people in the United States no longer need elaborate explanations of this idea. "Owing to the development of a cybernetic technology, the notion of a toil-less mode of life has become an article of faith to an ever-increasing number of young people" (1971, pp. 130–131, 93–94).[16]

Bookchin's rapture over new technologies seizes on such items as "self-regulating control mechanisms," "sensory devices" such as "thermocouples, photoelectric cells, X-ray machines, television cameras and radar transmitters." He rhapsodizes over a new "electronic 'mind' for coordinating, building and evaluating most . . . routine industrial operations"; "properly used," he explains, these devices "are faster and more efficient than man himself." Basic principles of efficiency "can be applied virtually to every area of mass manufacture—from the metallurgical industry to the food processing industry, from the electronics industry to the toymaking industry, from the manufacture of prefabricated bridges to the manufacture of prefabricated houses" (1971, pp. 101–102).

Machines can take over mining and agriculture. "We could operate almost any machine, from a giant shovel to an open-strip mine to a grain harvester . . . either by cybernated sensing devices or by remote control with television cameras." Perhaps there would be no need for human involvement at all. "It is easy to foresee a time, by no means remote, when a rationally organized economy could automatically manufacture small 'packaged' factories without human labor . . . most maintenance tasks would be reduced to the simple act of removing a defective unit . . . a job no more difficult than pulling out and putting in a tray. Machines would make and repair most of the machines required to maintain such a highly industrialized economy" (1971, p. 105).[17]

His reveries are paralleled by a grotesque resourcism. The postscarcity society will derive energy from many sources, he claims, including "solar, wind, hydroelectric and geothermal energy . . . heat pumps, vegetable fuels, solar ponds, thermoelectric converters, and, eventually, controlled thermonuclear reactions" (1971, p. 119). He thrills over immense tidal power installations and solar ponds, to which some ten thousand square miles might be committed (1971, p. 126). Farming can occur from the comfort of the air-conditioned cabs of giant tractors, and livestock can be fed by means of one "of the most promising technological advances in agriculture made since World War II": the "augermatic feeding" technique, which "by linking a battery of silos with augers," can allow the "farmer" to mix and send feed to animal pens "merely by pushing some buttons and pulling a few switches."

Bookchin describes such technology as an example of "a cardinal principle of rational farm mechanization"—the use of machines to eliminate "arduous farm labor" (1971, pp. 115–116). This technological apparatus will, in Bookchin's view, leave only "a gentle, human imprint on nature." Agriculture, now an integral part of society, will be engaged with the same playfulness and creativity that people bring to gardening, and will renew "the sense of oneness with nature that existed in humans from primordial times." We are told in a formulation evocative of Orwell's *1984*: "The region will never be exploited, but it will be used as fully as possible" (1971, pp. 118–119).

One must be forgiven for wondering if a "grim fatalism" about technological calamity is more realistic than this bizarre mix of futurism and pastoralism. It is hard to imagine anyone feeling at one with nature in the midst of all that technological what-have-you—tidal dams, giant solar grids, prepackaged factories and houses, ad nauseam—while sitting, say, in the air-conditioned cab of a giant tractor or at a bank of remote-control television monitors. Bookchin, on the contrary, effuses about new patterns of production bringing about a "new animism" in which "sun, wind, waters, and other presumably 'inorganic' aspects of nature . . . would cease to be mere 'resources' . . . and would become manifestations of a larger natural totality, indeed as respiritized nature, be it the musical whirring of wind generator blades or the shimmer of light on solar-collector plates" (1980, p. 93).

It gets worse: "I have no compunction in using esthetic metaphors to describe what might ordinarily be dismissed as 'noise' and 'glare' in the vernacular of conventional technology," he continues, adding another dose of Orwellian language manipulation. "If we cherish the flapping of sails on a boat and the shimmer of sunlight on the sea, there is no reason why we cannot cherish the flapping of sails on a wind rotor and the reflection of sunlight on a solar collector. Our minds have shut out these responses and denied them to our spirit because the conventional sounds and imagery of technology are the ear-splitting clatter of an assembly line and the eye-searing flames of a foundry" (1980, pp. 93–94).

Bookchin's ebullience can be traced to a long tradition, the ideology of the "technological sublime," which replaced the ideology of the pastoral sublime in North American culture in the last century. His sentimentality about technics echoes earlier modes; the sound of assembly machines and the fires of blast furnaces and forges once evoked as much aesthetic delight as his solar panels do for him. The early idea of "incredible abundance," writes Leo Marx, which became arguably "the most important single distinguishing characteristic of American life," now "is less closely associated with the landscape than with science and technology." By the mid-nineteenth century, Marx comments, "The idea that machine power is fulfilling an ancient mythic prophecy evokes some of the most exuberant writing." Only the incentives of "comfort and status" can explain the enormous success of American technology, he explains; inventions have come to be thought of as "vehicles for the pursuit of happiness." Thus, "Americans have seized upon the machine as their birthright."[18]

Bookchin is carried into almost shamanistic ecstasies only when describing machinery. And assertions that an "early" Bookchin evolved into a "mature" Bookchin do not hold up. Apart from a few details, for example his reversal on nuclear power, his perspective remains more or less the same.[19] He continues to stress that the "ability to manipulate nature and to function actively in natural and social history is a desideratum, not an evil" (1982, p. 307). He disparages labor-intensive techniques demagogically, saying he's never known "any of [his] fellow workers to applaud" labor-intensive options, "much less the lofty 'purity' produced by fasts on vision-quests." Replying to criticism of his mid-1960s technophilia, he rejoins, "I would gladly open a surface [strip] mine with a 'great shovel' if an ecological society truly needed new ores and fuels" rather than send miners into the earth (1989a, p. 18).

One wonders if Bookchin wouldn't *enjoy* this chance to be at "one with nature," whether an ecological society required it or not. At any rate, he apparently thinks that the giant shovel and its fuel will not be produced by miners, steel and petroleum workers, and all the other operatives of a mass technological grid, but by machines making machines making machines in prefab factories, somewhere out of sight of the municipal communes. He also continues to laud "the new material possibilities created by technology after the Second World War" and to insist that in his ecotopia "a high premium would be placed on labor-saving devices—be they computers or automative machinery" (1990, pp. 151, 196).[20]

Myths of Accumulation and the Accumulation of Myths

The incentives to which Leo Marx refers above correspond to what Lewis Mumford called the Megamachine's "immense bribe, which is bound to become bigger and more seductive as the Megamachine itself proliferates, conglomerates, and consolidates."[21] In the absence of what he might consider a rational realization of plenitude, Bookchin settles for capital's; the most pressing task of technology will be to produce "a surfeit of goods with a minimum of toil." Thanks to "liberatory" technology, "Free communities would stand at the end of a cybernated assembly line with baskets to cart the goods home" (1971, pp. 130, 133). Instead of a redeemed relation to being and the object itself, he presents the fantasy of industrial cornucopia.

To be sure, Bookchin argues in places against a world of "limitless needs" and "the mindless abundance of goods" of industrial–capitalist consumer society, but his reasoning is vague, to say the least. Postscarcity, he tells us, demands the "material possibility of choosing" one's needs, which will only work "if the individual [has] the autonomy, moral insight, and wisdom to choose rationally" (1982, p. 69). The idea that people might choose to do with less from the starting point of this society—to live simply not only so that others may simply live, as radical egalitarian Christians put it, but for its own rewards—is anathema to Bookchin. The neurosis of scarcity is so great, he contends, that people "may well require a superfluity of goods so immense in quantity that the prevailing fetishism of needs will have to be dispelled *on its own terms*" (1982, p. 71; emphasis added).

Thus, he wishes to resolve a neurotic fixation, a fetishism of accumulation, by *enabling* it. "Society," he says, "may well have to be overindulged to recover its capacity for selectivity." (He doesn't reveal exactly *who* will overindulge society.) "To lecture society about its 'insatiable' appetites, as our resource-conscious environmentalists are wont to do, is precisely what the modern consumer is not prepared to hear" (1982, pp. 71–72). Nor are very many modern consumers—poor or rich—"prepared" to hear Bookchin's "moral insight and wisdom"—they want speed boats!

His solution is wishful thinking based on partial truths. "The existing technics of the western world—in principle, a technics that can be applied to the world at large—can render more than a sufficiency of goods to meet everyone's *reasonable* needs," he writes. Because the underprivileged will never accept arguments that address the possibility of a need for social and ecological limits, social ecology must demonstrate, "and not merely on theoretical or statistical grounds alone[,] . . . that affluence can ultimately be made available to all—but *should* be desirable to none. It is a betrayal of the entire message of social ecology to ask the world's poor to deny themselves access to the necessities of life on grounds that involve long-range problems of ecological dislocation, the shortcomings of 'high' technology, and very specious claims of natural shortages in materials, while saying nothing at all about the artificial scarcity engineered by corporate capitalism" (1982, pp. 261–262).

Here Bookchin shifts from an argument to provide a Western-style affluence—so that everyone will somehow see its irrationality—to accusing those who suggest the need for limits of wanting to deny people their basic *necessities,* and finally to implicit denial of natural shortages altogether. He typically dismisses anyone who raises the problem of consumer society as a malthusian elitist pushing a "hunger politics" and lifeboat ethics (1994a, pp. 4, 10). But there's a lot of ground between starvation and industrialized affluence. It is possible, in fact, to speak of artificial shortages manufactured by capitalism while recognizing the ecological destruction and shortages a high-energy-consumption, mass-production society must inevitably generate.

Bookchin is too trapped in vestigial progressivism to realize, when he states that a free society's affluence "would be transformed from a wealth of things into a wealth of culture and individual creativity" (1982, p. 69), that this latter state of affairs was already possible in aboriginal, classical and vernacular societies, which did not need to go through a "stage" of the "wealth of things" to achieve authentic plenitude (if that is what his ambiguous formulation means). If he means such a society would pass through the "wealth of things" of modern society, what exactly differentiates, from the point of view of the modern consumer, his idea from the sackcloth and ashes image he portrays in his disparaging reference to environmentalists? If it implies a kind of two-stage theory of benign indulgence of accumulation neurosis, the argument that a stable "wealth of things" is ever achieved is far from persuasive.

As Christopher Lasch has written, the consumer's relation even to things is abstract, compulsive, and far less rich than that achieved in nonindustrial societies. "It is misleading to characterize the culture of consumption as a culture dominated by things," Lasch argues. "The consumer lives surrounded not so much by things as by fantasies . . . a world that has no objective or independent existence and seems to exist only to gratify or thwart his desires." Thus cultural analysis must decide "whether the invasion of culture and personal life by the modern industrial system produces the same effects that it produces in the social and political realm: a loss of autonomy and popular control, a tendency to confuse self-determination with the exercise of consumer choices, a growing ascendance of elites, the replacement of practical skills with organized expertise."[22]

At any rate, what are we to make of the proposal to develop mass technics and a combination consumer-producer utopia *in order to reject them*? " 'High' technology must be permitted to exhaust its specious claims as the token of social 'progress' and human well-being," Bookchin says, "all the more to render the development of ecological alternatives as a matter of *choice* rather than the product of a cynical 'necessity' " (1982, p. 262; emphasis in original). For the victims of affluence, he recommends social ecology pieties; for the victims of envy, a commodity bulimia.

Ultimately, this convoluted scenario fails to acknowledge what may be the greatest problem for a future sane society, that the industrial bribe has everywhere—even where its dubious benefits have proved the most meager—tended to undermine the capacities of human beings to resist it, to choose another way, another kind of plenitude. Instead, Bookchin offers the dogma of progress: not only must we pass through the crucible of class and hierarchic society, but we must even pass through consumerism to recover Eden and ecological wisdom. Of course, neither ecological wisdom nor the planet can wait for this grotesque overindulgence to have its curative effect.

A liberatory theory cannot be based on what the modern consumer, citizen, or proletarian is "prepared to hear." The urgent need to critique affluence from within this society, and not in some future time, is not necessarily a "hunger politics"; rather, it is intimately connected to our present social, spiritual, and ecological crisis. The recognition that less might become more could come from a radical rejection of the fetishism of artifactual abundance without having to go through Bookchin's transitional surfeit. Transformation isn't a question of "better delivery," of much, much more of the same, but rather a new relationship to the phenomenal world—something akin to what anthropologist Marshall Sahlins has called "a Zen road to affluence, departing from premises somewhat different from out own."[23] If anything will save humanity and the world we inhabit, it is surely insight into "premises somewhat different from our own" rather than a contrived satiety at industrialism's vomitorium.

The Machine against the Garden

Bookchin's writings on technics and especially on agriculture starkly reveal the inadequacy of his "nature-based" ethics and social ecology in determining liberatory (or even appropriate) ecological choices. His idea of agriculture—giant tractors, remote control, pulling a few switches—does not even remotely resemble what attracts people to gardening or what keeps farmers on the land. He evinces no intuitive appreciation of the activity or its spiritual rewards, or even any noticeable desire for direct relation with the phenomenal world of sun and soil, plants and animals. Bookchin, who in one memorable passage said that "the factory floor must yield to gardening" (1971, p. 72), prefers instead to turn the garden into a factory.

Yet the garden is a context where small scale, the "softness" of technics, labor-intensiveness, and technical limits all crucially matter, and where technological transformation is bound to have more than simply quantitative utilitarian results. It is a context that so-called labor-saving devices tend to uproot, unravel, and destroy, and where people might rightly "sit down and decide," if they had any choice in the face of technological invasion, to *choose* labor-intensive community over mechanization.

Bookchin's appeals for an abundance based on efficient technological development—embellished however they may be by claims to "an ethical context of virtue" (1982, p. 307)—parallels the logic of capitalism: where capital seeks to employ technics and instrumental reason to maximize returns, Bookchin proposes the same essential strategy to maximize yields. In contrast to short-term capitalist plunder, for example, he recommends protecting the soil through "minimum tillage" agriculture, using "planters which apply seed, fertilizers and pesticides (of course!) simultaneously" to reduce soil compaction from machines (1971, p. 116).

The parenthetical exclamation is Bookchin's; whether he regrets it now or not is unimportant. What is significant in retrospect is that he was arguing for technicized agricultural efficiency right when capitalist and state socialist planning bureaucrats, following utilitarian-analytic reason, were promoting not so very different efficiency schemes and a green revolution. Despite the libertarian rhetoric, Bookchin was only able to envision *what capital itself was bringing about*. So much for dialectics.[24]

In contrast with the fevered dreams of dialectical gadget fetishism, in the real world of agricultural progress whole societies and age-old forms of life and subsistence have been and continue to be pulverized by such schemes. The objection that economics is the source of the problem, and not *also* industrial technicization, ignores the myriad effects a reductionist orientation toward maximum yields (following an ideology of abundance, comfort, rationalization, and efficiency) inevitably has on subtle, organic processes and the web of meaning in which they are embedded. Bookchin's fervent advocacy of pesticides, tractors, massive energy installations, and the like is itself proof that the profit motive is not the sole source of ecological "accidents" and megatechnic disasters; certainly *he* cannot be accused of being motivated by greed for profits in his support for them.

At the center of the ideology shared by Bookchin and development technocrats is the mystique of the labor-saving device. But as a whole, these technologies have actually expanded work. Winner has observed that "the very nature of advanced technologies . . . demands much more of the human being than any previous productive arrangement. . . . Under the relentless pressure of technological processes, the activities of human life in modern society take place at an extremely demanding cadence. Highly productive, fast-moving, intensive, precision systems require highly productive, fast-moving, intense, and precise human participants."[25]

Bookchin defends such devices not only for a future revolutionary society but within the present context. "Modern working women with children could hardly do without washing machines to relieve them, however minimally, from their daily domestic labors—before going to work," he writes (1995, p. 29). Of course, in the present context, these women also need nuclear reactors to power their washing machines, gas for their cars to get to work, shopping malls, and slave labor in Asia and Latin America (and Los Angeles) to provide the clothes they'll wash, too.

Compare his one-dimensional idea with Mumford's comment on the washing machine as a microcosm of progress: "In the workshop and the household there were plenty of tedious tasks," Mumford writes, "but they were done in the company of one's fellows at a pace that allowed for chatting and singing; there was none of the loneliness of the modern housewife presiding over a gang of machines." The wide diffusion of hand labor, apart from its superiority to machine products, reflected "the tool-user's essential autonomy and self-reliance," he says. "Had this craft economy, prior to mechanization, actually been ground down by poverty, its workers might have spent the time given over to communal celebrations and church-building on multiplying the yards of textiles woven or the pairs of shoes cobbled." Certainly, he adds, a society enjoying the innumerable festivals and holidays of traditional societies "cannot be called impoverished."[26]

Anyone who has lived both in the modern urbanopolis and in a small village understands this passage. As the celebrated natural farmer Masanobu Fukuoka observes, "A life of small-scale farming may appear to be primitive, but in living such a life, it becomes possible to contemplate the Great Way." He notes that a Japanese one-acre peasant formerly enjoyed a three-month respite at the end of each year, spent

hunting rabbits and writing haiku poems. Gradually, the new year holi-
day dwindled to a brief break, and now television has colonized what
little leisure is left. "There is no time in modern agriculture for a farmer
to write a poem or compose a song."[27]

Lasch writes that modern industrial technologies "have been defended, like mass culture, on the grounds that although they may have
taken some of the charm out of life, they have added immeasurably to
the comforts enjoyed by ordinary men and women. . . . But it is precisely the democratizing effects of industrial technology that can no
longer be taken for granted. If this technology reduces some of the
drudgery of housekeeping, it also renders the housekeeper dependent
on machinery—not merely the automatic washer and dryer but the
elaborate energy system required to run these and innumerable other
appliances—the breakdown of which brings housekeeping to a halt."
Consequently, "modern technology undermines the self-reliance and
autonomy both of workers and consumers. It expands man's collective
control over his environment at the expense of individual control; and
even this collective control, as ecologists have pointed out again and
again, is beginning to prove illusory as human intervention threatens to
provoke unexpected responses from nature."[28] Thus, to return to Marshall Sahlins's excellent phrase, a "Zen road to affluence" might mean
wash your own clothes. Even the idle rich are known to work hard
without complaint at their avocations, especially gardening.

Those who see food production in technological terms alone,
writes farmer-poet Wendell Berry, oversimplify "both the practicalities
of production and the network of meanings and values necessary to define, nurture, and preserve the practical motivations. . . . A healthy
farm culture can be based only upon familiarity and can grow only
among a people soundly established upon the land; it nourishes and
safeguards a human intelligence of the earth that no amount of technology can satisfactorily replace." This culture is precisely what has been
destroyed not only by capitalist economic relations but also by the
companion ideology of technological progress and the dependencies it
has engendered. However difficult to initiate, Berry observes, a reverse
movement is necessary. "It will probably require several generations—enough to establish complex local cultures with strong communal memories and traditions of care." Where Bookchin apparently sees
only arduous, mindless toil, Berry, who has some real experience with
farming, sees opportunity for transformation.[29]

"It is difficult to conceive of higher social effectiveness with
lower industrial efficiency," Ivan Illich has written. "To recognize the
nature of desirable limits to specialization and output, we must focus
our attention on the industrially determined shape of our expecta-

tions." Rather than simply retooling under different political arrangements, "we must radically reduce our expectations that machines will do our work for us. . . . The only solution to the environmental crisis is the shared insight of people that they would be happier if they could work together and care for each other. Such an inversion of the current world view requires intellectual courage for it exposes us to the unenlightened yet painful criticism of being not only antipeople and against economic progress, but equally against liberal education and scientific advance."[30]

Here, too, Bookchin's error resides in his Marxism. For Marx, the workers become appendages of the machine because the machines and labor process are owned and controlled by the capitalists. The former confront the material products of their labor—machines and industrial apparatus as well as commodities—as an "alien power" because it all "belongs to some *other man than the worker*."[31] This schema does not take into account the life processes involved as cultural and epistemological contexts in their own right. Alienation is not limited to a problem of who owns or who directs mass technics. Commenting on Marx's passage, Winner argues that the governance imposed by this "other man" is not decisive; "the steering is inherent in the functioning of socially organized technology itself," which is to say that the owners and bosses must steer at the controls their technology provides. As the monster says to Victor Frankenstein, "You are my creator, but I am your master."[32]

Technology socializes those who operate it because mass industrial technics require that they operate *within* it. While people may think of the vast webs of instrumental and economic relations as simple tools to be either used properly or abused, one does not simply apply an Archimedean lever to a global petrochemical grid, or a communications-informatics grid. We are increasingly enclosed in them, functioning as cogs within them. As Ellul argues, "What is being established is no longer the subordination of man to technology, etc.," but far more deeply, "a new totality."[33]

Modern industrial society and its technics, Winner points out, with their "enormous size, complex interconnection and systemic interdependence," demand "precise coordination" of conceptual technique, organization and apparatus. Hierarchy, specialization, and stratified, compartmentalized organizational structures are inescapable—including the need for vast training institutes, hierarchies of command, record-keeping, resource exploitation with its natural sacrifice zones on a global scale, complex backup systems, police, and an army of skilled, semiskilled, and unskilled operatives to carry out the required procedures. The resulting web of human and technical relations, this "total order of networks," is not so much tool-like as it is "a technical ensemble that demands routinized behavior," becoming, ultimately, a "way of life": "We do not *use* technologies so much as *live* them."[34]

Thus mass society is not simply a reflection of organized commodity production and consumption, as Bookchin has suggested, with its technics subject to "more fundamental economic factors." Rather, the organic structures of society are dissolved and reconstituted not only by new economic forms and forces but by the social organization and dependencies generated by mass technics. Production for profits certainly brings a significant factor of cruelty and irrationality to the formula, but a world system mediated by machines brings its own objective and subjective content to human life whether "use value" or "exchange value" has generated it. As Cornelius Castoriadis has put it, all tools (and by this he means technics generally) are themselves "institutions as well as embodiments of meaning."[35]

For example, because they are fabulously efficient vehicles for the transmission of market values, computers and television have helped shape our expectations and culture to conform to market society; but our very sense of reality has also been transformed by these instruments and the webs of meaning they themselves establish. Mass technics have dramatically furthered a kind of psychic numbing and the fragmentation of knowledge, undermining and complicating (though not entirely suppressing) human agency and responsibility. They have created an opaque and dangerous technological structure that undermines biological foundations on a daily basis, and which, in its present configuration, is impossible to control.

To his credit, Bookchin has little regard for the present organization of affairs. But he never explains how free communities can expropriate the technology capital generated to use it for communal, libertarian ends. His views in this regard are, in fact, utterly naive. "Is society so 'complex' that an advanced industrial civilization stands in contradiction to a decentralized technology for life?" Bookchin asks. "My answer to this question is a categorical *no*" (1971, pp. 136–138; emphasis in original). The apparent complexity comes from bloated bureaucracy, needless administration, and wasteful capitalist endeavors.

"I do not wish to belittle the fact that behind a single yard of high quality electric wiring lies a copper mine, the machinery needed to operate it, a plant for producing insulating material, a copper smelting and shaping complex, a transportation system for distributing the wiring—and behind these complexes other mines, plants, machine shops and so forth," he says. But even if copper could be provided only by national distribution, "[in] what sense need there be a division in the current sense of the term? There need be none at all." The idea "that modern industry has become too complex," he writes, is misleading. In fact, only "a stupendous, often meaningless, social machinery" stands in the way. "But it is not the complexity of the machinery that inhibits our ability to deal with the imperatives," he reassures us, but "a system of industrial clientage" that is to blame (1982, p. 311).

It's astonishing that Bookchin thinks the complexity of the apparatus itself—and logically, then, the ensemble of networks that comprise this apparatus—has so little to do with externally and internally restructuring our lives. Outside my window as I type this, a line-worker is working on a telephone line. Her job is a mystery to me, and countless other jobs are mysteries to us both. There is no way she and I can know how a vast system of industrial production that Bookchin favors could or would work—even if the factories are small and prefabricated. The "trays" that will be effortlessly replaced to repair or adapt the machine-making machines in Bookchin's utopia, for example: someone would have to be trained to know how each one functions, but it is simply impossible for any human being to know how they *all* do or how they all fit together. Many of the dislocations and system breakdowns in industrial society are the result of this very opacity and information overload.[36]

Our inevitable compliance with the opinions of experts—even if we could discuss and vote on their findings in the assemblies of a Bookchinesque municipalist utopia—would of necessity be based on persuasion and faith—a faith, I might add, that appears after a century or more of rapid scientific-technological expansion, accommodation and acculturation, in which the mysteries of life have been supplanted by the miracles of science, the technological environment has been naturalized in the minds of most people, and much of culture has been reduced to "a collage of specialized bodies of knowledge," as Roszak has put it, "in which the thinking of ordinary people has been rendered worthless."[37]

From Prometheus to Epimetheus

The subjectivity of technological civilization is, of course, rooted in economic–instrumental values—the dominant ideology nearly everywhere on the planet, except for a few quixotic holdouts and the displaced or contaminated victims of technology who might have seen through the facade. According to the myth of the machine, however, technological development is not only inevitable but ultimately beneficial. This is the ideology of progress that Bookchin feels compelled to defend from skeptics with flimsy platitudes (1994b). Not only can we not stop prog-

ress, the idea of stopping it is considered crazy, irresponsible, immoral—perhaps fascistic (1995, pp. 29–30, 61). Even technology's disasters are now employed as justifications for its continued maintenance.

Bookchin dramatically reveals himself to be an acolyte of this myth
when he argues for advanced technology to protect nature from itself—for example, from "ice ages, land desiccation, or cosmic collisions
with asteroids." NASA will apparently be turned into a municipalist organization—and with no division of labor, either. "If there is any truth
to the theory that the great Mesozoic reptiles were extinguished by cli-
matic changes that presumably followed the collision of an asteroid
with the earth," he explains, "the survival of existing mammals might
well be just as precarious in the face of an equally meaningless natural
catastrophe unless there is a conscious, ecologically-oriented life-form
that has the technological means to rescue them" (1990, p. 38).

Of course, it probably won't be a "meaningless natural catastrophe" that extinguishes mammal life, but a series of "meaningful" catastrophes set off by the very megatechnic civilization Bookchin portrays
as nature's only hope. His projection is a Rube Goldberg nightmare
filled with lurid delusions of grandeur and scientific hubris. Not only
would we need a massive missile system (reminiscent of Ronald Rea-
gan's Star Wars fantasy) to deflect asteroids, but a complex technics advanced enough to deflect *entirely unimagined threats*—suggesting,
among other things, a genetic engineering arsenal of colossal proportions. Bookchin fails to notice that our defense systems, antibodies, and
fail-safe backups will likely do us in long before the cosmic threats arrive.

Equally significant is his comment that it would hardly be anthropocentric, except under exploitive capitalist conditions, of course, "to
turn the Canadian barrens—a realm that is still suspended ecologically
between the highly destructive glacial world of the ice ages and the
richly variegated, life-sustaining world of temperate forest zones—into
an area supporting a rich variety of biota." He continues: "I frankly
doubt that a case can be made against a very *prudent, nonexploitative,*
and *ecologically guided* enterprise of this kind . . . unless we put blink-
ers on our eyes that narrow our vision to an utterly dogmatic and
passive-receptive 'nature-oriented' outlook" (1992, p. 170; emphasis in
original). Presumably, this is what he means when he postulates a
"more advanced interface with nature" (1982, p. 39) and "a new, emi-
nently ecological function: the need to create more fecund gardens than
Eden itself" (1982, p. 303). One swoons imagining the Eden Bookchin
might make of the Canadian barrens.

One thing is certain: Bookchin's view is anything but passive.
While we are still in the "prehistory" of such capacities, we are told (a

classic Marxian formulation of progress), humans "can intervene in [nature], even try to manage it consciously, provided they do so in its own behalf as well as society's" (1982, pp. 34, 39; 1980, p. 271). Indeed, "human intervention can be as creative as natural evolution itself" (1987b, p. 40). Not only does he argue that we have no choice but to intervene to some degree in the natural world (a reasonable assumption of inevitable responsibility), but he tells us that humanity has a "right to intervene in the natural world, *to do even better than 'blind' nature in fostering variety and natural fecundity*" (1982, p. 268; emphasis added).

This species of box puzzle paradox is nothing but a paean to the power of technology. We can surmise only general patterns in social and natural history; the whole, and the endless minute workings of the whole, remain mysteries. Humility about such matters is called for, especially given our capacity to alter human and nonhuman nature alike by wrecking them beyond recognition. Instead, Bookchin's dialectic of freedom turns out to be only a variant of the ideology of bourgeois progress and human mastery, a mastery exercised by a "life-form . . . that expresses *nature's* greatest powers of creativity" (1990, p. 36; emphasis in original), which is, in fact, "nature itself rendered self-conscious" (1982, p. 315).

Elsewhere, it is true, he qualifies—for example, his contrast of the "Promethean quest . . . to 'dominate nature'" technologically with "the ecological ethic of using technology to harmonize humanity's relationship with nature" (1980, p. 109). Of course, it should be clear by now that Bookchin gives little indication of recognizing the difference between domination and harmony, between Promethean meddling and a humble sense of human limits. Historically, the notion of controlling nature has typically been legitimated by claims to a higher rationality. Prometheus, representing foresight, has always been the hero of the rationalists, who have forgotten that he comes accompanied by his dull-witted brother, Epimetheus, representing hindsight and miscalculated catastrophe. After approving Charles Elton's "sensitive" comment that "the world's future has to be managed," through a process like "steering a boat," Bookchin adds: "What ecology, both natural and social, can hope to teach us is the way to find the current and understand the direction of the stream" (1982, p. 25). Of course, it is one thing to try rationally to steer humanity's paper boat on the swells, squalls, and currents of nature's great sea, another entirely to think we could steer *the sea itself,* an idea Bookchin takes to hallucinatory extremes, as we have seen.

For a writer whose ideas are based on a notion of potentiality, Bookchin's static idea of technology fails dismally to see technics in

their full development—not only the dubious potentiality of their evolution into a liberatory society, but other potentialities that do not fit
his schema. We do not yet fully know the real meaning of industrialism; it is still being played out in our very being, somatically and genetically, and in the myriad ripples and feedback loops now traveling
through both human societies and the natural world. In Bookchin's
simplistic view of technology, "free municipalities" will one day somehow stand with shopping bags at the end of their cornucopic assembly
line, picking and choosing only the technics and products they rationally desire, while somehow avoiding the accompanying "accidents,"
side effects, and toxic residues.

Bookchin recognizes that "potentiality must not be mistaken for
actuality. The great bulk of humanity is not even remotely near an
understanding of its potentialities . . . [and] a humanity unfulfilled is
more fearsome than any living being" (1982, p. 237). Yet if we remain that unrealized after so much History, Civilization, and Progress, then a more skeptical view of technology might be called for,
since neither Bookchin nor anyone else has demonstrated a capacity
to unlock what he only *intuits* is humanity's potential to master a
vast, complex industrial system and place it in its "proper context."
There might in fact be an *inherent* potentiality toward alienated consciousness and totalitarian control in the modern technology so inspiring to Bookchin that far outweighs its potential for liberatory
practice. It is, after all, the product of capitalism. Bookchin nowhere
addresses this possibility.

Yet for those who have the courage to look clearly at life today,
the claims of mass technics are already dramatically eroded by decades,
even centuries, of catastrophe, imperial plunder and war, the unprecedented dislocation of human communities, and the ongoing eclipse of
the human spirit. A new perspective now haunts the industrial capitalist
necropolis. As inchoate and embryonic as it may now be, this "epistemological luddism," as Winner has called it, is where the real possibility for social ecology must lie. It does not propose "a solution in itself
but . . . a method of inquiry" that, instead of focusing on obfuscatory
notions of "use" and "misuse," "insists that the *entire structure* of the
technological order be the subject of its critical inquiry."[38]

Different values are emerging because many people have now seen
enough of the artificial paradise—in fact, more than they care to see—
and are seeking pathways to abandon it. Too late, perhaps—we may
only be seeing the proverbial twilight flight of Minerva's owl of understanding. But whatever the outcome, this sensibility is a natural consequence of the conditions in which we presently find ourselves. Bookchin himself once wrote that "wherever possible, we must 'unplug' our

'inputs' from a depersonalized, mindless system that threatens to absorb us into its circuitry" (1982, p. 335). "We can no longer retain techniques that wantonly damage human beings and the planet" (1990, p. 188). It might even be a reasonable idea to "turn our backs on the entire heap," he says. But since we are too mired in industrialism's debris to do so, he continues, we must instead "tread cautiously—seeking firm ground wherever we can in the real attainments of science and engineering" (1982, pp. 245–246).

Yet the "real attainments" of science and engineering are hardly solid ground, braided as they are into the instrumental reason central to the exterminist system. (It's also difficult to understand how or why we should turn our backs on the very "heap" that we've been told has laid the preconditions for a postscarcity paradise; our dialectician will have to sort that out.) Unsure whether or not there is a driver to the Megamachine, as we have seen, Bookchin wavers between the poetry of the past and the future, finally choosing capital's future—with its computers, augermatics, and even giant shovels if historical necessity and progress demand them—over the otherness of the past. Science, he says, should "make room for other metaphysical presuppositions" to illuminate what science cannot know (1982, p. 239), yet ultimately, in the face of the greatest crisis in our species' history, we can find "solid ground" only in the epistemologies of domination itself.

Bookchin's failure to resolve such contradictions paralyze him, and social ecology with him. To redeem social ecology requires returning to a starting point that comes from insights provided neither by science nor by engineering. As Mumford observed, a "reliance upon mechanical solutions" to our present exigencies would only solidify the Megamachine, "when what is actually required is mechanical simplification and human amplification."[39] This view hardly signifies the "return to the Stone Age" that Bookchin claims critics of technology desire (1995, p. 36); *any* way out is going to demand a careful negotiation with technics, a process which Roszak has called "the selective reduction of industrialism," requiring "an economics which uses neither people nor nature as its proletariat."[40]

We have no choice but to face the legacy that modernity has given us. We cannot evade the responsibility to think critically and rationally about the crisis we face. *But reason is whole.* A future social ecology, worthy of its desire for redemption and renewal, would recognize that it is not in scientific rationality and technological mastery but in other domains—starting from an *authentically* dialectical understanding that reorients life around perennial, classic, and aboriginal manifestations of wisdom we have yet to address fully—where firm ground, if any, must be found. Revolution will be a kind of return.

Acknowledgments

Parts of this essay appear in another version in my book, *Beyond Bookchin: Preface for a Future Social Ecology* (New York: Autonomedia/Detroit: Black & Red, 1996). I wish to thank Marilynn Rashid, Lorraine Perlman, John Clark, Kathleen Rashid, and Steve Welzer for their editorial advice and encouragement throughout the process of writing both texts.

Notes

1. Here Bookchin is using the word *technics* to refer to technology in a general way, but also to distinguish the reductive instrumentality of modern technology from the matrix of craft, art, and ethical values embodied in the Greek root *techne*. The pervasive confusion about the word *technology* and related terms such as *technics* and *technique* is illuminated by Langdon Winner's introduction to his indispensable *Autonomous Technology: Technics-Out-of-Control as a Theme in Political Thought* (Cambridge, MA: MIT Press, 1977), pp. 8–10. Winner describes the gradual and eventually dramatic transformations in our sense of the word *technology*, from a "specific, limited and unproblematic meaning" as referring to a "practical art" or the study of the practical arts, to "an unbelievably diverse collection of phenomena—tools, instruments, machines, organizations, methods, techniques, systems, and the totality of all these and similar things in our experience" (p. 8). See also my *Against the Megamachine: Essays in Empire and Its Enemies* (New York: Autonomedia, 1988), pp. 119–120.

2. "Although it is all too easy to blame on technics what is really the result of bourgeois interest, technics, when divested of any moral constraints, can also become demonic under capitalism" (1990, p. 188). Bookchin never explains what might become demonic about technics separate from capitalist "abuse." It's hard to imagine it happening unless technics has *some* measure of autonomy.

3. See "We All Live in Bhopal," in my *Against the Megamachine*, pp. 42–47.

4. E.g., see Jacques Ellul's comment that "technology is inevitably part of a world that is not inert. It can develop only in relation to that world. No technology, however autonomous it may be, can develop outside a given economic, political, intellectual context" (*The Technological System* [New York: Continuum, 1980], p. 31; see also pp. 18, 303–304).

5. Lewis Mumford, *Technics and Civilization* (1934; reprint, New York: Harcourt, Brace & World, 1962).: "The fact is that in Western Europe the machine had been developing steadily for at least seven centuries before the dramatic changes that accompanied the 'industrial revolution' took place" (p. 3). In *The Myth of the Machine, Vol. 1: Technics and Human Development* and *Vol. 2: The Pentagon of Power* (New York and London: Harcourt Brace Jovanovich, 1967 and 1970, respectively), he develops the idea of the ancient megatechnic precursors of modern technological civilization.

6. Jacques Camatte, *The Wandering of Humanity* (Detroit: Black & Red, 1975).

7. One is reminded of anthropologist Eleanor Burke Leacock's criticism of Lévi-Strauss, who "does not conceive of structures at different levels as interacting dialectically, i.e., influencing and ultimately transforming each other. Instead, his view of relationships among levels is functionalist" (*Myths of Male Dominance* [New York: Monthly Review Press, 1981], p. 218). In Lévi-Strauss's case, the mythic communications system makes up the said structure; in Bookchin's the economy and abstract relations of domination. Thus a hierarchic society produces an industrialism of domination, and a libertarian socialism gives a liberatory industrialism, sort of the way Stalinism provided socialist fallout from its nuclear tests to counter capitalist fallout from those in the West.

8. See his response to a questionnaire in the Fall 1977 issue of *CoEvolution Quarterly*. Rightly criticizing the appropriate technology movement's "mechanical substitutes for a new spiritual community based on mutual aid [and] libertarian communism," he raises the idea of "a liberatory (not morally neutral 'appropriate' and 'intermediate') technology," without ever making clear what would constitute it.

9. See Langdon Winner's excellent survey, *Autonomous Technology*, esp. pp. 306–313, and Joseph Weizenbaum, *Computer Power and Human Reason: From Judgment to Calculation* (San Francisco: Freeman, 1976).

10. Ivan Illich's book *Shadow Work* (Boston and London: Marion Boyars, 1981) is a valuable source of insight into this history. What Lewis Mumford called "the new coalition" of "money power with political power" arose along with the state's military machine, with its dependence on technical innovation and the mass production of weapons, leading to an "economic dynamism" linked to "bellicose destruction"—the basic features of today's national military megastates, with their ultimate weapons. See Mumford, *The Pentagon of Power*, p. 242.

11. Jacques Ellul, *The Technological Society* (New York: Knopf, 1964), p. 63.

12. See Karl Polanyi's *The Great Transformation: The Political and Economic Origins of Our Time* (Boston: Beacon Press, 1971) for useful discussion of the emergence of market society. Carolyn Merchant shows the interconnections between science, capitalism and technics in creating not only a "new economic and scientific order" but a new alienated relation to nature and ourselves—keys to understanding the origins and the character of this phenomenon. "A slow but unidirectional alienation from the immediate daily organic relationship that had formed the basis of human experience from earliest times was occurring. Accompanying these changes were alterations in both the theories and experiential bases of social organization which had formed an integral part of the organic cosmos" (*The Death of Nature: Women, Ecology and the Scientific Revolution* [New York: Harper & Row, 1980], p. 68). Theodore Roszak's *Where the Wasteland Ends: Politics and Transcendence in Postindustrial Society* (Garden City, NY: Doubleday/Anchor, 1973) is an excellent survey of this transformation.

13. Mumford, *Pentagon of Power*, p. 327.

14. Mumford's *Pentagon of Power* and Bookchin's *Post-Scarcity Anarchism* both appeared in 1970–1971.

15. Thus the fallacy of John Clark's claim that "Bookchin rejects the view . . . that technology can be looked upon instrumentally, as a means toward either liberation or domination, depending on how it is used." In fact, the upshot of Clark's own discussion of this question in Bookchin makes the opposite conclusion clear. See 1980, pp. 24–25, 255–256, and Clark, *The Anarchist Moment* (Montreal and Buffalo: Black Rose Books, 1984), pp. 207–212.

16. Bookchin made this claim in 1965, reaffirmed it in 1970, and has never reconsidered it. Mumford had a less sanguine view of this "faith": "The notion that automation gives any guarantee of human liberation is a piece of wishful thinking" (*Pentagon of Power,* p. 191). Bookchin's enthusiasm about computers was ironically accurate, though. As *Whole Earth Catalogue* founder Stewart Brand told the April 1994 issue of the magazine *Outside,* "Computers won out. Everything else failed. Communes failed, dope failed, politics failed. The main thing that survived and thrived was the computer."

17. In *Forces of Production: A Social History of Industrial Automation* (New York: Knopf, 1984), David Noble points out that efforts of total automation have proven to be "a flight from reality." As Theodore Roszak notes in a discussion of Noble's argument, a "workerless production process loaded with 'sensors, monitors, counters, alarms, self-actuated repair devices' . . . is more and more prone to breakdown or malfunction," reaching "a level of complexity and delicacy that finally overwhelms the ability of those who must use them. 'The greater complexity required to adjust for unreliability,' Noble comments, 'merely adds to the unreliability.'" See Roszak's useful and readable *The Cult of Information: A Neo-Luddite Treatise on High Tech, Artificial Intelligence, and the True Art of Thinking* (Berkeley and Los Angeles: University of California Press, 1986), pp. 128–129.

18. Leo Marx, *The Machine in the Garden: Technology and the Pastoral Ideal in America* (London and New York: Oxford University Press, 1973), pp. 40 and 198–205.

19. He has changed his mind about nuclear power, it is true. It is now "intrinsically evil" (1990, p. 188). But even hydroelectric and geothermal power schemes have proven disastrous, and wind is highly problematic, and not purely because of corporate avarice. As far as his "evolution" goes, Bookchin has practiced singularly bad faith. He is not above touting the prescience of his "new approach," his early attempt to "go beyond . . . traditional trends in environmentalism," explored in his own "pioneering articles," citing those quoted above (1990, pp. 154–155). When a letter to the Canadian anarchist magazine *Kick It Over* raised questions about his views in the 1960s, however, he called the use of quotations from his work "immoral and cynical." The essays, he fumed, could not be taken out of the context of their time, "when mechanization, and even controlled thermonuclear reactors, were seen as desirable by me—and given [their] time, I have *no regrets whatever for having held this view*" (1989a, p. 18; emphasis in original).

20. It was *capitalism,* of course—the kind of society that according to Bookchin himself must create inherently antiecological technics—that created

such "material possibilities." Mumford argued that it was in the crucible of World War II and the postwar period that the modern Megamachine was forged. Weizenbaum, in fact, points to the timeliness of cybernation after World War II—in averting "catastrophic crises" for American capitalist and military elites in managing an increasingly complex global empire. See *Computer Power and Human Reason*, pp. 27—28.

21. Mumford, *Pentagon of Power*, p. 330. Bookchin provides only a radical veneer to another kind of surrender to this bribe. Despite interesting parallels in their work, this failure in Bookchin illuminates one of the critical differences between them. Even with its many limitations and contradictions, Mumford's work is more profound, more suggestive, and ultimately more useful to a future social ecology than Bookchin's.

22. Christopher Lasch, *The Minimal Self* (New York and London: Norton, 1984), pp. 46, 30, 41. See also Stuart Ewen, *Captains of Consciousness: Advertising and the Social Roots of the Consumer Culture* (New York: McGraw-Hill, 1976).

23. Marshall Sahlins, *Stone-Age Economics* (New York: Aldine, 1972), p. 2. Bookchin confuses the objective conditions of consumerism that compel people to buy products created for the market by corporations (which means that needs aren't so much false as they are artificial), with the subjective conditions that bring people to consider as basic necessities irrational products that they could easily do without. He dismisses criticism of consumerism as a middle class liberal environmentalist shibboleth (see, e.g., 1994, pp. 26–28; 1990, pp. 71–72, 1989b). There is never any complicity with the "grow-or-die" economy in his scenario; consumers are always the passive victims of corporations, never the willing participants in demonstrations to maintain the production of "Classic" Coca Cola. Populist rhetoric does nothing to illuminate the subjective conditions of modern capitalism among either those who benefit relatively from them or those who go without.

24. "To sense and comprehend after action is not worthy of being called comprehension," we read in an ancient Taoist text. "Deep knowledge is to be aware of disturbance before disturbance, to be aware of danger before danger, to be aware of destruction before destruction, to be aware of calamity before calamity" (quoted in Thomas Cleary's Introduction to Sun Tzu's *The Art of War* [Boston and London: Shambhala, 1988], p. 3).

25. Winner, *Autonomous Technology*, pp. 204–205.

26. Mumford, *Pentagon of Power*, pp. 137–138.

27. Masanobu Fukuoka, *The One-Straw Revolution: An Introduction to Natural Farming* (Emmaus, PA: Rodale Press, 1978), pp. 110–111.

28. Lasch, *The Minimal Self*, p. 43.

29. Wendell Berry, *The Unsettling of America* (New York: Avon Books, 1978), pp. 43–45; emphasis in original.

30. Ivan Illich, *Tools for Conviviality* (New York: Harper & Row, 1973), pp. 49–50. See also Illich's *Toward a History of Needs* (New York: Pantheon Books, 1978), p. 67. Illich is also a problematic and contradictory writer—a discussion far beyond the scope of this essay.

31. Karl Marx, *Economic and Philosophic Manuscripts of 1844* (Moscow: Progress Publishers, 1974), p. 70; emphasis in original.

32. Winner, *Autonomous Technology*, pp. 36–40; Mary Shelley's *Frankenstein*, quoted by Winner, p. 311. The word "ultimately" must be stressed here; Winner is not arguing that elites do not make critical decisions concerning the direction of technics. Yet while capitalist economic imperatives shape and direct technological development, technological synergism also determines the direction of economic imperatives. See Winner, *Autonomous Technology*, pp. 202, 227–228; Ellul, *Technological Society*, pp. 255–265; and C. Wright Mills, *The Causes of World War Three* (New York: Simon and Schuster, 1958).

33. Ellul, *Technological System*, p. 203.

34. Winner, *Autonomous Technology*, pp. 200–203.

35. Cornelius Castoriadis, "The Crisis of Marxism and the Crisis of Politics," in *Society and Nature*, Vol. 2, September–December, 1992, p. 205. Bookchin consistently commits errors in this regard. One glaring example is his treatment of the clock, no labor-saving device at all, since it does no work. Its "triumphal invention," he says, "lessened the need for arduous toil and greatly increased the effectiveness of craft production" (1982, p. 258). The clock is, of course, the quintessential autonomous machine, bringing a completely new totality into operation. "The invention and perfection of the mechanical clock," writes Mumford, "was the decisive move toward automation" (*Pentagon of Power*, pp. 176–177). The Benedictines, who developed the mechanical clock, were "perhaps the original founders of modern capitalism," he says, "for the clock is not merely a means of keeping track of the hours, but of synchronizing the actions of men." See *Technics and Civilization* (New York: Harcourt, Brace & World, 1963), pp. 13–14.

36. In *The Cult of Information*, Roszak discusses a form of "computer Alzheimer's" that is rendering many programs—including in very sensitive and potentially dangerous systems—unreadable and erratic because of the inevitable layered changes that many different programmers have made on them over time. See pp. 193–195.

37. Roszak, *Where the Wasteland Ends*, pp. 236–237.

38. Winner, *Autonomous Technology*, pp. 325–235, 226.

39. Mumford, *Pentagon of Power*, p. 286.

40. Roszak, *Where the Wasteland Ends*, pp. 385–386.

Bibliography of Bookchin's Works Cited in This Chapter

1971. *Post-Scarcity Anarchism.* San Francisco: Ramparts Press.

1980. *Toward an Ecological Society.* Montreal: Black Rose Books.

1982. *The Ecology of Freedom.* Palo Alto, CA: Cheshire Books.

1986. *The Modern Crisis.* Philadelphia: New Society Publishers.

1987a. *The Rise of Urbanization and the Decline of Citizenship.* San Francisco: Sierra Club Books.

1987b. "Thinking Ecologically: A Dialectical Approach." *Our Generation* 18, no. 2 (Spring–Summer 1987): 5–40.

1988. "When the Earth Comes First, People *and* Nature Suffer." *Guardian,* August 3, 1988.

1989a. "Environmentalism and Class: A Letter from Murray Bookchin." *Kick It Over,* March 1989.

1989b. "Death of a Small Planet." *Progressive,* August 1989.

1990. *Remaking Society: Pathways to a Green Future.* Boston: South End Press.

1991. with Dave Foreman. *Defending the Earth: A Dialogue between Murray Bookchin and Dave Foreman.* Edited and with an introduction by Steve Chase. Boston: South End Press.

1992. "Recovering Evolution: A Reply to Eckersley and Fox." *Nature and Society* 2 (September–December 1992).

1994a. *Which Way for the Ecology Movement?: Essays by Murray Bookchin.* Edinburgh and San Francisco: AK Press.

1994b. "History, Civilization, and Progress: Outline for a Criticism of Modern Relativism." *Green Perspectives,* No. 29, March 1994.

1995. *Social Anarchism or Lifestyle Anarchism: An Unbridgeable Chasm.* Edinburgh and San Francisco: AK Press.

Chapter 8

" 'Small' Is Neither Beautiful nor Ugly; It Is Merely Small"

Technology and the Future of Social Ecology

ERIC STOWE HIGGS

Introduction

That technology is a central theme in the work of Murray Bookchin is an inescapable conclusion for any reader.[1] It features prominently in almost every one of his texts. For example, in introducing the idea of material abundance in the opening essay of *Post-Scarcity Anarchism,* he writes:

> This technological revolution, culminating in cybernation, has created the objective, quantitative basis for a world without class rule, exploitation, toil or material want. The means now exist for the development of the rounded man, the total man, freed of guilt and the workings of authoritarian modes of training, and given over to desire and the sensuous apprehension of the marvelous.[2]

A substantial portion of *Toward an Ecological Society* is taken up with describing the liberatory possibilities of alternate technology. Two of twelve chapters of his major theoretical work, *The Ecology of Freedom,* concern technology. In the final chapter of *Remaking Society,* "technology and decentralization" is the penultimate section. Clearly, a critical evaluation of technology has an important function in Bookchin's overall program of social ecology.

A review of writings by and about Bookchin, however, reveals no critical interpretation of his ideas about technology.[3] George Bradford, in his blistering criticism of deep ecology that is also sympathetic to Bookchin's social ecology, writes in the first footnote,

> While Bookchin responds powerfully [in his essay "Thinking Ecologically: A Dialectical Approach"] to the entire discussion on the question of humanism and the dangerous mixture in deep ecology of sentimental mysticism and a harshly instrumental scientistic methodology, his notion that "human intervention into natural process can be as creative as that of natural evolution itself" suggests the very technological hubris that deep ecology confuses with humanism. Given Bookchin's view of creative human intervention (and his naive position on technology in *Post-Scarcity Anarchism*), what is his attitude toward biotechnology, which uses the same essential argument in legitimating its destructive meddling into the fundamental structures of nature? Bookchin's perspective needs a thorough critique which hopefully will be undertaken soon.[4]

It is ironic that a writer as influential and provocative as Bookchin would be subject to (almost) no critical scrutiny of a main feature of his work. Several key questions can be raised concerning this theoretical lapse. Why is there so little secondary use of his writings on technology? Why has Bookchin not been regarded as an important commentator by scholars of technology? What is it about his view of technology that occludes such reflection? Does the structure of his theory of technology limit the potential for social ecology as an emancipatory social and political project? If this is the case, can the theory be reformed such that it provides a solid ground for the flourishing of social ecology (or at the very least, appropriate recognition for its accomplishments)?[5]

In this essay I will argue that Bookchin promotes a kind of technological pluralism in theory that provides little purchase on the larger political and philosophical questions central to his work: the dissolution of hierarchy, the development of a "free" nature, and so on. Significant aspects of his view of technology reflect a common instrumental view of technology. According to this view, technology has no moral value apart from that which is vested in it by the users of those devices and machines. This kind of instrumentalism is now widely regarded by scholars of technology as unhelpful in forming a coherent political understanding of the significance of technology. My purpose is to provide commentary on Bookchin's use of technology, criticism that I hope will strengthen what is surely one of the most coherent and hope-inducing sets of radical environmental reforms in the contemporary period. This criticism would be much easier if Bookchin displayed only a naïve instrumentalism. In fact, his account of technology is complicated by concepts such as "managerial radicalism" and his typically incisive ideas about institutional life and state authority. Indeed, several different strains of technology theory operate, at times simultaneously, in his work. For this reason I present three different readings—pluralism as confusion, pluralism as evolution, and pluralism as pragmatism—that together provide a more coherent account of Bookchin's views about technology and also point to the need for a coherent philosophy of technology within social ecology.

In sorting through Bookchin's concepts of technology, I intend to accomplish three tasks: first, to explicate his understanding of the function of technology; second, to suggest why this view of technology is inadequate when judged against other more sophisticated views of technology; and third, to propose a theoretical stance concerning technology that is more useful to his ultimate goal of "remaking society." More than this, I believe that it is important to articulate a theoretical position that can advance social ecology more generally; it just so happens that this goal is coterminal with a critical treatment of Bookchin. In the last section of this essay, I will appeal to the work of a contemporary communitarian philosopher of technology, Albert Borgmann. All of this has the practical end of advancing social ecology as a philosophical, political, and social program.

If it is not already clear, this criticism is offered in a spirit of respect and solidarity. When I studied at the Institute for Social Ecology in 1986 it was apparent to most of my fellow students that social ecology offered a coherent, radical political alternative to various nascent environmental theories such as deep ecology. The polarization and eventual marginalization of social ecology following a series of virulent exchanges between Bookchin and various deep ecologists in the late 1980s,[6] coupled with Bookchin's detachment from many of his former collaborators, correspondents, and colleagues, was difficult to imagine a decade ago. This volume is one attempt to revive an appreciation for the liberatory possibilities of social ecology, to share some of the burden of responsibility for developing its theoretical positions, and to straighten out some of the misinterpretations appearing now in the literature. Despite the hybridization of deep ecology, it remains unfocused theoretically, essentialist in its view of nature, and indiscriminate in assigning responsibility for contemporary environmental crises.[7] It was Murray Bookchin's *Toward an Ecological Society* that inspired me as a graduate student to seriously address the role of technology in thinking through political change. It came as a welcome antidote to ceaseless enthusiastic reports from the alternate technology front: "Devices will set us free!" The title for this chapter is borrowed from a passage in *The Ecology of Freedom* that itself was apparently borrowed from Tony Mullaney.[8] "Small is beautiful," of course, is the expression coined by E. F. Schumacher and which became a virtual chant for appropriate technologists in the 1970s and 1980s. No slight is intended to this inspired work, but as it became absorbed in the mass culture of environmentalism in the 1970s, anything small in the way of technology became something of value. Bookchin exploded this notion by showing how small technologies could be just as destructive as large ones for social and ecological communities.[9] To understand technology requires peering into the background of social and political arrangements. Unfortunately, this point is largely lost, and my essay is an attempt to bring this kind of analysis once again back into the common conversation and struggle for the future of social ecology.

Theoretical Pluralism

What is an accurate and concise characterization of Bookchin's theory of technology, indeed the theory of technology that currently supports social ecology? The answer, I believe, is a kind of technological pluralism that grants gives priority to instrumentalism. There are several ways in which one can read such pluralism: as an indication of profound theoretical confusion; as an evolution from a naïve instrumentalism in the early Bookchin to a much more sophisticated reading of technology in his more recent writings; or as a pragmatic response to various social and theoretical demands. A plausible case can be made for any of these readings, but no individual perspective captures the essence of the entire program of technology within social ecology. More important, these readings, either alone or in combination, are insufficient to comfort an underlying concern that social ecology is stymied by the lack of a coherent theory of technology.

Bookchin argues openly for what I term a pluralist general theory of social ecology, best typified in this passage from *The Ecology of Freedom*:

> It has become clear to me that it was the *unity* of my views—their ecological holism, not merely their individual components—that gave them a radical thrust. That a society is decentralized, that it is farmed organically, or that it reduces pollution—none of these measures by itself or even in limited combination with others makes an ecological society. . . . Combined in a coherent whole and supported by a consistently radical prac-

tice, however, these views challenge the status quo in a far-reaching manner—in the only manner commensurate with the nature of the crisis.[10]

His unified theory is an amalgam of concepts bonded together by the
adhesive of his central organizing principle: a pathology of domination.
The main principle, which is detailed exhaustively as the "emergence
and dissolution of hierarchy" in *The Ecology of Freedom,* is supported
as he suggests by individual components drawn from the radical Left, a
combination of nineteenth-century anarchism, orthodox Marxism, and
the neo-Marxism of the Frankfurt School. What is crucial for any inter-
preter of Bookchin's social ecology is not the individual theories, which
he would prefer left aside for his privileged interpretation, but the unity
of their expression.[11] The dutiful social ecologist is asked to accept
wholeheartedly the master reading of Greek philosophy, social and political philosophies of the eighteenth and nineteenth centuries, and
modern radical theory, and concentrate on the genius of the combination. That so many have been swayed by this particular unity is a testament to Bookchin's radical vision. However, in order for social ecology
to gain a wider audience and realize the revolutionary potential latent
in its broadest outlines, constructive critics must concentrate (and be allowed to concentrate) on retracing and reconfiguring some of the steps
taken by Bookchin.

His theoretical eclecticism does not require a consistent theory of
technology to integrate seamlessly with the core. In its place we find
an exotic assortment of theories and observations about technology,
some of which bear connection with one another and others that are
by any account incommensurable. This is fertile ground for planting
multiple interpretations and may well explain, as I mentioned earlier,
why it is that Bookchin has received so little attention for his views
on technology by other theorists of technology and why social ecology is missing an incisive account of technology. He proceeds with
little regard for contemporary philosophers and theorists of technology.[12] Passing reference is made to Lewis Mumford's substantial radical libertarian theories of technology and to Langdon Winner's socialist criticisms. Jacques Ellul, arguably the single scholar who
launched North American critical interpretation of technology in the
1960s, is scarcely mentioned. Even Hans Jonas's work on technology,
The Imperative of Responsibility, which was published in English in
1984, is missing, despite Bookchin's demonstrated respect for this
critical theorist. At the very least, a consideration of these theoretical
positions would enrich social ecology by providing it with coherence
and wider intellectual and popular support.

Pluralism as Confusion

The first of the three readings of Bookchin's pluralist views on technology is articulated by David Watson (Chapter 7, this volume). Watson, a regular contributor to *Fifth Estate*, an anarchist broadside, is clearly sympathetic to social ecology. However, in "Social Ecology and the Problem of Technology," he dismantles Bookchin's position on technology by arguing that Bookchin has done a disservice to social ecology by perpetuating an incoherent theory of technology. In his detailed and uncompromising exegesis, which covers much of Bookchin's published work, Watson exposes several views of technology. The most distinctive shows Bookchin as an unapologetic champion of a hodgepodge of so-called alternative, appropriate, and labor-saving technologies. On this view, technology is a means to reach the proper end of an ecological society. This objective involves the replacement of human labor with machine labor, a point that Watson suggests echoes a familiar Marxist view. In *Remaking Society*, Bookchin writes "that a high premium would be placed on labor-saving devices—be they computers or automatic machinery—that would free human beings from needless toil and give them unstructured leisure time for their self-cultivation as individuals and citizens."[13]

In 1989 Bookchin's instrumentalism reached an embarrassing crescendo with his call "to turn the Canadian barrens—a realm that is still suspended ecologically between the highly destructive glacial world of the ice ages and the richly variegated, life-sustaining world of temperate forest zones—into an area supporting a rich variety of biota."[14] This outrageous proposal demonstrates a lack of understanding of the cultural and ecological realities of the circumpolar arctic. Watson does a good job of showing how this technological optimism grows to an oppressive scale when the fragments from Bookchin's various writings on technology are placed together. What was once enticing to me as a student—a radical political program that incorporated an elaborate description of communal ecotechnologies—is rendered as an uncritical instrumentalism.

However, to engrave Bookchin's writings with a crude instrumentalism is unfair. Watson articulates another competing strain that offers an antidote to instrumentalism, namely, the view that technology does more than merely amplify social realities: technology is a social construction. Throughout his work Bookchin voices two apparently diametric opinions, one that holds technology to be a collection of instrumentalities and the other that suggests that the technical imagination is "*never* socially neutral."[15] In places, he is clear about the social and political content of technologies, so clear that one almost forgets his instrumentalism. He writes, for instance:

> Solar power, wind power, methane, and geothermal power are merely *power* insofar as the devices for using them are needlessly complex, bureaucratically controlled, corporately owned or institutionally centralized. Admittedly, they are less dangerous to the physical health of human beings than power derived from nuclear and fossil fuels, but they are clearly dangerous to the spiritual, moral and social health of humanity if they are treated merely as *technologies* that do not involve new relations between people and nature and within society itself.[16]

On first reading *The Ecology of Freedom*, I was struck by this searing observation:

> The historic problem of technics lies not in its size or scale, its "softness" or "hardness," much less the productivity or efficiency that earned it the naive reverence of earlier generations; the problem lies in how we can *contain* (that is, absorb) technics within an emancipatory society. In itself, "small" is neither beautiful nor ugly; it is merely small. Some of the most dehumanizing and centralized social systems were fashioned out of very "small" technologies; but bureaucracies, monarchies, and military forces turned these systems into brutalizing cudgels to subdue humankind and, later, to try to subdue nature.[17]

Watson argues that the essential problem with Bookchin's view of technology is his willingness to accept the idea that the margins of a capitalist economy can shift easily to the core of an ecological society: it is confused to think that the odious aspects of capitalist technology can be converted to the ends of liberation. Bookchin is caught in a Marxist mode of believing that technologies can be converted from bad to good, from "swords" into "ploughshares" in the Christian idiom, and it is this view that exposes his instrumentalism. In *Post-Scarcity Anarchism* and *Toward an Ecological Society*, where we witness the most programmatic descriptions of an ecotechnological society, it is clear that for Bookchin the radical challenge of technology is the appropriation of devices for liberatory ends, in spite of their being a product of capitalist production. This criticism can be blunted as well. As Watson notes, Bookchin, especially in his earlier work, focuses on showing how modern technologies constitute an entirely new character and potential for revolutionary change. In the end, Watson makes a convincing case that Bookchin's theory of technology is confused, representing an unreasonable combination of Marxist theory and technological optimism.

Pluralism as Evolution

Another way of interpreting Bookchin's pluralism is through an evolutionary explanation. Early in his writing career, in the late 1950s, he viewed technology from an anarcho-syndicalist perspective, arriving at the realization of pervasive environmental damage caused by industrial agriculture (the subject, largely, of his first book, *Our Synthetic Environment*). Following Watson's argument, Bookchin also shows a strong strain of an older Marxism, one in which technology is seen through the eyes of an industrial worker.[18] During the 1960s and 1970s, a period characterized by massive, at times progressive, cultural upheaval in North America, Bookchin developed in full form the program of social ecology. By 1982, with the publication of *The Ecology of Freedom,* he had established a new way of thinking about ecological questions in a political context and had gained a reputation as an uncompromising radical thinker. The two chapters on technology in *The Ecology of Freedom*—"Two Images of Technology" and "The Social Matrix of Technology"—display a much richer understanding of the social and political context of technology than his previous work.

Alan Rudy and Andrew Light detected this evolution in their analysis of Bookchin's minimal treatment of social labor in producing a radical social ecology. They suggest that "Bookchin's hopes for modern technology were initially extraordinarily high" and that his subsequent "ecotopic visions have become increasingly low-technology affairs."[19] They attach this change in viewpoint to other aspects of his theory: "As Bookchin's views on technology have changed, so have his views on revolutions in developed countries and on the political tactics demanded by the modern situation."[20]

Watson notes some development, but is much less sanguine that it constitutes a profound realignment, arguing against the idea of a maturity of Bookchin's views on technology over the years. This pluralism-as-evolution reading bears recognition, but we should also take Watson's argument seriously. Even if we were to strip away Bookchin's early writings on technology, an obviously misguided hermeneutic practice, we are left with the inconsistencies and confusions in his later work.

Pluralism as Pragmatism

The third reading, a more charitable view of Bookchin's pluralism, suggests that the confusion evident in his account of technology is a pragmatic strategy to achieve a theoretical "unity" from various and not always easily connected positions. This strategic pastiche of existing theories has one obvious benefit: it is assured of attracting the attention of committed anarchists and libertarian socialists, and also a wider revolutionary audience. This kind of strategy has traditionally received poor billing by activists as opportunism and by theorists, as noted earlier, as simply confused. Andrew Light and Eric Katz, however, have provided us with a way of interpreting such a strategy: "The call for moral pluralism, the decreasing importance of theoretic debates and the placing of practical issues of policy consensus in the foreground of concern, are central aspects of our conception of environmental pragmatism."[21]

Recall that Watson believes that Bookchin's theory of technology is a muddle that "paralyze[s]" Bookchin "and social ecology with him."[22] But what if we read Bookchin as an environmental pragmatist? Instead of receiving his instrumentalism as a throwback to an old-style Marxism, we can instead consider his multiple tacks with technology as an attempt to salvage the best of capitalism (e.g., sophisticated ecotechnologies) and graft these on to his radical municipalism. Similarly, we could imagine his approach to technology on two levels. The first is his conversion of capital production to serve his "nonauthoritarian Commune of communes,"[23] and the second is his awareness of the social and political construction of technology. At the first level we find his extensive proposals for "ecotechnologies." The second level provides a more abstract design, based on an understanding of the political and social contexts of technology including the principles of emancipation and ecological justice, for the practical alternatives to flourish. The problem with this reading, as Watson would no doubt point out, is that it cannot be verified through a close reading of Bookchin's work. For such a reading to be successful, Bookchin would have to acknowledge such an approach. His commitment to theoretical "unity" is not strong enough evidence of a premeditated pragmatism. As a pragmatist, Bookchin should have no difficulty articulating a coherent *theory* of technology that would operate at one level in relation to, and in support of, a more popular instrumentalism. This is simply not the case.

In this section I have described three possible readings of the pluralism that I attribute to Bookchin's view of technology. The first, bolstered by Watson's argument, shows that Bookchin's pluralism is evidence of a profound theoretical confusion. The second reading illustrates how his views about technology have shifted over the past four decades; the pluralism is inherent in the way he has developed his thinking from a naïve instrumentalism to a more sophisticated constructivism. This is no doubt true, but observation must not obscure the inconsistencies that linger in his later works. The third reading construes the pluralism according to a fresh view of environmental pragmatism. Adhering to Light and Katz's strictures about pragmatism, especially the need to produce coherent theories to explain one's more practical moves, Bookchin's view of technology falls back to the first reading. In sum, each of the three readings offers a helpful way of understanding what Bookchin is up to regarding technology.

In the end, however, the pluralism-as-confusion reading makes the most sense. Pluralism, in Bookchin's case with a strong dose of instrumentalism, works against a popular understanding of the *patterns* of life created by *systems* of modern technology. The widespread belief that technology is politically inert has given rise to a disturbing lack of radical ambition and activism around issues such as reproductive technologies, industrial agriculture, corporate distribution systems, franchise marketing, so-called ecofriendly products, electronic networking, and computerization of the workplace. While Bookchin is acutely aware of these issues, and while his writings are peppered with indictments of these resistant strains of modern capitalism, he is not able to offer up much more than "alternative technology" as an antidote. Without a theoretical position that shows how technology connects to politics, technology will remain a neutral conduit for various social and political activities. Instead, technology needs to be recognized properly as a set of contemporary patterns that define contemporary notions of passivity, consumption, alienation, and domination. Once technology is truly understood, oppositional movements based on social ecology will have a much better chance of achieving decentralized nonhierarchical "free" communities well integrated with nature. The next steps in the reconstructive process are to identify the various genera of theories of technology, locate an appropriate metatheoretical location for Bookchin's viewpoint, and propose a coherent and compatible philosophy of technology for social ecology.

The Limits of the Social Ecological Theory of Technology

Where does Bookchin's theory of technology fit within the philosophy of technology? Albert Borgmann suggests that current and recent theories of technology can be described by three "essential types: the substantive, the instrumentalist, and the pluralist views of technology."[24] Those who had a substantive view of technology behold a world in which artifacts and procedures utterly define the operation of social life. Typically, a substantivist regards technology as a debilitating force, one that acts in its own right against the best intentions of its makers. From this we derive the well-known and controversial idea of technological determinism.[25] Jacques Ellul is certainly the best example of the substantivist view. His work portrays a spiritually denuded society composed of individuals who have given up their own autonomy to the autonomy of machinery. His exhaustive social studies of modern life, especially his *The Technological Society*, leave little room for reform. Borgmann suggests that "the substantive view is theoretically inviting because of its ambition and radicality. It seeks to give a comprehensive elucidation of our world by reducing its perplexing features and changes to one force or principle. That principle, technology, serves to explain everything, but it remains itself entirely unexplained and obscure."[26] The difficulty is that once one passes by the initial allure of the description of life in a technological society, one is left standing beside a theory that merely demonizes technology.

By far the most common view of technology belongs to the instrumentalist, who believes technology constitutes the material amplification of human action. The instrumentalist comprehends contemporary high technology as a seamless extension of all previous tools and machinery; the difference is one of quantity, not of quality. The most significant aspect of this view is the moral separation of user and tool. Tools themselves are regarded as morally inert or neutral. All consequence or value is shifted to the tool maker or tool user. The instrumentalist view of technology is, as Borgmann suggests, "congenial to that liberal democratic tradition which holds that it is the task of the state to provide means for the good life but wants to leave to private efforts the establishment and pursuit of ultimate values."[27] He continues by observing that radical critics of social life often find a variant of instrumentalism acceptable. For them, it is naïve to ignore the ends of a political–technological order:

> A penetrating inquiry of technology must inevitably be a social critique. This approach, sometimes called "politicized technology," is an important kind of instrumentalism. Indeed, if one is persuaded that the political dimension is decisive in human endeavors, any analysis of technology can be evaluated as to its political salience, and it *becomes possible to give an array of prominent analyses from left to right*.[28]

The promiscuous political function of instrumentalism is one reason for its popularity, and it also helps to explain why a radical theorist such a Bookchin would find it so congenial. Instrumentalism is difficult to criticize because in a sufficiently local and short-term way all artifacts and procedures are morally neutral. However, this ignores the social, ecological, and political context in which these things reside. A means is never independent of its ends.

A third view—pluralism—opens up through the combination of substantivism and instrumentalism. Substantivism provides a dark and constrained picture of reality which we know to be both true at a general level when considering the interconnectivity of global industrial systems, and false at a local level when we are able to provide direct counterexamples to the power of technology from our own experience. The latter, instrumentalism, explains much at the local level but is hopelessly inadequate to account for the magnitude of technological effects on social life and politics. A pluralist admits many kinds of explanation and accounts of technology into his or her view. When explanatory problems arise—for example, the inadequacy of instrumentalism in accounting for political realities—the pluralist provides counterexamples that solve or challenge the particular problem. Such an ad hoc approach permits a remarkably wide range of viewpoints to operate simultaneously, and if taken far enough, lends the impression that there is no adequate account of the character of technology. Borgmann counsels against this perspective:

> Technology, in fact, does not take shape in a prohibitively complex way, where for any endeavor there are balancing counterendeavors so that no striking overall pattern becomes visible. It is intuitively apparent that in modern technology the face of the earth is transformed in a radically novel way; and that transformation is possible only on the basis of strong and pervasive social agreements and by way of highly disciplined and coordinated efforts. These crucial matters escape the pluralist's minute and roving scrutiny.[29]

This account of pluralism captures precisely the way in which Bookchin uses theories of technology. His instrumentalism is tempered with what I earlier called a social constructivist view, that is, the belief that the value of technology is shaped by the context of social practices. These are freely admixed such that the result is more like an agglomeration than an alloy. Bookchin rarely resorts to a substantivist position except as a rhetorical device to emphasize a specific and foreboding aspect of technology. Most of the time, he is remarkably cheerful about the potential of alternative technologies.

The question that springs directly from this analysis is: Why is a coherent theory necessary? I have hinted earlier in this chapter that coherence is vital for social ecology to prosper and that raiding various theories of technology without proper regard for the work of contemporary theorists of technology will hamper the long-term agenda of social ecology. More specifically, the current pluralism of social ecology, in the absence of a metatheory of technology that would orient its various views, provides no clear description of the social and political function of technology. If social ecology is to prosper, it requires a diagnostic political theory—one capable of articulating clearly how the various features of a technological society operate in relation to political ends.

There is no theory of technology in Bookchin's work that provides such a means. Social ecology, with its emphasis on communalism, freedom, and ecology, should have a theory of technology that enlarges these features and provides at least the hint of how to proceed toward these ends. Watson calls for such a view, I think, at the end of his essay, suggesting that we must search for politically progressive pathways not along the traditional routes of science and technology, but along the paths blazed by an enduring, community-based knowledge. The search for such a theory, then, must step outside the constructs of scientific and technological rationality, resist the debilities (and temptations) of contemporary capitalism, provide diagnostic interpretation, and set a sense of direction for the future of social ecology. Does such a theory exist? The answer, as one would expect, is a tentative yes. Borgmann describes what a theory should adhere to if it is to avoid the problems of substantivism, instrumentalism, and pluralism:

> Clearly, the theory of technology that we seek should avoid the liabilities and embody the virtues of the dominant views. It should emulate the boldness and incisiveness of the substantive version without leaving the character of technology obscure. It should reflect our common intuitions and exhibit the lucidity of the instrumentalist theory while overcoming the latter's superficiality. And it should take account of the manifold empirical evidence that impresses the pluralist investigations and yet be able to uncover an underlying and orienting order in all that diversity.[30]

This provides the broad criteria for a theory, but we must ensure also that it fits with the political sensibilities of social ecology. One obvious starting place would be the Marxist theories of technology advanced, for example, by Feenberg and Jonas. They share a radical political commitment with Bookchin, but they are hampered by Bookchin's uneasy relationship with Marxist theory. In any case, in spite of their crucial contributions, neither Feenberg nor Jonas provides a diagnostic theory of technology, that is, a theory that explains the operation of contemporary technological patterns.[31] Another starting point is a communitarian philosophy of technology, and I turn to this view for unconventional inspiration.

Toward a Renovated Theory of Technology for Social Ecology

The philosophical investigation of technology has blossomed in the past two decades to the point where many major departments of philosophy in North America can lay claim to specialists in the area. A conversation about theory has opened among philosophers, social scientists, natural scientists, humanists, and engineers. The results of this work are impressive and controversial. A new constructivist method has gained prominence to show how practices and beliefs in science and technology are elementally social and political activities, and as such are subject to critical interpretation.[32]

Technology studies have tended to be largely descriptive, and it is this descriptive base that has provided as impetus for theoretical development. A variety of important philosophical positions have been articulated, covering epistemological, ontological, and moral issues.[33] However, until the publication, in 1984, of Albert Borgmann's *Technology and the Character of Contemporary Life* (hereafter cited as *TCCL*), no comprehensive theory of technology had been put in place. His interconnected notions of the "device paradigm" and "focal things and practices" provided the central theoretical concepts around which a descriptive and prescriptive viewpoint could work. Attention came slowly to *TCCL,* and only now are scholars beginning to recognize its theoretical importance.[34]

Borgmann's theory of technology begins with the observation that everyday life is punctuated by events and activities to which we attach significance; they provide focal points from which we evaluate and orient the rest of our lives. These focal activities, and other activities to varying degrees, are embraced by contexts that include, depending on individual circumstances, social, cultural, political, economic, spiritual, and ecological factors. Borgmann argues that there is a contemporary pattern, which we can call technology, that threatens the stability and viability of these practices and their contexts:

> This pattern is visible first and most of all in the countless inconspicuous objects and procedures of daily life in a technological society. It is concrete in its manifestations, closest to our existence, and pervasive in its extent. The rise and the rule of this pattern I consider the most consequential event of the modern period. Once the pattern is explicated and seen, it sheds light on the hopes that have shaped our times, on the confusions and frustrations that we have suffered in our attempts to realize those hopes, and on the possibilities of clarifying our deepest aspirations and of acting constructively on our best insights.[35]

This pattern functions in the following way. A "thing," which can be any object or practice and exists within a recognizable social context, is decomposed into two separate components: machinery and commodity. A "commodity" refers to the arrangements and artifacts with which we interact in the foreground of experience. The "machinery" constitutes the myriad systems of production that give rise to the foreground, but that is generally cloaked from normal observation. The rise of a clear separation between machinery and commodity, foreground and background, is at the heart of the pattern. Take the following example:

> Surely a stereo set, consisting of a turntable, an amplifier, and speakers, is a technological device. Its reason for being is well understood. It is to provide music. But this simple understanding conceals the characteristic way in which music is procured by a device. After all, a group of friends who gather with their instruments to delight me on my birthday provide music too. A stereo set, however, secures music not just on a festive day but at any time, and not just competent flute and violin music but music produced by instruments of any kind or any number and at whatever level of quality. To this apparent richness and variety of technologically produced music there corresponds an extreme concealment or abstractness in the mode of its production. Records as unlabeled physical items do not bespeak, except to the most practiced of eyes, what kind of music they contain. Loudspeakers have no visible affinity to the human voice, to the brass or the strings whose sound they reproduce, I have little understanding of how the music came to be recorded on the disk and by what means it is retrieved from it.[36]

Here, the music that reaches us in our living rooms constitutes the commodity, and the various means of its production—the stereo set, the recordings, the production system for the recordings and the stereo set, the electrical supply—are its machinery. The divide that opens between foreground and background is present in many contemporary activities and continues to separate as devices become more pervasive. The sheer magnitude of the pattern tends to obscure our view of its operation. It is, in a sense, too mundane and common to attract much critical attention.

This problem has two implications. First, it means that scholars and citizens typically focus attention on large or glamorous aspects of technological culture: automobiles, aircraft, computers, and so on. Outside of these charismatic devices lie myriad technologies and procedures that constitute daily life. Second, it deflects an understanding of how technology spreads widely into activities, procedures, and artifacts that we do not traditionally associate with technology. Elsewhere, I have termed this the process of commodification.[37] The example of insurance illustrates this point:

> In a pretechnological setting, security in the face of catastrophes is had from the goodwill and charity of parents, siblings, or neighbors. But such

security was sometimes unreliable and always burdensome to giver and receiver. Insurance technology first reduces security to the guarantee of a cash payment, then decomposes the resourcefulness and precariousness of society by mathematical means, and finally institutes a financial and legal machinery to insure the collection and distribution of money. The machinery of this device, especially at first, was not primarily a physical entity but a network of computations, contracts, and services. But it had from the start the concealed, inaccessible, and disburdening character of technological machinery. Accordingly, the commodity, though it was a regrettable necessity more than a final consumption good, represented commodiously available security, support that did not require asking, imposing, or begging but could be claimed through a call to the insurance agent.[38]

This expansion to "nontechnological" areas is what provides the device paradigm significant explanatory power. It helps interpret the rise and tenacity of a culture of consumption and provides a direct connection to social life and politics through the removal, reduction, or transformation of these local contexts. As a pattern, it allows us to understand technology not as a force in its own right, as technological determinists would argue, but as a deeply imbedded social and political pattern, or what Ursula Franklin terms "a mindset."[39] With any deeply imbedded pattern, there are possibilities for its redirection or reconstitution, and this is what makes the device paradigm especially useful for social ecology.

The commodification of experience will strike most radical and critical scholars as a problem prima facie. But what kind of finer points can be made to support this intuition? Borgmann argues that the device paradigm decomposes traditional approaches to work and leisure. This occurs in several ways. First, skillful engagement with bodily and social practices is transformed to solitary or passive activity. Second, traditional activities are simply displaced by more alluring and disburdening forms of leisure (television is a prime example). Third, work becomes increasingly divided from other forms of related work and production more dependent on deskilled or deskilling labor. Fourth, the separation between action and context described above reflects a simultaneous separation between action and consequence. There is a tendency to be less aware of the ecological and social consequences of one's work and leisure activities. These concerns fall directly from the device paradigm. As the pattern of technology becomes wider spread, it is more difficult to comprehend normality; the rule of the device paradigm becomes the new norm. We are inured to a new political arrangement that is constructed through our relations with devices. This, I believe, is one of the main points that Bookchin misses in his theory of technology. He has not accounted for the close interplay between politics and technology.

Borgmann suggests that the device paradigm is strongly compatible with modern liberalism. Liberalism is constituted by three main forces: liberty, equality, and self-realization. The first two are plainly evident in the negative definition of liberalism as "an enterprise of breaking fetters and throwing off burdens."[40] Self-realization helps stabilize equality by promoting equal access to the development of "common and various talents" and by providing a center for the attainment of freedom.[41] Self-realization easily converts to a fetishism of the personal, especially when it becomes possible to amplify the personal through appeal to commodities. A person's attainment of freedom and equality is funneled through the disburdening character of technology. This is the deep-rooted connection that Borgmann draws between liberalism and technology. It is not merely the case that the rise of modern technology coincided conveniently with liberalism, or that technology became a servant to liberalism, but that the particular pattern of commodity relationships that we termed the device paradigm became a decisive component of contemporary politics.

The main implication for social ecology is that the underlying conditions of technology make the reform of capitalism difficult. In effect, the reform of capitalism is inextricably linked to the reform of the device paradigm. This is a point that I argued in previous writings about the commodification of planning and nature.[42]

Borgmann proposes a three-stage program of reform. First, at an individual level, he suggests the centering of one's life on things that truly matter, that is, artifacts and practices that provide bodily and social engagement. "Engagement" refers to skillful practice and the particularities of family and community life. Second, while reform necessarily begins in individual awareness, it must be supported and strengthened by public focal practices. There must be sufficient flexibility and openness in institutional life, however this is construed, to permit regular "communities of celebration." The final reform strategy, one that appears rather weak alongside the radical and revolutionary perspective of social ecology, is a proposal for a two-sector economy. This involves the elevation of a local community-based economy to a point where large-scale industrial activity serves the interests of local economies. This latter proposal is staked out in coarse detail in one of the closing chapters of *TCCL*. The apparent naïveté of this view is matched by the apparent innocence of Bookchin's proposals for flourishing alternative technologies.

Borgmann's theory of the device paradigm, including his proposals for reform, provide a coherent and reasonably complete theoretical account of contemporary technology. It carries the necessary theoretical weight to support a political program. Not only is it expansive in scope, but it directly confronts the connection between technology and politics. The potential fit of the device paradigm with social ecology is marred by one rude question: Are the two ideologically compatible? In the same way that Bookchin had neglected contemporary philosophy of technology, Borgmann has largely ignored social ecology. There is, at present, no direct link between the two. Borgmann is a small "r" republican in the American tradition. He identifies with conservative political traditions and, by and large, eschews radical political change. In this sense, his theories seem uneasy bedmates for social ecology. However, his republicanism connects him politically with Bookchin's New England republicanism (recycled through Bookchin's proposals for "radical municipalism"), and philosophically to the communitarian tradition in recent social and political philosophy.[43] It is the focus that Borgmann gives to locally bounded social context in understanding technological patterns that makes his work attractive to social ecologists. Of course, the obverse is true. Borgmann's conservative reform proposals would be strengthened through appeal to more radical and far-reaching political and economic stratagems, the kind provided via social ecology.

In this chapter I have provided the barest outlines of a coherent theory of technology, namely, Borgmann's device paradigm, that offers a much-needed bolster to Bookchin's pluralist account of technology. To reach this point required a critical interpretation of Bookchin's views on technology against contemporary philosophy of technology, a task that was made much easier by David Watson's careful commentary. This is, however, a modest beginning. What lies ahead is a systematic reconfiguration of a theory of technology such that one of the acknowledged "pillars" of social ecology is made coherent and is connected to a wider body of contemporary critical and radical scholarship.

Notes

1. By his own admission, "technics" lines up with ecology and urbanism as one of three main elements in writings that span forty years: "That my writings in ecology, urbanism, and technics have not always been celebrated by my colleagues on the Left can, in my view, be attributed to one reason: my commitment to anarchism" "(Introduction," in *Toward an Ecological Society* [Montreal: Black Rose Books, 1980], p. 22).

2. Murray Bookchin, "Post-Scarcity Anarchism," in *Post-Scarcity Anarchism* (Berkeley, CA: Ramparts Press, 1971), p. 33.

3. Janet Biehl, *A Bibliography of Published Works by Murray Bookchin in Chronological Order* (Burlington, VT: Social Ecology Project, 1993). Biehl

prepared this helpful source on the occasion of Bookchin's seventieth birthday, on January 14, 1991. The revision I am using is dated September 13, 1993.

4. George Bradford, "How Deep Is Deep Ecology? A Challenge to Radical Environmentalism," *Fifth Estate*, Vol. 22, Fall, 1987, p. 23. I am indebted to David Watson (a.k.a. George Bradford) for his thorough and critical review of Bookchin's view of technology, the first of its kind, which appears in this volume. My speculations on the role of technology theory in the future of social ecology are made much easier thanks to this base of scholarship.

5. There is a continual problem with the accuracy of the portrayal of the positions of social ecologists that must be addressed by social ecologists themselves. For example, George Sessions's recent comment that social ecology is "a spinoff from the Marxist social justice movement but concerned, as well, with urban pollution problems" ignores the political organization of social ecological theory around anarchism, and reduces the sweep of social ecology by portraying it as an arcane, urban-focused splinter organization. See Sessions, "Reinventing Nature: The End of Wilderness? A Response to William Cronon's *Uncommon Ground*," *Trumpeter*, Vol. 13, 1996, p. 35.

6. See, e.g., Murray Bookchin and Dave Foreman, *Defending the Earth: A Dialogue between Murray Bookchin and Dave Foreman*, ed. Steve Chase (Boston: South End Press, 1991); Robyn Eckersley, "Divining Evolution: The Ecological Ethics of Murray Bookchin," *Environmental Ethics*, Vol. 11, 1989, pp. 99–116 (first part of Chapter 2, this volume); Bookchin, "Recovering Evolution: A Reply to Eckersley and Fox," *Environmental Ethics*, Vol. 12, 1990, pp. 253–274.

7. As a veteran of the social ecology–deep ecology struggles of the late 1980s, I find it too easy to slip back into sectarian debate. Various aspects of deep ecology proved attractive to a mass movement, and it has become a vernacular environmental position. Deep ecology continues to be used in a wanton fashion to cover a disturbingly wide spectrum of belief. The fact that a corporate environmental relations specialist can have works of deep ecology prominently displayed on office bookshelves and can liberally sprinkle deep ecological epithets throughout corporate planning documents is testament to its plastic character. There is an important distinction required between a mass popular movement based on a coherent set of principles and a mass movement. It is clear also that deep ecology, at least various accounts, has become better developed over the past five years, including elements that point to similar radical political positions advocated by social ecologists. Does this mean either that deep ecology can subsume social ecology, or that social ecology will be represented as a subcategory of deep ecology? I think not, but this requires a metatheoretical view about the general drift of radical environmentalism.

8. Tony Mullany, "If Big Is Not Good, Small Is Not Beautiful (2 parts)," *Peacework*, December 1975 and January 1976, cited in Murray Bookchin, "The Concept of Ecotechnologies and Ecocommunities," in *Toward an Ecological Society* (Montreal: Black Rose Books, 1980), p. 111.

9. Langdon Winner, *The Whale and the Reactor: A Search for Limits in an Age of High Technology* (Chicago: University of Chicago Press, 1986), pp. 61–85, provides a critical view of alternative technology.

10. Murray Bookchin, "Introduction," in *The Ecology of Freedom* (Palo Alto, CA: Cheshire Books, 1982), p. 3.

11. Alas, Bookchin has been notoriously unbending to commentators and critics. All criticism, it would appear, is taken very hard, and this behavior has tended to marginalize him from larger intellectual conversations in the past half decade.

12. Philosophy of technology has been established as an identified subfield in professional philosophy in North America at least since the early 1970s. See Carl Mitcham, *Thinking through Technology: The Path between Engineering and Philosophy* (Chicago: University of Chicago Press, 1994). Not many members of this intellectual community, best represented by the Society for Philosophy and Technology, share Bookchin's anarchism, but neo-Marxists such as Andrew Feenberg and communitarians such as Albert Borgmann have much to offer.

13. Murray Bookchin, "From Here to There," in *Remaking Society* (Montreal: Black Rose Books, 1989), p. 196.

14. Murray Bookchin, "Recovering Evolution: A Reply to Eckersley and Fox," *Nature and Society,* Vol. 2, September–December, 1992, p. 170 (cited in Watson, Chapter 7, this volume).

15. Murray Bookchin, "Two Images of Technology," in *The Ecology of Freedom* (Palo Alto, CA: Cheshire Books, 1982), p. 226; emphasis in original.

16. Murray Bookchin, "Open Letter to the Ecology Movement," in *Toward an Ecological Society* (Montreal: Black Rose Books, 1980), p. 78. Of course, such a statement seems in direct contradiction to a main claim he makes about scale: "Without variety and diversity in technology as a whole, solar energy would merely be a substitute for coal, oil, and uranium rather than function as a stepping stone to an entirely new way of dealing with the natural world. . . . What is no less important, "alternate energy"—if it is to form the basis for a new ecotechnology—would have to be scaled to human dimension." See "Energy, 'Ecotechnology,' and Ecology," in *Toward and Ecological Society* (Montreal: Black Rose Books, 1980), p. 92.

17. Murray Bookchin, "The Social Matrix of Technology," in *The Ecology of Freedom* (Palo Alto, CA: Cheshire Books, 1982), pp. 240–241.

18. This is clearly evident in "Toward a Liberatory Technology," published originally in 1965 in *Comment,* and reprinted many times, most noticably in *Post-Scarcity Anarchism.*

19. Alan Rudy and Andrew Light, "Social Ecology and Social Labor: A Consideration and Critique of Murray Bookchin," *Capitalism, Nature, Socialism,* Vol. 6, June, 1995, pp. 83–84.

20. Ibid., p. 85.

21. Andrew Light and Eric Katz, "Introduction," in Andrew Light and Eric Katz, eds., *Environmental Pragmatism* (London: Routledge, 1996), p. 5. Light, in his essay in the same volume, "Compatibilism in Political Ecology," illustrates pragmatism in practice by finding metatheoretical compatibility between materialists such as Bookchin and ontologists such as Naess.

22. David Watson, "Social Ecology and the Problem of Technology," Chapter 7, this volume. Watson could also stand to temper his criticisms

slightly to allow for the fact that, in spite of his profound shortcomings, Bookchin has placed technology much closer to the center of a radical program than most other theorists. And, when one considers the breadth of his larger liberatory project and the promise that many of us believe social ecology holds, it is appropriate to nod (very slightly, as necessary) gratefully in Bookchin's direction.

23. Murray Bookchin, "Introduction," in *The Ecology of Freedom* (Palo Alto, CA: Cheshire Books, 1982), p. 2.

24. Albert Borgmann, *Technology and the Character of Contemporary Life* (Chicago: University of Chicago Press, 1984), p. 9.

25. For an account of the revival of interest in technological determinism, see Leo Marx and Merritt Roe Smith, *Does Technology Drive History?* (Cambridge, MA: MIT Press, 1994).

26. Borgmann, *Technology,* p. 9.

27. Ibid., p. 10.

28. Ibid, emphasis added.

29. Ibid., p. 11.

30. Ibid., pp. 11–12.

31. At the time of writing I have not had an opportunity to appraise Feenberg's new book, *Alternative Modernity* (Berkeley and Los Angeles: University of California Press, 1995), which may overcome my criticism about the diagnostic power of his earlier work.

32. Joseph Rouse provides a lucid summary in "What Are Cultural Studies of Scientific Knowledge?," *Configurations,* Vol. 1, Winter, 1993, pp. 1–22.

33. By no means an exhaustive list, main proponents include Carl Mitcham, Paul Durbin, Albert Borgmann, Larry Hickman, Don Idhe, Langdon Winner, and Andrew Feenberg.

34. Andrew Light and I coorganized a workshop with many of the prominent scholars in philosophy of technology, in Jasper, Alberta, in September 1995, to discuss the significance of *Technology and the Character of Contemporary Life*. Along with David Strong, we have prepared a book based on the workshop presentations, *Philosophy in the Service of Things,* which is currently under contract to the University of Chicago Press.

35. Borgmann, *Technology,* p. 3.

36. Ibid., pp. 3–4.

37. "Commodification and Naturalization: The Recreation of Nature as Technology," in Joseph Pitt and Elena Lugo, eds., *The Technology of Discovery and the Discovery of Technology* (Blacksburg, VA: Society for Philosophy and Technology, 1992), pp. 301–318.

38. Borgmann, *Technology,* p. 117.

39. Ursula Franklin, *The Real World of Technology* (Concord, ON: House of Anansi Press, 1991), p. 12.

40. Borgmann, *Technology,* p. 89.

41. Ibid., p. 89.

42. Eric Higgs, *Planning, Technology, and Community Autonomy* (unpublished Ph.D. diss., Department of Philosophy and School of Urban and Regional Planning, University of Waterloo, Ontario, Canada,1988). An abridged

version appears as "The Landscape Evolution Model: A Case for a Paradigmatic View of Technology," *Technology in Society*, Vol. 12, 1990, pp. 479–505; see also Jennifer Cypher and Eric Higgs, "Colonizing the Imagination: Disney's Wilderness Lodge," *Capitalism, Nature, Socialism*, Vol. 8, December, 1997, pp. 107–130.

43. I am thinking here of the brand of communitarianism embraced by Charles Taylor, especially in his *Sources of the Self* (Cambridge, MA: Harvard University Press, 1989).

Part III

Historical Considerations and Comparisons

Chapter 9

Ecology and Anthropology in the Work of Murray Bookchin
Problems of Theory and Evidence

ALAN P. RUDY

Introduction

Murray Bookchin's doctrine of social ecology consists of two more or less equal parts: his interpretation of the anarchist or anarcho-communalist (communist) tradition, on the one hand, and his reading of evolution and the science of ecology, on the other. Like some other social critics past and present, Bookchin claims to have discovered an objective foundation for social liberation in evolution, that is, in nature, itself. Processes of ecological and biosocial evolution, according to Bookchin, have a predominantly mutualistic and symbiotic (as contrasted with commensualistic, competitive, or predatory) character. Bookchin's work is inextricably wound around a core critique of empiricist and structuralist approaches to nature and society; his vision opposes traditional theories of competitive determinations of social and ecological relations with a theory of mutualistic cooperation. For Bookchin, the mutualistic and cooperative evolution of stable and complex ecological, social, and ecosocial relations points to domination (perhaps *the* key word in Bookchin's works) and hierarchy, which oppose free and cooperative modes of ecosocial relations, as the root cause of modern ecological and social problems.

This chapter investigates Bookchin's account of the preliterate societies, out of which he asserts that domination and hierarchy have emerged, and the ecological claims upon which these anthropological positions are built. The chapter's first part is an account of Bookchin's arguments about social and ecological evolution. The second part criticizes Bookchin's theses and the positions toward which these theses take him. While Bookchin's work is undeniably exciting in its totalizing critique and wide historical scope, I argue that his arguments dualistically oppose deterministic natural and social science rather than provide analysis of real contradictions and then generate a new synthesis. Highly contested natural scientific categories and widely debated anthropological theories are forced into a predetermined anarchist form.

In order to advance radical social and ecological movements, a more sophisticated analysis of the relationships within and between the many forms of mutualism and competition—and the many modes of cooperation and hierarchy—is needed. While not advanced here, such a position must be predicated on material and historical investigations of the enablements, constraints, and contingencies of diverse ecological and social modes, forms, and processes at interrelated levels of analysis. Bookchin's opus advances the possibility of, and the opportunity for, such a synthesis by insisting on alternative visions of ecological and social history rooted in mutualism and cooperation. The next step is to synthetically transcend dualistic alternatives to generate new syntheses.

Bookchin on the Production of Domination

The central concepts of Bookchin's social ecology are freedom and spontaneity, and domination and hierarchy. The first two terms are generally opposed to the second two. Since natural evolution is predominantly driven by freely spontaneous forms of mutualism and symbiosis (according to Bookchin), the most important moment in social ecological history lies at the point where hierarchy and domination emerge and are institutionalized. In contradistinction to traditional liberal and Marxist accounts of protohumanity, Bookchin argues that organic communities of *Homo sapiens* (symbiotically integrated into ecological conditions and evolutionary patterns) unevenly produced and gradually instituted forms of social domination. From these relations of domination then sprung ideologies and practices oriented to the domination of nature. The components of this section address Bookchin's writings on the social evolution of domination following the natural evolution of society.

Preliterate Organic Societies Rooted in Natural Evolution before Domination

Bookchin argues that prior to the emergence and institutionalization of hierarchy and domination, organic tribal societies were primarily organized along gender lines. While Bookchin has characterized the functional and (re)productive relations within preliterate societies somewhat differently in different writings, he has always insisted on a strict differentiation of activities and values between the male *civil sphere* and the female *domestic sphere*. In *Remaking Society,* Bookchin argues that blood ties, gender, and age (in that order) are the biologically defining moments of association and self-identity within preliterate organic societies.[1]

> Women exercised full control over the domestic world: the home, family hearth, and the preparation of the most immediate means of life such as skins and food. Often a woman built her own shelter and tended to her own garden as society advanced toward a horticultural economy.

> Men, in turn, dealt with what might be called "civil affairs"—the administration of the nascent, barely developed "political" affairs of the community such as relations between bands, clans, tribes, and intercommunal hostilities.[2]

He suggests that this division of social functions generated distinct, though complementary, subcultures within organic societies. In *The Modern Crisis,* Bookchin argues that women operated within separate, sororal realms that ran the "home, garden, cleaning, food preparation, parenting and many other functions." The male, fraternal realm embraced "hunting, 'politics,' and, where it exists, the men's house, into which all the males withdr[e]w after puberty."[3] These two subcultures were different within organic communities, but there was a balance between them. The relationship between male and female realms contained no relations of hierarchy or domination/submission. According to Bookchin, our "earliest institutions were based on blood ties, age groups, and gender functions—all biological facts, yet distinctly social in that [these] natural affinities are given structure and stability, cohered by ideologies, and expanded to include seemingly 'alien' groups through marital exogamy and the exchange of gifts."[4]

In *The Ecology of Freedom,* Bookchin's account stresses the importance of age, and includes an account of the frequent shamanistic involvement of elderly members of both sexes in the masculine civil sphere. Nevertheless, until "less traditional forms of differentiation and stratification" began to appear, the civil sphere of elders and males "was simply not very important to the community." It was "counter-balanced by the enormous significance of the woman's 'domestic' sphere."[5] Stratification within sororal and fraternal societies and within the organic communities at large was derived from the idiosyncratic characteristics and skills of individuals, as opposed to biological relations associated with blood, gender, and age. While Bookchin argues that the traditional domination of the field of anthropology by men has resulted in a scholarly exaggeration of the importance of men in preliterate societies, he does suggest that tensions existed within the egalitarian relations between individuals and gender-based societies in organic communities. He writes that

> certainly tensions in *values,* quite aside from social relationships, must have simmered within primordial hunting and gathering communities. To deny the very existence of the latent attitudinal tensions that must have existed between the male hunter, who had to kill for his food and later make war on his fellow beings, and the female foodgatherer, who foraged for her food and later cultivated it, would make it very difficult to explain why patriarchy and its harshly aggressive outlook ever emerged at all.[6]

Though male roles as "specialists in violence" complemented female roles as "specialists in nurturing," Bookchin suggests that this biologically based specialization contributed to the tensions that in turn contributed to the emergence and institutionalization of domination.[7]

For Bookchin, the socially and ecologically balanced complementarity of the blood-, gender-, and age-based institutions within organic preliterate communities was a result of the grading of these relations out of nature and teleological evolutionary processes. For social ecology, the link between the evolution of external nature and social nature is profound. The "very natural processes that operate in animal and plant evolution along the symbiotic lines of participation and differentiation reappear as social processes in human evolution, albeit with their own distinctive traits, qualities, and gradations or phases of development."[8] Symbiotic, participatory, and differentiating processes not only appear in human evolution but in the social institutions of preliterate societies.

The social ecological key to Bookchin's understanding of the egalitarian and complementary relations of prehistory lies in the derivation of those relations from the driving processes behind evolution. Evolution, for him, is predominantly governed by gradations of participatory and mutualistic processes that generate ever greater differentiation and complexity in nature. Further, increasing differentiation and diversity of nature is associated with an indeterminate—though teleological—striving by nature for subjectivity and self-consciousness.[9] Searching for "really dialectical ways of process-thinking that seek out the potentiality of a later form in an earlier one, that seeks out the 'forces' that impel the latter to give rise to the former, and that absorb the notion of process into truly evolutionary ways of thought about the world,"[10] Bookchin constructs his own social ecological theory of evolution. For him,

> Social ecology "radicalizes" nature, or more precisely, our understanding of natural phenomena, by questioning the prevailing marketplace image of nature from an ecological standpoint: nature as a constellation of communities that are neither "blind" nor "mute," "cruel" nor "competitive," "stingy" nor "necessitarian," but, freed of all anthropocentric moral trappings, a *participatory* realm of interactive life-forms whose most outstanding attributes are fecundity, creativity, and directness, marked by complementarity that renders the natural world the *grounding* for an ethics of freedom rather than domination.[11]

Perhaps more than any other sentence in Bookchin's work, this last quotation summarizes the foundation upon which social ecology is constructed.

The central issue for Bookchin is that ecological and social evolution is participatory, interactive, fecund, creative, direct, and complementary. This "fact" demands a new social ecological imperative, the acceptance or rejection of which will determine the future ecological and social sustainability of the world. While individual and collective human "selfhood, reason, and freedom" emerge from nature teleologically, he insists that this process is not inexorable.[12] "Our notion of teleology need not be governed by any 'iron necessity' or unswerving self-development that 'inevitably' summons forth the end of a phenomenon from its nascent beginnings."[13] Nevertheless, "there seems to be a kind of intentionality latent in nature, a graded development of self-organization that yields subjectivity and, finally, self-reflexivity in its highly developed human form."[14] Understanding human beings to have been generated by processes flush with ecological forms of subjectivity and intentionality, social ecology insists that these processes establish the necessary ethical ground for sustainable social and natural activities.

In order to illuminate these processes, and the resulting ecological ethic, Bookchin argues that "it is the *logic* of differentiation that makes it possible to relate the mediations of nature and society into a continuum" rather than the particulars of plant-animal communities and natural evolution.[15] The "logic" of evolutionary processes in nature generates the uneven historical accumulation of differentiated, complex, and ever more self-reflexive life forms that culminates in humanity.[16] He ties the logical products of evolution to the internal processes of natural and social evolution by connecting ecological and social participation, differentiation, complexity, and reflexivity.

> Complexity, a product of variety, is a crucial factor in opening alternative evolutionary pathways. The more differentiated the life-form and the environment in which it exists, the more acute is its overall sensorium, the greater its flexibility, and the more active in its own evolution. . . . The greater the differentiation, the wider is the degree of participation in elaborating the world of life. . . . Participation unites the biotic ecocommunity with the social ecocommunity by opening new evolutionary possibilities in nature and society. Differentiation yields richer possibilities for the elaboration of these ecocommunities and adds the dimension of freedom, however nascent in nature or explicit in society.[17]

The mediation of evolutionary processes and products is further developed in Bookchin's emphasis on the dialectics of plant-animal "contexts," as opposed to species development. It is "not only the environment which 'chooses' what 'species' are 'fit' to survive but species themselves, in mutualistic complexes as well as singly, that introduce a dim element of 'choice'—by no means 'intersubjective' or 'willful' in the *human* meaning of these terms."[18]

The contextual participation at the heart of Bookchin's evolutionary theory leads him to stress symbiosis, mutualism, and wholeness over competition, predation, and adaptation within social ecology's evolutionary theory. "Mutualism, not predation, seems to have been the guiding principle for the evolution of [the] highly complex aerobic life forms that are common today."[19] Wholeness, "the *relative* completion of a phenomenon's potentiality, the fulfillment of latent possibility as such, all its concrete manifestations aside,"[20] represents

> a dynamic *unity of diversity*. In nature, balance and harmony are achieved by ever-changing differentiation, by ever-expanding diversity. Ecological stability, in effect, is a function not of simplicity and homogeneity but of complexity and variety. The capacity of an ecosystem to retain its integrity depends not on the uniformity of the environment but on its diversity.[21]

As a result, in "contrast to biotically complex temperate zones, relatively simple desert and Arctic ecosystems are very fragile and break down easily with the loss or numerical decline of only a few species."[22]

For Bookchin, "unity in diversity" is the fount of evolutionary potential and the determinant of ecosystemic stability.

The "unity of diversity" in the social ecological "logic of differentiation" represents an essential moment in Bookchin's understanding of materiality: of the physics, chemistry, and genetics underlying biology and sociality. For him, the "universe bears witness to an ever-striving, *developing* not merely 'moving' substance, whose most dynamic and creative attribute is its ceaseless capacity for self-organization into increasingly complex forms."[23] Within the realm of bioevolution,

> life *is* a counteracting force to the second law of thermodynamics, an "entropy-reduction" factor. The self-organization of substance into ever more complex forms, indeed, the importance of form itself, as a correlate of function, as a correlate of self-organization, implies the unceasing activity to achieve stability; and finally, that complexity is a paramount feature of organic evolution and of the ecological interpretation of biotic interrelationships. All these concepts taken together are ways of understanding nature as such, not mere mystical vagaries.[24]

In fact, interpreting the work of Lynn Margulis and others, Bookchin argues that the emergence of eukaryotic cells (and thereby subsequent multicellular organisms) is rooted in "a symbiotic process that integrated a variety of microorganisms into what can reasonably be called a colonial organism."[25] Independent single cellular organisms are suggested to have symbiotically self-integrated themselves to generate a more complex, organismal unity in diversity with greater evolutionary potential. This represents, for Bookchin, more evidence that the central moment in evolution lies in mutualism, contextual ecological relations, and complementarity over competition, independent activity, and predation.

These points are made as components of social ecology's critique of reductionist and empiricist ecological science. Absolutely central to this argument is Bookchin's insistence that nature is inherently flush with fecundity and a limitless capacity for differentiation. Criticizing conventional and traditional Marxist approaches to nature, he argues that social ecology no longer looks at nature as predominantly a limitation, barrier, or constraint to human reproduction and social development.

> More than any other single notion in the history of religion and philosophy, the image of a "blind," "cruel," "competitive," and "stingy" nature has opened up a wide, often unbridgeable chasm between the social world and the natural world, and in its more exotic ramifications, between mind and body, subject and object, reason and physicality, technology and "raw

materials," indeed the whole gamut of dualisms that have fragmented not only the world of nature and society but the human psyche and its biological matrix.[26]

As we have seen, Bookchin argues that preliterate humanity existed in participatory and fecund communities of complementary blood-, gender-, and age-based societies generated by and graded out of nature. Thus, to understand nature as a stingy realm of necessity over which humanity must struggle to gain mastery is to deny our own natural and biological roots: to ideologically and practically contradict the processes that generated us in the first place. As proof of his point, Bookchin appeals to his anthropological presentation: "An [accurate] image of nature as 'stingy' would have produced 'stingy' communities and self-seeking human participants"[27] in preliterate communities, which he claims did not occur.[28]

The stable, mutualistic, and interdependent communities of preliterate humanity were biosocially rooted in the graded inheritance of natural evolution and the particularity of a "protracted period of development"[29] during childhood. Thus, the

> emergence of society is a *natural* fact that has its origins in the biology of human socialization. . . . It is when *social* parents and *social* siblings, that is, the human community that surrounds the young, begin to participate in a system of care, that is normally undertaken by biological parents, that society begins to truly come into its own. . . . Society thereupon advances beyond a mere reproductive group towards institutionalized human relationships, and from a relatively formless animal community into a clearly structured order.[30]

The qualitative breaking point from prehuman communities to human societies arrives when human beings establish "*institutionalized* relationships, relationships that living things literally institute or create but which are neither ruthlessly fixed by instinct on the one hand nor idiosyncratic on the other."[31] It is the institutionalization of socially reproductive relationships, then, that coordinates the social and biological roots of "instinctual maternal drives, . . . the sexual division of labor, age-ranking, and kin-relationships."[32]

By these means Bookchin differentiates animal communities and human societies from each other while linking them through bioevolutionary processes.[33] He elaborates a fully social ecological theory of whole individuals participating in the social context that situates personal development. "The making of a human being, in short, is a collective process, a process in which both the community and the individual *participate*. It is also a process which, at its best, evokes by its own variety of stimuli the wealth of abilities and traits within the individual that achieve their full degree of *differentiation*."[34] The unity of nature and society in diversity is coherent and differentiated and the individual and collective levels of society can "*embody* the *creativity* of nature" as the "great achievements of human thought, art, science, and technology serve not only to monumentalize culture, *they serve also to monumentalize natural evolution itself*."[35]

These processes, unified and differentiated, generated organic human societies with mutualistic and participatory relations to their ecological contexts in practice, ideology, and language.[36] Rituals of ecological persuasion saturate relations with nature. Humanity was, for Bookchin, "no less a part than animals in this complementary orbit in which human and nonhuman were seen to give of themselves to each other according to mutual need rather than 'trade-offs.' "[37] He argues that the "animism and magic" by and through which preliterate rituals approached the subjectified natural world furthered the use of reason as a means to mutualistically "unite and create" nature and society. Bookchin explicitly opposes his account to those that suggest natural objectification, division, and destruction as the predicate of the extension of reason and society. Objectification is understood by Bookchin as a "repressive abstraction" serving "manipulative forms of human predation" that directly contradict ecological subjectification and the participatory unity of animistic and conciliatory forms of human provisioning.[38]

Further, despite parochial xenophobia, "principles if the 'irreducible minimum,' substantive equality, the arts of persuasion, and a conception of differentiation as complementarity" are said to structure resource distribution within these communities. These relations also generate a deep "commitment to *usufruct*."[39] Here, Bookchin argues that individuals freely appropriated resources simply "by virtue of the fact that they were using them."[40] Given his position on the fecundity of nature and of relations between nature and organic society, resources are assumed to be sufficiently available so as to make this practice relatively sustainable and unconscious. As both social and natural "fecundity originates primarily from growth, not from spatial 'changes' in location,"[41] the internally and ecologically participatory, mutualistic, and usufruct-based preliterate societies prospered, differentiated, and expanded.

Of course, Bookchin is not primarily concerned with generating a coordinated theory of natural and preliterate social evolution. His primary focus is to establish a social and ecological ground for contemporary politics: to examine and explain "those junctures in social evolution where splits occurred which slowly brought society into opposition to the natural world, and explain how this opposition emerged from its inception in prehistoric times to our own era."[42] The primal juncture, of which all subsequent junctures are elaborations, lies at the point(s) when freely participatory and organically differentiated preliterate societies began to institutionalize relations of domination. Accordingly, Bookchin's "greatest single concern has been with the interplay between the evolution of domination and that of freedom"[43] and the "seeming conflict between the 'realm of necessity' and the 'realm of freedom.'"[44] Having emphasized the integrated complementarity of mutualism, symbiosis, participation, and differentiation in natural and social evolution, Bookchin must explain the historical decline of these processes.

The Institutionalization of Domination

Bookchin grounds the alienation of humanity from nature, and thereby the destruction of stable ecological conditions by human beings, in "the domination of the young by the old, of women by men, and of men by men."[45] Relations of hierarchy and domination, for Bookchin, are inherently *social* and

> must be viewed as *institutionalized* relationships, relationships that living things literally institute or create but which are neither ruthlessly fixed by instinct on the one hand nor idiosyncratic on the other. By this, I mean that they must comprise a clearly *social* structure of coercive and privileged ranks that exist apart from the idiosyncratic individuals who seem to be dominant within a given community, a hierarchy that is guided by a social logic that goes beyond individual interactions or inborn patterns of behavior.[46]

Hierarchy and domination, as institutions, can only be found in human *societies*, and cannot be said to exist in animal *communities*.[47] Bookchin defines hierarchy and domination as complex "cultural, traditional and psychological systems of obedience and command . . . in which elites enjoy varying degrees of control over their subordinates without necessarily exploiting them."[48] The combination of domination's necessary roots in singularly human institutionalized social relations, together with the cultural, traditional, and psychological focus of Bookchin's position, lay the groundwork for his insistence that the domination of human by human precedes attempts by humans to dominate nature:

> As a historical statement [social ecology] declares, in no uncertain terms, that the domination of human by human *preceded* the notion of dominating nature. Indeed, human domination of human gave rise to the very *idea* of dominating nature. . . . Men did not think of dominating nature until they had already begun to dominate the young, women, and, eventually, each other.[49]

This position explicitly rejects materialist perspectives that he claims argue that "the 'domination of man by man' emerges from the need to 'dominate nature,' presumably with the result that once nature is subjugated, humanity will be cleansed of the 'slime of history' and enter into a new era of freedom."[50] Since nature is understood by Bookchin to be primarily predicated on fecund, participatory mutualisms, and since human individuals and societies are evolved from nature, domination and hierarchy (which oppose participation and mutualism) must be extra- or unnatural. The prevalence of social domination and hierarchy, and ecological depletion and pollution, in the modern world leads Bookchin to "ask what humanity's 'place' in nature may be" and leads him to argue that this question "has now become a moral and social question . . . one that no other animal can ask of itself. . . . And for humans to ask what is their 'place' in nature may be is to ask whether humanity's powers will be brought into the *service* of future evolutionary development or whether they will be used to *destroy* the biosphere."[51]

The initial conditions of domination are biosocial in character. He suggests that the frailty and insecurity of the elderly, the different value orientations within the gender-based civil–domestic division of labor and functions, and the increase in tribal intercourse as populations grew are the three keys to the historical production of domination. Emphasizing different components of these relations in different texts, Bookchin insists that these processes advanced in an uneven and nonlinear fashion. Nevertheless, perhaps because domination does exist, there appears a certain ineluctability to the process in the multiple accounts he provides.

In organic preliterate societies, while the "sexes compliment each other economically, the old and the young do not. . . . The need for social power, and for hierarchical social power at that, is a function of [the elderly's] loss of biological power."[52] Gerontocracy is the first form of institutionalized hierarchy Bookchin addresses. Despite his assertion of the institutionalization of usufruct and the "irreducible minimum" within these societies, Bookchin argues that gerontocracy emerges as a means by which elderly individuals (of both sexes) manipulate younger members of organic societies in order to assure that they be provided for materially. As they become unable to "pull their own weight,"[53] and especially during "periods of difficulty,"[54] the non- or underproductive existence of the elderly makes their infirmities and vulnerabilities a source of great insecurity. Bookchin asserts that these insecurities lie behind the development of shamanism and other claims by the elderly to be able to manipulate nature for social benefit. "In the tension between extreme personal vulnerability on the one hand and the embodiment of the community's traditions on the other hand, they may have been more disposed to enhance their status, to surround it with a quasi-religious aura and a social power, as it were, that rendered them more secure with the loss of their physical power."[55] By these means, Bookchin claims that the elderly "slyly" use their "fictive" manipulation of nature to become "specialists in fear," comparable to male specialization in violence and female specialization in nurture.[56]

The establishment of gerontocracy, however, did little to displace the gender equality of these early preliterate societies; yet the treatment of nature as something "out there," something that must be manipulated, is asserted to have had pernicious results throughout history.

> More than any other single notion in the history of religion and philosophy, the image of a "blind," "cruel," "competitive," and "stingy" nature has opened up a wide, often unbridgeable chasm between the social world and the natural world, and in its more exotic ramifications, between mind and body, subject and object, reason and physicality, technology and "raw materials," indeed the whole gamut of dualisms that have fragmented not only the world of nature and society but the human psyche and its biological matrix.[57]

Before the male domination of men, the young, and women, gerontocratic mindsets laid the ideational groundwork for the institutionalization of command and obedience, as well as provided the foundation for dualistic conceptions of nature and society.

The potential for the failure of nature manipulation by shamans, Bookchin claims, engendered alliances between elderly shamans and the idiosyncratic hierarchies associated with "big men" within the tribes, thereby extending the realm of domination through selective intratribal cooperation.[58] Shamans, fearing the violent retribution of the tribe, allied themselves with top-ranking specialists in violence as a means of protection should their claims to control nature prove false. These associations fed on the differential value orientations between the male "civil sphere" and the female "domestic sphere" at about the same time during the social evolution that Bookchin asserts horticultural societies emerged, multiplied, and expanded. This geographic expansion generated scarcities of "cultivable land" and the initiation of intertribal warfare.[59] "With increasing intercommunal conflicts, systematic warfare, and institutionalized violence, 'civil' problems became chronic. They demanded greater resources, the mobilization of men, and they placed demands on woman's domain for material resources."[60] In this account, successful younger warriors then began sharing sociopolitical power with the elderly shamans and other big men as the "civil sphere" increased in importance in relation to that of women's "domestic sphere." By these processes, dominant men increasingly forced servility on lesser men and women, younger members of the societies, and the older members of society through the replacement of blood, gender, and age identification with fealty oaths to individuals—oaths that eventually mutated into inherited positions of social domination.

Domination, by these routes, develops within preliterate societies as a moment "latent within organic society itself,"[61] as the societies differentiate internally under changing social and ecological conditions. The emergence of the practical fact of social domination and the ideologies of social and ecological domination derives from the "distinctly social interests" of specific groups within the early communities. It is in this setting, however, that Bookchin asserts that social domination—and ideologies of the social domination of nature—emerge simultaneously with the idea of freedom.[62] This puts a radically different spin on the emergence of domination than that which is generally present in Bookchin's account of the issue. A correlation between the actuality of social domination, ideologies of natural domination, and the concept of freedom suggests that these processes may be more "natural" than Bookchin asserts, and also that the institutionalization of domination may not lie at the root of our present ecological and social crises.

Critique

My view is that Bookchin's work on the emergence of social domination and processes of natural evolution is internally contradictory. His approach—in direct opposition to his own theoretical and methodological positions—inflexibly forces social and ecological processes into prefabricated anarcho-ecological boxes. And his claims to having discovered dialectical processes of participatory ecological and social emergence are not represented in his surprisingly static, and frequently dualistic, schema. I will first address problems with his anthropology of preliterate societies, and then show how these problems are rooted (or

have parallels) in his perspectives on nature, ecology, and the graded
evolution of humanity from nature.

Critique of Preliterate Anthropology in Social Ecology

The central contradiction in Bookchin's account of the origin of domination lies in his foundational insistence that the domination of human
by human historically precedes ideologies of, and attempts to practice,
the domination of nature, or that struggles within society generate
struggles between society and nature. Accepting, for the moment,
Bookchin's edenic view of the participatory, mutualistic, and complementary character of nature–society relations within preliterate organic
societies, it is clear that parallel insistence on social ecological fecundity
and differentiation is denied by his own account.

First, the position that hierarchy and domination are primordially
rooted in the insecurities of the elderly, especially "during periods of
difficulty," suggests that social domination is at least rooted in struggles
associated with human biological nature. To understand this approach
any other way demands the acceptance of dualistic accounts of the evolution of human bodies as opposed to the evolution of external nature.
Nature can never be understood as something "out there" within a dialectical ecology and yet Bookchin leaves the naturalness of human ag-
ing—the personal and social struggles associated with the declining
physical condition of the elderly—out of his account.

Further, and possibly more importantly, Bookchin never investigates the consequences of his approach to the emergence of "periods of
difficulty" for his prior positions regarding the unity in diversity of society and nature within organic communities. Traditionally, "periods of
difficulty" have been understood as forms of ecological scarcity
whether rooted in drought, flood, fire, plague, or some other vagary of
nature. Bookchin, rejecting these positions, denies that the fecund nature from which humanity evolved ever presented early human societies
with such forms of scarcity, or, if it did, such situations had no systematic effect on social institutions. Ecological scarcity is understood in a
manner parallel to idiosyncratic hierarchy; neither is presented as sufficiently sustained to have lasting effects. In fact, however, Bookchin provides no climatological or epidemiological data—because none exist—to support such an assertion of perpetual social ecological
fecundity in prehistory. His claims, like those of the people he criticizes,
are ideologically determined and materially groundless. The only difference is that Bookchin chooses to fall on the opposite side of the fence
from his critics. This oppositional approach is not dialectical but rather
dualistic. Those he criticizes say ecological scarcity generates hierarchy;

he says social domination generates ecological scarcity; and no mediation or synthesis emerges.

Bookchin, by rejecting extrasocial ecological scarcity, must suggest some source for "periods of difficulty" and so he asserts that such conditions emerge only following the relatively extensive spatial spread and population growth of horticultural societies (leaving out altogether the wide geographic reach of pastoral societies widely accepted within anthropology to have emerged at the same time as horticulturalism). Bookchin's position that, from relatively isolated societies, organic human communities expand, meet, and subsequently compete for horticultural resources and space, is deeply problematical.

First of all, at what point (uninvestigated by Bookchin) within the growth of organic society does the local population experience conditions in which fecund local ecologies no longer supply sufficient calories and shelter and therefore must colonize new spaces and ecologies? Put another way, doesn't the geographic and population expansion of preliterate communities suggest that local resources were being exhausted, or were being found insufficient? If this suggestion is correct, it seems that ecological scarcity may be understood in these cases to have been generated by social practices, as Bookchin asserts. But they must be understood as having been generated by social practices Bookchin himself describes as cooperative and mutualistic in their relation to surrounding natures. Something is amiss. Either practices understood by such societies to be participatory and mutualistic were, in fact, not reproducably participatory or mutualistic, *or* Bookchin's assertions about the character of society–nature relations within such organic societies is wrong. The problem lies in trying to pin scarcity, and thereby domination, on nature or on society, rather than associating scarcity with the contingent relations inherently within and between these two moments in our world. Bookchin is quite correct to criticize theorists who dualistically ascribe domination to natural scarcity. On the other hand, his positions are no better than those he so stridently attacks because he, equally dualistically, pins domination on social ideology and fear.

Second, given that Bookchin asserts that the ideological roots of institutionalized domination, in the form of shamanism, emerge from the insecurity of elderly members of organic societies at about the same time as horticultural societies begin to compete with one another, it seems obvious that shamanism is understood to be derived from conditions of ecological scarcity, despite the claim that such conditions are socially generated as a result of the expansion of organic horticulturalism. If the tenuous position the elderly feel themselves to be in during "periods of difficulty" generates "fictive," shamanistic claims about the

ability to control nature in order to control younger members of societies, it could be argued that the social competition for unintentionally
scarce resources generated the "need" within the elderly for institutionalized social domination. A large chicken-and-egg question emerges,
but only if we dualistically separate social and ecological contributions
to what can be called ecological scarcity. Independent of the utilitarian
or animistic understanding that preliterate societies brought to bear on
their relations with nature, and independent of the diverse ecological
cycles of abundance, flood, drought, and climactic temperance, the key
point is that Bookchin, no different from those theorists he attacks,
cannot escape attributing domination's ideological and material roots
to anything other than ecological scarcity.

Bookchin cannot have his cake and eat it, too. What is needed is a
dialectical theory of the *enablements* and *constraints* generated by ecological and social conditions across diverse spatial and temporal locations. The stereotypic strawmen that Bookchin batters for grounding
social domination and exploitation in struggles with nature are, in fact,
wrong. However, Bookchin's alternative to flawed Enlightenment and
structural Marxist theories is to produce a dualistic opposite rather
than a dialectical synthesis. As a result, his work is as thoroughly contradictory as (and no more nuanced than) the work of the writers and
thinkers he criticizes.

Another source of deep problems in Bookchin's account of the rise
of domination lies in the rigid gender division of labor he insists on
within preliterate societies. A good part of the problem comes from
Bookchin's preference for two (and only two) anthropological studies
from the 1950s that have been long superseded by the flowering of anthropology since the early 1960s.[63] Further, the two studies Bookchin
cites are investigations of North American tribes generated during the
mid-twentieth century. To extrapolate from two studies of modern,
long post-European contact, tribal entities to a unified theory of social
organization within preliterate societies is to practice an extreme form
of unevolutionary thinking. Contemporary forms of social relations can
only be assumed to represent historical predecessors in highly problematic fashion, assumptions as problematic as the idea that modern chim-
panzees and gorillas provide linear information about the common an-
cestor we share with those two species.[64] Unlike the extremely
confident tones of Bookchin's work, while some anthropologists use
contemporary ethnographic research to develop postulates and theories
about preliterate and protohuman communities, such work universally
notes the paucity of actual data and the speculative character of the
work presented.

Perhaps even more critically, anthropological data generated since the late 1950s and early 1960s has shown that gender, age, and power relations within tribal societies (preliterate and otherwise) vary to extraordinary extents. To argue, as Bookchin does, that rigid "domestic" and "civil" spheres, deterministically rooted in biological characteristics that encourage masculine specialization in violence and feminine specialization in nurturing, were maintained across the manifold savannah, forest, riverine, oceanic, mountain, and desert peoples, without any associated ecological and social differentiation is to reject his own position on social ecological evolution. If biosocial evolution and differentiation are predicated on species–context dialectics, and if social participation actively situates institutional complexity within organic societies, then Bookchin's model of organic societies does not fit his own social ecological theory.

Further, given that Bookchin must be writing about the preliterate societies spread across most of the globe around the time of the horticultural revolution, extensive contemporary research indicates that the concept of "isolated tribe" has little or no meaning, despite the more sedentary characteristics of modern tribal societies. Intertribal contact across ecosystems was common, while frequency depended on the mobility of communities and their neighbors. To assume geographic isolation and homogenous spatial development of horticulturalism is difficult to defend given the gatherer-hunter/fisher reproductive characteristics of human societies before the emergence of horticulturalism and pastoralism. Bookchin's argument implicitly suggests relatively even geographic isolation among organic societies, the coherence and complementarity of which collapses at the point when concentric circles expanding from centrally isolated points meet and competition for resources begins.

However, *Homo erectus* had resided throughout Africa, Europe, and Asia for more than 1.5 million years prior to the (proposed) migration of *Homo sapiens* from Africa approximately 100,000 years ago. Prehistoric peoples were by no means sedentary or evenly spread out, and the social and ecological diversity of neighboring hunter-gatherers, fisherfolk, horticulturalists, and pastoralists suggests anything but generalizable, simple, and solely gender-based divisions of labor. In fact, whether one accepts a multiregional or an "out of Africa" model of human evolution, the geographic expansion of *Homo sapiens* demanded the development of "social, and especially trade, networks to supply needed materials from distant places."[65] Along these lines, not only does the "civil sphere" necessary for trade and social contact relative to the "domestic sphere" have more importance than Bookchin suggests, but the importance of trade, ecological diversity, and different divisions of labor within such societies indicates that his strong, dualistic separation of the two spheres makes no sense. The point is not to deny gender divisions of labor, or tendencies within such cross-cultural divisions, but to suggest that social divisions of labor are certainly as complex and uneven as the diverse ecologies within which they are found.

Not only that, but the strict separation of a "civil" from a "domestic" sphere is predicated on a form of biological essentialism that reproduces long-outdated culture–nature dualisms. Anthropological investigation of, and models for, the role of women in evolution increasingly reject universalizing assumptions that posit the kind of woman:domestic::man:hunter relationships that Bookchin profers. Increasingly, paleoanthropologists believe that gathering and scavenging by both sexes were practiced long before hunting, an activity that was not necessarily a solely male occupation.[66]

> Critical re-examination of the public/private (or the public/domestic) dichotomy in recent anthropological writings on kinship draw attention to the fact that the distinction between a public sphere and a private, or domestic, sphere is characteristic only of certain types of societies, notably those that have separated the workplace from the household; that the distinction can take a variety of different forms; and that it cannot be taken at face value even in societies for which it is relevant, since it should be understood more as an ideological construct that masks certain features of the social system than as a description of how that system is structured.[67]

Whereas Bookchin is critical of the historical biases and blindness of male anthropologists with respect to their failure to investigate women's roles and the "domestic sphere," he is uncritical of the same bias as expressed by the dualistic assumption that women's domestic/nurturing social roles are determined by their biological capacity for reproduction and men's hunter/violent roles are determined by their greater size.

The gender dualisms in Bookchin's work aside, his ideas about ritualized egalitarian animism are problematic. His preliterate cultures are said to personify "animals, plants, even natural forces and perfectly inanimate things as well as human beings."[68] Treating all the world as subjects precludes objectification, an ideological and technical practice he terms "repressive abstraction"[69] and the roots of which he finds in the original "fictive" manipulation of nature claimed by insecure elderly shamans. However, without objectification, means/ends distinctions are impossible and, as a result, so is the situational, reflexive, and participatory nature of free will. Individuals, under these conditions, must always treat themselves and all other things as subjects, that is, ends in themselves. This precludes treating one's own (individual and social) life as an object and thereby disables the establishment of institutionalized activities.[70] Social and personal activity, under such conditions, must be reconstituted at every meeting of subjects, whether natural or social. The objectification Bookchin rejects is, in fact, a necessary condition for self-reflexivity and the ideational and memorial recognition of historical time, that is, a significant part of what it is to be human.

It is possible that Bookchin would want to distinguish between "Objectification," as instrumental and reified abstraction, and "objectification," as a dialectical moment (with subjectivity) in the evolution of mind. However, if this is the case, he fails to do so. Without this distinction, Bookchin's theorization of an "organic" preliterate ecological phase in human society comes into question. His grounding of human sociality in subjective animism, active mutualism, and ecosocial creation precludes the very institutionalization of social relations that he insists differentiate human *societies* from animal *communities*.[71] Again, rather than dialectically synthesize a complex relation between conventional wisdom and its dualistic opposite, Bookchin chooses opposition. Conventional science and structural Marxism overemphasize the importance of objective, scientific mentality in human evolution, while Bookchin overemphasizes the importance of subjective animistic mentality in the same processes.

Here, the parallel with his insistence that the domination of nature is linearly and historically rooted in social domination, in direct opposition to conventional accounts (from both the left and the right) which argue that the domination of society derives from the need to dominate nature, is striking. In neither case does Bookchin take a step forward (so to speak). By rejecting all forms of objectification, it appears that Bookchin's argument demands that organic preliterate *societies* be only prehistoric *communities*. His communities cannot be societies because they can have no social institutions, or established systematic relations objectively and abstractly understood by their participants, given their subjective animistic orientation. It seems that, by Bookchin's own account, social history emerges from community development only with the objectification of nature and the protoinstitutionalization of shamanistic domination he so abhors. It just may be that Bookchin's own account at least partially supports components of the theoretical and empirical research he is criticizing.[72] If social history only begins with the institutionalization of domination, then much of his later accounts (not addressed in this chapter) of the struggle between evolutionarily oppressive institutions of domination and the volitionarily progressive institutions of freedom begins to fall apart.

For Bookchin, the interactive evolution of human mind, body, and society is in no way contradictory; it is embedded in an indeterminate "logic of differentiation" until human mentality and physiology are evolutionarily complete. The argument here is that such a view of human nature is untenable. Bookchin insists on a mutualistic dialectic of evolution rather than a dialectic that integrates competition and mutualism, and all the gradations in between, into a more nuanced vision of the combined and uneven development of subjectivity and objectification, of the complex mediations of evolving social relations and ecological conditions, and of the diversity of social and ecological determinations. Ecosocially enabling relations cannot be abstracted from the social and ecological forces that constrain socioecological transformation.

It is important to emphasize this last point. Human beings have made their lives individually and collectively within social and environmental conditions that they did not make or choose, and within a world full of unintended consequences, since before the emergence of the hunter-gatherer societies. If this is the case, what is demanded is an investigation of early societies that strives to see how coordination and command, organization and hierarchy, and agreement and domination are creatively embedded *in* one another, that is, are each other's content and context, rather than an investigation that asserts, without support, that such dialectically related relations preclude each other.[73]

Given social ecological "unity in diversity," a less dualisitic view of the evolution of human subjectivity and society would more comprehensively investigate the diversity of preliterate institutions and their manifold ecological milieus. Bookchin overemphasizes "unity" and underplays "diversity" despite his insistence that participatory mutualism generates diversity and complexity. His emphasis on ecological fecundity stresses the bounty "nature" freely provides but omits analysis of the daily social organization of gathering, scavenging, and hunting demanded by ecological constraints (which are not only or always only enablements). Bookchin stresses ecological normalcy but never explains ecosocial "periods of difficulty." His work is one-sided, perhaps as a polemical strategy, but one-sided nevertheless. Bookchin's early preliterate societies are subjectively animistic, cooperative, free, and agency-driven until domination emerges; then are introduced objectified technopractices, hierarchy, power, and structural constraints, both ecological and social, to the social world. A more nuanced approach to human evolution generates a much less idealized vision of our past and would look more closely at the importance of the diversity of different forms of social organization in hunting–gathering, production, and reproduction. Such an approach would not demand a totalizing origin story, and therefore a story of the Fall, in order to account for what must be, in Bookchin's view, "unnatural" domination.

What is needed is an empirically supported theory of human biosocial evolution that looks at the complex and contingent enablements and constraints within and between forms of ecological and social organization. All forms of ecological constraint need not be tied to scarcity, as has been historically done by liberal scientists and Marxist theorists. Equally, ecological constraints need not be tied to the modes of social and ideological practice that linearly suppress cultural differentiation as a result of their hierarchical character. Examples of social and technical creativity in the face of ecological constraints abound. Some of these innovations have generated highly participatory and ecologically sustainable forms of social ecological practice, others have produced hierarchical and destructive modes of life. Further, examples of extraordinarily hierarchical and yet ecologically sustainable societies are also part of the historical record.

No dialectical theory of ecological and social evolution and change can afford linear statements such as "social domination is derived from the need to control nature" or "ideologies of the domination of nature are derived from ideologies, and the practice, of social domination." Materially, insufficient scientific and anthropological data exists to support any such linear claim about prehistory. Also, the ecological and social diversity associated with the historical record argues strongly against such statements. Philosophically, contradiction and negation is not about conflictual oppostion or mutualistic generation but rather the complex unity that is conflictual generation and mutualistic opposition. Fecundity and scarcity, and domination and freedom, are inherently connected sides of the same coin. What we need is to understand the general patterns and specific forms of the multivalent interactions associated with these and other terms, rather than the linear promotion of one term versus the promotion of another.

Critique of Evolutionary Biology in Social Ecology

The problems with Bookchin's origin story for domination are rooted in his overemphasis of mutualism in nature and evolution. While appropriately critical of the reductionist, empiricist, and adaptationist moments within the sciences of evolution, he gives only lip service to (i.e., effectively rejects) competition, predation, and contingency. Moreover, Bookchin fails to address the diverse and differentiated forms of ecological relations associated with the terms "mutualism" and "symbiosis."[74] Mutualism, the beneficial interaction between species, and symbiosis, the close association of interspecies life patterns, remain incompletely understood as a result. For example, mutualistic relations range from those that are characterized by direct beneficial

interaction to those that are indirectly beneficial; direct mutualistic relations can be symbiotic, closely amd immediately associated, or nonsym-
biotic, distantly amd contingently associated. Symbiosis can be facultative (nonnecessary) or obligate (necessary), and can be comprised of
monophylic (two), oligophylic (less than five), or polyphylic (many) relations. Among these diverse relations only direct symbiotic mutualisms
are generally coevolved and obligate—yet this is the only form of mutualism Bookchin ever addresses.

Further, there is little scientific evidence that mutualism and symbiosis are the predominant forces in evolution as Bookchin claims, even
among those aware of the politicized history and meanings of the biological subdiscipline.

> On the one hand, an enormous number of ecologically and economically
> important interactions, found throughout the biosphere, would seem to
> be mutualistic. On the other hand, few studies have actually demonstrated
> increases in either fitness or population growth rate by both of the species
> in an interaction. Interactions have generally been shown to be mutual-
> isms by describing what is exchanged. Mutualism may be everywhere, but
> its existence remains practically unproven.[75]

In fact, in the review article on mutualism, symbiosis, and evolution
from which the last quote was taken, the authors find the two key
questions associated with theories of mutualism: "(a) When will mutualisms develop (note that we do not say 'evolve') and in what sorts of
species and environments will they be found? (b) When will a community involving mutualists persist (again, we do not say 'be stable')?"[76]
The complexity of evolved, facultative, and monophylic relations
within ecologies generate diverse and contingent interaction effects between mutualists (+/+), commensualists (+/0), predators (+/–), and competitors (–/–) such that changing relations between sets of species or individual species and their environment can facilitate or exacerbate
other sets of relations in equally diverse fashion.

Bookchin asserts that mutualism made possible the stable evolution of complex ecologies under harsh conditions and suggests that the
greatest amount of mutualism, species differentiation, and ecological
differentiation occurs in temperate and equatorial zones. While this
may be correct, the greater species richness, productivity, and biomass
of such zones may also bring about greater predation and competition—at this point no one knows.[77] What is clear is that Bookchin's appropriate rejection of monofocal competition- and scarcity-based theories of evolution led him to produce an equally monofocal theory, on
the other side of the competition–mutualism dualism, which overem-

phasizes mutualism and fecundity as much as the others overemphasize competition and scarcity.

This point is especially clear in the example at the end of *The Ecology of Freedom* where Bookchin argues that the evolution of aerobic eukaryotic cells represents a foundational mutualism at the evolutionary source of complex life forms on earth. Appealing to the pathbreaking work of Lynn Margulis, he goes so far as to suggest that the processes by which single-cell aerobic (oxygen-utilizing) organisms combined with other single-cell anaerobic organisms, to generately evolve multicellular organisms capable of aerobic and anaerobic respiration, represent a critical mutualistic juncture in the history of evolution. This juncture is certainly pivotal within evolutionary history; however, the processes by which single cellular organisms combined were anything but mutualistic, though they clearly produced obligate mutualisms.

Single cellular combination resulted from two processes, phagocytic absorption (without digestion) of one organism by another, or invasion of one organism by another. At root these were predatory acts under conditions of high ecological stress and scarcity that simultaneously generated selection pressures for (1) resistance to digestion and (2) resistence to invasion. Bookchin fails to note that this moment in ecological history represents the moment when single cellular anaerobic respirants had so "polluted" their environment with oxygen by-products that cellular metabolites were available in increasingly short supply and competition for those metabolites was intense. These conditions certainly contributed to the generation of increasingly sophisticated forms of aerobic respiration at the same time that they contributed to the predatory processes of cellular absorption and invasion.

Bookchin is certainly correct when he argues that mutualism, symbiosis, and commensualism are central to the story of the evolution of aerobic life on this planet. However, and equally certainly, these extraordinarily diverse forms of species interaction can be neither analytically nor materially separated from competition, predation, and contingency. A truly dialectical theory of evolution would go beyond the competition–mutualism dualism, rather than further it from the other side as Bookchin does.

Without advancing an alternative theory of evolution it is possible to criticize even more deeply Bookchin's theoretical preoccupation with participatory differentiation, species complexity, and ecological stability.[78] Bookchin asserts that increasing ecological complexity, derived from nature's teleological movement toward self-organization and self-consciousness, generates ever more stable ecologies populated by fairly generalized genera. These genera, which are said to evolve during sporadic or rapid periods of change, are regarded as more ecologically flexible than more specialized genera, and are thereby understood to evolve more slowly than their specialized neighbors. Bookchin argues that this generalized and flexible species–environment relations may even produce "long periods of [evolutionary] stasis."[79]

Here, however, there is an apparent contradiction in Bookchin's account. He finds that the most stable ecologies are those with the greatest differentiation and interaction of flexible symbiotic genera; yet as species differentiation increases within bounded space and time there must necessarily be greater specialization or competition, given the energetic constraints of any particular ecosystem. Bookchin has a choice: either he must accept that his differentiation-complexity theory suggests greater specialization or that it suggests greater competition among generalized species. He fails to recognize the necessity of that choice, preferring to assert teleological moves toward differentiation, mutualism, *and* generalization. There is another option left for Bookchin, to argue that evolution produces ever smaller populations of generalized species, though he cannot make this choice given his insistence that ecological "fecundity originates primarily from growth, not from spatial 'changes' in location."[80]

The central problem here relates to levels of analysis, most specifically in terms of levels of complexity. Levins and Lewontin note that the widespread assumption of an "increase in complexity and information during evolution does not stand on any objective ground and is based on several confusions."[81] The first confusion relates to the measurement of the complexity of an organism or the complexity of organism–environment relations. Are mammals more internally differentiated or biochemically complex than bacteria, than coral, than angiosperms, than fish? Are mammalian interactions with other members of their species or their "parasites, predators, competitors, and symbionts"[82] more complex than those of "lower" species? How could we measure such internal or interactional complexity?

The second confusion identified by Levins and Lewontin is associated with the character of ecosystemic complexity. While more complex body-plan structures evolved later in evolutionary time from species with less complex body plans, there is no clear trajectory by which the less complex species can be found to become extinct nor a trajectory that suggests the more complex species resist extinction, nor that the intraspecies ecological interactions later in evolutionary time are more complex than those earlier in time. Third, since "earlier" species survive, evolve, and become extinct just as "later" ones do, suggesting that general ecological complexity increases is deeply problematic. Fourth,

Levins and Lewontin find that the common assumption of a linear correlation between increasing ecological complexity and increasing systemic information (increasing self-organization in Bookchin's case), is
not borne out by material investigation. This assumed linear relation is
predicated on another assumption that posits complexity and information to be parallel concepts. For example, Bookchin repeatedly asserts
that the latent intentionality of nature generates graded self-
organization, subjectivity, and "finally, self-reflexivity in its highly developed human form."[83] The claim here is that, prior to *Homo sapiens*,
nature's self-organization was less developed than after the point when
modern humans evolved: that ecological subjectivity was less developed. At the level of species–environment interactions at which Bookchin insists ecologists must study evolution, there is no means to assess
or investigate this claim, whether quantitatively or qualitatively. Independent of whether science is dialectical or empiricist, the strong assertion of irrefutable claims without acknowledged self-criticism is unac-
ceptable.

Finally, Bookchin uncritically embraces deeply problematic organicist assumptions about nature and society. As Levins and Leontin, put
it, the position that "form and function together with strong interdependence of the diverse elements are the components of complexity,
which in turn leads to stability through greater homeostasis . . . has no
apparent basis in fact or in theory. . . . No apparent increase in overall
taxonomic or ecological diversity has occurred for the last 150 million
years."[84] Depending on evolutionary patterns and ecological contingency, different groups and species have increased or decreased in diversity as have particular equatorial, arid, polar, terrestrial, and aquatic
ecologies. There is no clear indication that the more social and self-
reflexive mammalian species have been ecologically more successful
than have other taxonomic groups, nor that the ecosystems within
which such mammals reside are more complex than those in which
they do not. In fact, "If complexity of a community is defined as the
number of species interactions multiplied by the strength of the interactions, it been shown that as this complexity increases, by adding
more species or by increasing the strength of the interaction[s], the
probability that the community will be stable to perturbation decreases
rather than increases."[85]

Stephen Jay Gould has recently argued that were the tape of life re-
wound "to the early days" and let "replay from an identical starting
point . . . the chance becomes vanishingly small that anything like human intelligence would grace the replay."[86] He emphasizes that the
evolution of life on Earth has been one of species differentiation under

conditions in which the number of phyla and genera has been "decimated" over the last 750 million years. Punctuated evolution, in this model, is rooted in the periodic proliferation of species from groups, species, and individuals that fortuitously survived devastating events at various levels of ecological strata.[87] Rather than the tendential movement toward complex, self-organized, participatory, and stable mutualistic differentiation that Bookchin asserts, evolution embodies moments of enabling contingency, bio- and geostructural constraint, and participatory interdetermination, all of which mediate and interpenetrate with gradations of ecological symbiosis, predation, mutualism, parasitism, commensualism, and competition. The evolutionary moment of Bookchin's social ecology, out of which he explicitly derives his social histories, analyses, and critiques, is too linear and is clearly weighted to lean toward preconceived, politically motivated (if admirable) notions of social freedom, participatory democracy, and personal agency. Sharing his vision of the ecological potential of humanity does not demand accepting the route by which Bookchin reaches his conclusions or his undialectical account of the driving forces behind ecological and social evolution.

As with the previous section of the critique, what is needed and what contemporary environmental sociologists, evolutionary biologists and anthropologists, and some political economists are working on are theories that assess the complex interdeterminations (enablements, constraints, and contingencies) of ecological, personal, and communal conditions in their association with historical production of increasingly international political, economic, and cultural forces and relations.

Conclusion

Despite Bookchin's polemical attacks on the monolithic and simplistic character of Marxist and scientific accounts of social and ecological evolution, they are far more complex, nuanced, and focused on participation and mutualism than he suggests. It is certainly the case that the evolutionary, historical, and contemporary relation of society and nature must be reconceptualized, and it is certain that such an exercise will subsequently demand new forms of social and evolutionary theory and data collection. The key, however, is that the new approach to theorizing and researching the intertwined evolution of life, society, domination, exploitation, and ecological destruction must be more sophisticated than previous forms. For all Bookchin's effort to generate a theory that goes beyond scientistic and structuralist theses of natural and social evolution, what he has done is generate a dualistic antithesis rather than a new synthesis.

Any new synthesis will have to move beyond the opposition of deterministic versus spontaneous and structuralist versus participatory approaches. In this way, the universalizing assumptions historically associated with anarchism and socialism must be sublated and overcome. A new synthetic view of ecological and social evolution must take into account both general trends in, and specific forms of, natural and human processes. Bookchin's account of the early stages of ecological and social evolution and domination implicitly asserts a lack of diversity in ecological relations and social environments that undermine his whole approach. Without an approach that assumes diversity and complexity at the start, and one that takes into account contingent changes from "above," "beside," and "below" chosen levels of analysis, Bookchin will continue to produce contradictory analyses.

Iwould argue that a more dialectical approach to evolution and history would suggest that evolutionary and institutionalized relations and products are never ethically clear-cut. Whether predicated on mutualism, predation, contingency, or some combination of the three, ecological evolution enables some forms of diversity and complexity and constrains others just as social processes rooted in combinations of institutional cooperation and competition and personal agency and shortcomings situate and foster different forms of social conservation and revolution.

Bookchin never makes explicit how the domination of humans by humans produces the ecological problems and crises beyond his unsupported assertion that at the root of shamanism is a consciously fictive claim to be able to control nature. We'll never know at what physiological and social moment idiosyncratic forms of social domination were first institutionalized, nor will we ever know what the ideological or practical basis for that institutionalization was. What we can be assured of, however, is that at that point in prehistory, just as in the present, human beings made their own lives individually and collectively in the face of both enablements and constraints associated with social ecological relations. The project, today, must be to analytically and practically understand the particular forms and general structures associated with contemporary enablements and constraints so as to produce ecologically and socially appropriate responses to social and ecological crises. The key to this process is to understand the graded mediations of exploitation and domination, mutualism and competition, and local democracy and national bureaucracy rather than to continue to generate polemical excurses in favor of one-sided approaches to unproductive dualism.

Notes

1. Murray Bookchin, *Remaking Society: Pathways to a Green Future* (Boston: South End Press, 1990), pp. 51–52.

2. Ibid., p. 52.

3. Bookchin, *The Modern Crisis* (Philadelphia: New Society Publishers), pp. 17–18.

4. Ibid., p. 26.

5. Bookchin, *Ecology of Freedom* (Palo Alto, CA: Cheshire Books, 1982), p. 5.

6. Ibid., p. 6.

7. Ibid., p. 79.

8. Ibid., pp. 42–43.

9. "In my view, reason exists in nature as the self-organizing attributes of substance; it is the latent subjectivity in the inorganic and organic levels of reality that reveal an inherent striving toward consciousness. In humanity, this subjectivity reveals itself as self-consciousness" (*Ecology of Freedom*, p. 11).

10. *Modern Crisis*, p. 15.

11. Ibid., p. 55; emphasis in original.

12. Ibid., pp. 12, 13.

13. *Ecology of Freedom*, p. 355.

14. Ibid., pp. 353–354. "If nature provides the ground for an ethics that has an objective ancestry in evolution's thrust towards freedom, selfhood, and reason, so too nature provides the ground for the emergence of society" (*Modern Crisis*, p. 16).

15. Ibid., p. 60; emphasis in original.

16. "Natural history is a *cumulative* evolution toward ever more varied, differentiated, and complex forms and relationships. . . . The *evolutionary* development of increasingly variegated entities, most notably, of life-forms, is also an evolutionary development which contains exciting, latent possibilities. With variety, differentiation, and complexity, nature, in the course of its own unfolding, opens new directions for still further development along alternative lines of natural evolution. To the degree that animals become complex, self-aware, and increasingly intelligent, they begin to make those elementary choices that influence their own evolution. They are less and less passive objects of 'natural selection' and more and more the active subjects of their own development" (*Remaking Society*, pp. 36–37).

17. *Modern Crisis*, pp. 25–27.

18. Ibid., pp. 56–57. "This is a concept of evolution as the dialectical development of ever-variegated, complex, and increasingly fecund *contexts* of plant-animal communities as distinguished from the traditional notion of biological evolution based on the atomistic evolution of single life-forms, a characteristically entrepreneurial concept of the isolated 'individual,' be it animal, plant, or bourgeois—a creature which fends for itself and either 'survives' or 'perishes' in a marketplace jungle."

19. Ibid., p. 359. Bookchin does not say that predation, competition, and

adaptation to environmental change do not occur. But, as the quote indicates, he very strongly privileges mutualism and symbiosis as central evolutionary forces. "Social ecology is largely a philosophy of participation in the broadest sense of the word. In its emphasis on symbiosis as the most important factor in natural evolution, this philosophy sees ecocommunities as participatory communities. The compensatory manner by which animals and plants foster each other's survival, fecundity, and well-being surpasses the emphasis conventional evolutionary theory places on their 'competition' with each other—a word that, together with 'fitness,' is riddled with ambiguities" (*Modern Crisis,* p. 25).

20. Ibid., pp. 60–61.

21. *Ecology of Freedom,* p. 24.

22. *Modern Crisis,* p. 58. Further, evolution "seems to be more sporadic, marked by occasional rapid changes, often delayed by long periods of stasis. Highly specialized genera tend to speciate and become extinct because of the very narrow, restricted niches they occupy ecologically, while fairly generalized genera change more slowly and become extinct less frequently because of the more divesified environments in which they can exist" (*Ecology of Freedom,* p. 360).

23. Ibid., p. 357.

24. Ibid., pp. 356–357.

25. Ibid., p. 359. "Eukaryotic flagella, she hypothesizes, derive from anaerobic spyrochetes, mitochondria, from prokaryotic bacteria that were capable of respiration as well as fermentation; and plant chloroplasts, from 'bluegreen algae,' which have recently been reclassified as cyanobacteria. The theory, now almost a biological convention, holds that phagocytic ancestors of what were to become eukaryots absorbed (without digesting) certain spirochetes, protomitochondria (which, Margulis suggests, might have invaded their hosts), and, in the case of the photosynthetic cells, coccoid cyanobacteria and chloroxybacteria."

26. *Modern Crisis,* p. 52.

27. *Remaking Society,* pp. 50–51.

28. "The overwhelming mass of anthropological evidence suggests that participation, mutual aid, solidarity and empathy were the social values early human groups emphasized within their communities" (*Remaking Society,* p. 28).

29. *Modern Crisis,* p. 34.

30. *Remaking Society,* p. 26.

31. *Ecology of Freedom,* p. 29. For Bookchin, animal "communities are not societies . . . they do not form those uniquely human contrivances we call institutions . . . [they have] genetic rigidity . . . not contrived rigidity" (*Modern Crisis,* pp. 16–17).

32. Ibid., p. 62.

33. "The principles of social ecology, structured around participation and differentiation, thus reach beyond the biotic ecocommunity directly into the social one, indeed, into the nature of the ego itself and the image it forms of the other. An ecological ethics of freedom thus coheres nature, society, and the in-

dividual into a unified whole that leaves the integrity of each untouched and free of a reductionist biologism or an antagonistic dualism" (*Modern Crisis*, p. 36).

34. Ibid., p. 35; emphasis in original. "Society, in turn, attains its 'truth,' its self-actualization, in the form of richly articulated, mutualistic networks of people based on community, roundedness of personality, diversity of stimuli and activities, an increasing wealth of experience, and a variety of tasks" (*Modern Crisis*, p. 59).

35. *Remaking Society*, pp. 35–36; emphasis in original.

36. "This organic, basically preliterate or 'tribal,' society was strikingly nondomineering—not only in its institutional structure but in its very language. . . . No less striking than the substantive equality achieved by many organic societies was the extent to which their sense of communal harmony was projected onto the natural world as a whole. In the absence of any hierarchical social structures, the aboriginal vision of nature was also strikingly nonhierarchical" (*Remaking Society*, pp. 47–48).

37. Ibid., p. 49.

38. *Ecology of Freedom*, pp. 98–99.

39. *Remaking Society*, p. 50; emphasis in original.

40. *Ecology of Freedom*, p. 50.

41. Ibid., p. 357.

42. *Remaking Society*, pp. 31–32; see also p. 38.

43. *Ecology of Freedom*, pp. 350–351.

44. Ibid., p. 10.

45. *Remaking Society*, p. 39.

46. *Ecology of Freedom*, p. 29; emphasis in original.

47. "By reducing a complex society to a mere community, we can easily ignore how societies differ from each other over the course of history. . . . Indeed, we risk the possibility of totally misunderstanding the very meaning of terms like 'hierarchy' as highly organized systems of command and obedience—these, as distinguished from personal, individual, and often short-lived differences in status that may, in all too many cases, involve no acts of compulsion. We tend, in effect, to confuse the strictly institutional creations of human will, purpose, conflicting interests, and traditions, with community life in its most fixed forms, as though we are dealing with inherent, seemingly unalterable, features of society rather than fabricated structures that can be modified, improved, worsened—or simply abandoned" (*Remaking Society*, p. 29). The key is the transition from benign (natural) status differentiation based on skill, gift-giving, and experience and age to domination by gradual, incremental, and more-or-less invisible changes in institutional relations (*Remaking Society*, p. 62).

48. Ibid., p. 4. For Bookchin, that hierarchy exists is "an even more fundamental problem than social class[], that domination exists today as an even more fundamental problem than economic exploitation" (*Modern Crisis*, p. 67).

49. *Remaking Society*, p. 44; emphasis in original. Further, the "notion that man is destined to dominate nature stems from the domination of man by man—and perhaps even earlier, by the domination of woman by man and the

domination of the young by the old . . . [and] that to harmonize our relationship with the natural world presupposes the harmonization of the social world" (Bookchin, *Towards an Ecological Society* [Montreal: Black Rose Books, 1980], pp. 60, 66–67).

50. *Modern Crisis,* p. 51; emphasis in original.

51. *Remaking Society,* p. 40.

52. *Ecology of Freedom,* p. 81.

53. The lives of the elderly "are always clouded by a sense of insecurity . . . incremental to the insecurity that people of all ages may feel in materially undeveloped communities" (*Ecology of Freedom,* p. 82).

54. *Remaking Society,* p. 53.

55. Ibid.

56. *Ecology of Freedom,* pp. 83, 100.

57. *Modern Crisis,* p. 52.

58. *Remaking Society,* pp. 58–59.

59. *Ecology of Freedom,* p. 7. "As 'civil' society became more problematic because of invaders, intercommunal strife, and, finally, systematic warfare, the male world became more assertive and agonistic—traits that are likely to make male anthropologists give the 'civil' sphere greater prominence in their literature, especially if they have no meaningful contact with the women of the preliterate society" (*Remaking Society,* p. 56). See also *Modern Crisis,* p. 21.

60. Ibid., pp. 56–57.

61. *Ecology of Freedom,* p. 80.

62. Ibid.

63. These studies by Dorothy Lee (*Freedom and Culture* [Englewood, NJ: Prentice-Hall, 1959]) and Paul Radin (*The World of Primitive Man* [New York: Grove Press, 1960]) are the only anthropological citations in Bookchin's major publications since 1980, and are those to which he repeatedly appeals.

64. See Linda Marie Fedigan, "The Changing Role of Women in Models of Human Evolution," *Annual Review of Anthropology,* Vol. 15, 1986, pp. 25–66, and Thomas N. Headland and Lawrence A. Reid, "Hunter-Gatherers and Their Neighbors from Prehistory to the Present," *Current Anthropology,* Vol. 30, No. 1, February, 1989, pp. 43–66, for anthropological reservations and citations about this issue.

65. Stephen Jay Gould, "So Near and Yet So Far," *New York Review of Books,* October 20, 1994, p. 28.

66. Fedigan, "Changing Role of Women"; Carol C. Mukhopadhay and Patricia Higgins, "Anthropological Studies of Women's Status Revisited," *Annual Review of Anthropology,* Vol. 17, 1988, pp. 461–495; Headland and Reid, "Hunter-Gatherers and Their Neighbors"; Duane Quiatt and Jack Kelso, "Household Economics and Hominid Origins," *Current Anthropology,* Vol. 26, No. 2, April, 1985, pp. 207–222. In Quiatt and Kelso, the comments are of particular interest.

67. Judith Shapiro, "Gender Totemism," in Richard R. Randolph, David M. Schnieder, and May N. Diaz, eds., *Dialectics and Gender: Anthropological Approaches* (Boulder, CO: Westview Press, 1990), pp. 1–19.

68. *Ecology of Freedom,* p. 98.

69. Ibid., p. 99.

70. "Man makes his life activity itself the object of his will and of his consciousness, . . . his own life is an object for him. Only because of this is his activity free activity" (Karl Marx, "The Economic and Philosophic Manuscripts of 1844," in Karl Marx and Frederick Engels, *Collected Works* (New York: International Publishers, 1975 [1932]), vol. 3, p. 276.

71. I may want to note that Bookchin takes prescientific people at their word in terms of their "ecological" and "organic" relation with each other and the world whereas, and it is dangerous to overstate this, it is only now that we have a posttheological and postscientific understanding of molding matter to our social needs as a result of the breakdown of parochial relations and the resulant global consciousness of ourselves as a species and our world as a unity that we have a developed sense of the long-term consequences of our actions on our social relations and our regional and global ecologies. The formulations in this discussion are deeply rooted in communications I had with George Katsiaficas with respect to an earlier draft of this paper.

72. This makes a certain amount of sense, particularly given the anthropological focus on the biosocial evolution of gender relations during Paleolithic times. A criticism of Bookchin's work that will not be fully developed here can be generated out of his insistence that fully formed human beings and societies existed before the institutionalization of domination. The predominant view within anthropology is that the capacity for social institutionalization emerged dialectically with the biological evolution of *Homo sapiens* and that the vast diversity of the content of social institutions is relatively indeterminate, neither inherently mutualistic nor competitive and usually and necessarily flush with dialectical, rather than oppositional, moments of each.

73. These last contrasted relations are taken from Bookchin's discussion of Engels's technodeterminism in *Toward an Ecological Society,* pp. 126–127.

74. The following discussion of mutualism and symbiosis is drawn from an excellent review article by Douglas H. Boucher, Sam Jones, and Kathleen H. Keeler: "The Ecology of Mutualism," *Annual Review of Ecology and Systematics,* Vol. 13, 1982, pp. 315–347.

75. Ibid., p. 316.

76. Ibid., p. 325.

77. "Whithout data on *proportions* of taxa or individuals that are mutulaistic, we can say little" (Boucher, Jones, and Keeler, "The Ecology of Mutualism," p. 328).

78. To date, the most thoroughgoing approach to dialectical ecology remains the work of Richard Levins and Richard Lewontin (*The Dialectical Biologist* [Cambridge, MA: Harvard University Press, 1985]). The extraordinary work and reporting of Stephen Jay Gould, most thoroughly advanced in *Wonderful Life: The Burgess Shale and the Nature of History* (New York: W. W. Norton, 1989), provides another perspective on the many-leveled nonlinearity and nonprogressivity of structurally constrained and participatorally contingent evolution.

79. *Ecology of Freedom,* p. 360.

80. Ibid., p. 357.

81. Levins and Lewontin, *Dialectical Biologist,* p. 17.

82. Ibid.

83. *Ecology of Freedom,* pp. 353–354.

84. Levins and Lewontin, *Dialectical Biologist,* p. 21.

85. Ibid., p. 22.

86. Gould, *Wonderful Life,* p. 14.

87. For Gould, the "history of life is a story of massive removal followed by differentiation within a few surviving stocks, . . . the current earth may hold more species than ever before, but most are iterations upon a few basic anatomical designs" reduced from those represented in the pre-Cambrian fossil record (*Wonderful Life,* pp. 25, 47).

Chapter 10

Evolution and Revolution
The Ecological Anarchism of Kropotkin and Bookchin

DAVID MACAULEY

Introduction

Echoing Murray Bookchin's earlier explorations of the affinities between anarchism and ecology, Kirkpatrick Sale once asked, "What better understanding of the liberatory possibilities of humankind could the ecologist get than from the anarchist; what better understanding of the liberatory character of the natural world could the anarchist get than from the ecologist?"[1] In the following essay, I will examine some of the contributions to an ecoanarchist tradition that, I argue, antedates, influences, and presently remains coextensive with Bookchin's own work. It is to this tradition which Bookchin returns, from which he departs, and occasionally over and past which he looks. In so doing, I situate Bookchin's thought historically against the work of Peter Kropotkin in particular and try to illuminate some of the tendencies and tensions within both anarchist and ecological theory. By showing how closely related Bookchin's work is to Kropotkin's opus, I hope to help restore the importance of Kropotkin for critical social theory and to stress his relevance for the future of social ecology. If, as I claim, comprehending Kropotkin is central to understanding Bookchin, then it would bode well for future social ecologists to investigate this earlier body of work as they advance and refine their theory. This comparative and critical enterprise is valuable ultimately because it allows us to better understand the origins of, changes in, and complexities concerning social ecology, and to further develop informed responses to environmental and political challenges.

As John Clark maintains, the defining features of anarchism as a political theory are, first, a view of an ideal society that is noncoercive and nonauthoritarian; second, criticism of present conditions from the perspective of the ideal; third, a characterization of human nature (or, following Hannah Arendt, the human condition) that justifies hope in progressing toward the ideal; and fourth, a strategy for political change that involves immediate use of decentralist, nonauthoritarian alternatives.[2] This definition is broad enough to be inclusive of most forms of anarchism yet specific enough to exclude those views that are not radically libertarian in nature. An *eco*anarchist perspective is an anarchist view that locates its analysis in the close relation it finds between the domination of humans by each other and the idea or practice of controlling the natural world. In addition, it frequently challenges reformist, narrowly anthropocentric, patriarchal, and bureaucratic institutions and ideas within society and prevailing forms of "environmentalism." It offers, in turn, a radical ecological perspective based upon coexistence with the natural world, appropriate or liberatory technology, decentralization, and organic ways of living and thinking. The ideal set forth is typically one of cooperation rather than competition, autonomy as opposed to authority, and human scale instead of hierarchy. Generally, an ecoanarchist perspective presents a vision of social and ecological harmony based upon notions of local place, region, or community; mutual aid; voluntarism; and direct action. Like much anarchist philosophy, it combines utopian aspirations with practical proposals, attempting to transcend traditional left/right divisions and to preserve the most valuable insights of the conservative and the radical.[3] In the forthcoming sections, I will summarize, compare, and analyze the views of Kropotkin and Bookchin in terms of their shared or respective positions on ecological mutualism, nature and human nature, ethical naturalism, anarcho- communism, evolution and revolution, decentralism and regionalism, and radical agriculture. I conclude by briefly relating their work to the ecoanarchist tradition and its future.

Mutualism in Natural and Social Ecology

Bookchin's association of ecological and anarchist ideas appears initially and most clearly in his 1965 essay, "Ecology and Revolutionary Thought." In this essay, he sketches the similarities in outlook of the two perspectives, arguing that the science of ecology is both critical and reconstructive (and one is tempted to add, with Paul Sears, "subversive"). He remarks, for example, that

> both the ecologist and anarchist place a strong emphasis on spontaneity. The ecologist, insofar as he is more than a technician, tends to reject the notion of 'power over nature.' He speaks, instead, of 'steering' his way through an ecological situation, of managing rather than recreating an ecosystem. The anarchist, in turn, speaks in terms of social spontaneity, of releasing the potentialities of society and humanity, of giving free rein to the creativity of people. Both, in their own way, regard authority as inhibitory, as a weight limiting the creative potential of a natural and social situation. Their object is not to rule a domain but to release it. . . . To both the ecologist and the anarchist, an ever-increasing unity is achieved by growing differentiation. . . . Just as the ecologist seeks to expand the range of an ecosystem and promote a free interplay between species, so the anarchist seeks to expand the range of social experience and remove all fetters to its development.[4]

Here and elsewhere, Bookchin extends the comparison to include a fundamental opposition to hierarchy and an orientation toward wholeness, complementarity, unity-in-diversity, spontaneity, and decentralism. In developing this connection and in forging the links between natural and social ecology, Bookchin looks to nineteenth-century libertarian utopians and anarchists who conceived of freedom as rational, community as necessary, and nature as a ground for the ethical while challenging their scientism and technological optimism. He lays a new emphasis in his analysis upon hierarchy (rather than class), freedom (not only justice), and domination (as opposed to exploitation). He also returns to Aristotle, Hegel, and the organismic tradition of the West for his *Naturphilosophie,* but argues for a more open teleology and a more radical conception of democracy and political life. Finally, Bookchin plumbs the revolutionary traditions for living and viable ideas that might inform present-day situations.

Perhaps as much as (and maybe more so than) any other figure, the mutualistic, naturalistic, and communitarian ideas of Peter Kropotkin find expression, reflection, and extension in Bookchin's work, where they are illuminated often as much indirectly as directly.[5] Briefly, Kropotkin (1842–1921) was a nineteenth-century Russian field naturalist, geographer, and revolutionary anarchist who pioneered advances in the social and natural sciences, including the fields of geology and zoology.[6] Kropotkin contested the stress within evolutionary theory upon notions of competition and struggle, placing new emphasis on cooperation, spontaneity, and mutual aid in animal and human societies. Works such as *Fields, Factories, and Workshops* (1899), *Mutual Aid* (1902), and *Ethics* (1924) influenced not only the development of Bookchin's social ecology but also the thought of Gandhi, Martin Buber, Lewis Mumford, Patrick Geddes, E. F. Schumacher, Ebenezer Howard, and Paul Goodman, among others.

Mutual Aid, one of Kropotkin's most significant philosophical and scientific contributions, was written as a response and corrective to Thomas Huxley's essay "The Struggle for Existence: A Programme," in which Huxley had interpreted Darwin in terms that emphasized strife as a necessary and inevitable condition for progress.[7] Kropotkin looked for evidence to support the view held by most Darwinists (but not always Darwin himself[8]) that struggle between animals of the same species was the dominant factor in evolution. What he *found* through field studies in Siberia was evidence for intraspecific support and cooperation and the inspiration for a philosophical and political anarchism that could be based, in part, upon this discovery. What he *lost* was a belief in the sole primacy of strife and competition as characteristics in the progressive development of life, along with the faith that the state and its bureaucracy could be useful to the majority of people. Kropotkin saw how a version of Darwinism was "socialized" (and sociologized) to provide ideological support for a laissez-faire philosophy by validating it as a "law of nature" that applied as well to human communities (a phenomenon that is arguably recurring with the advent and development of sociobiology).[9]

Kropotkin recognized clearly that there is struggle against "inclement Nature"—adverse circumstances caused, for example, by snow, rain, heat, or drought—that serves to control overpopulation and in which most species must engage. However, he failed to find much evidence for struggle against other members of the same species. Contrary to some interpretations, Kropotkin considered mutual aid as only *one* of the chief factors in evolution. And although he often suggests that mutual aid may be the most important factor, it is not the only significant influence.[10] Some of the most lucid and explicit statements of the role of mutual aid occur early and repeatedly in the work:

> As soon as we study animals—not in laboratories and museums only, but in the forest and prairie, in the steppe and the mountains—we at once perceive that though there is an immense amount of warfare and extermination going on amidst various species, and especially amidst various classes of animals, there is, at the same time, as much, or perhaps even more, of mutual support, mutual aid, and mutual defence amidst animals belonging to the same species or, at least, to the same society. Sociability is as much a law of nature as mutual struggle.[11]

Shortly thereafter he remarks that

> mutual aid is as much a law of animal life as mutual struggle, but that, as
> a factor of evolution, it most probably has a far greater importance, inas-
> much as it favours the development of such habits and characters as in-
> sure the maintenance and further development of the species, together
> with the greatest amount of welfare and enjoyment of life for the individ-
> ual, with the least waste of energy.[12]

Kropotkin's observations and studies of animals included land
crabs (who cooperate in migrations and sometimes rescue overturned
companions), ants (who regurgitate food for fellow ants and combine
in nests), bees (who work in common), pelicans (who fish in bands),
and house sparrows (who share food). Among mammals, Kropotkin
found association and mutual aid to be the rule. He remarks upon dogs
and jackals (who hunt in packs), deer, gazelle and ibex (who watch
over the safety of the herd), seals and walruses (who are given to mutual attachment and a striking sociability), and buffalo and reindeer
(who migrate in herds instead of competing when food is in short supply). According to Kropotkin, as one ascends the "scale of evolution,"
one sees association becoming more conscious, less instinctive, and
more reasoned.[13]

Sociability proper, too, is said to be the distinctive feature of the
animal world.[14] "The fittest are thus the most sociable animals, and sociability appears as the chief factor of evolution, both directly, by secur-
ing the well-being of the species while diminishing the waste of energy,
and indirectly, by favouring the growth of intelligence."[15] Kropotkin
attests to instances of compassion between injured animals and claims
that such feeling is the first step in the direction of developing higher
moral sentiments. From this evidence, he adduces a double-edged injunction that is at once proscriptive and prescriptive:

> "Don't compete!—competition is always injurious to the species, and you
> have plenty of resources to avoid it!" That is the *tendency* of nature, not
> always realized in full, but always present. . . . "Therefore com-
> bine—practise mutual aid! That is the surest means for giving to each and
> to all the greatest safety, the best guarantee of existence and progress,
> bodily, intellectual, and moral."[16]

In his treatment of early and primal peoples, Kropotkin extends his
views to consider mutual aid in the lives of humans. He criticizes Hobbes's picture of natural man and the state of nature, and he notes the
continued recourse to Hobbes's ideas.[17] Kropotkin argues that the (nuclear) family arises late in human evolution, pointing out that few

"higher" mammals exist in small families or live alone. The message is, in short, that bands, tribes, communities, and the concomitant cooperation that these groupings entail have been significant in the evolutionary process. Kropotkin mentions or explores the Papuas of New Guinea who live without chiefs in a kind of "primitive communism," the Polynesians who exist in a state of near social harmony, and the Eskimos who are closely interdependent and who—like other tribes and races—periodically abandon debt or redistribute land in order to reestablish an equality of condition. Such people consider individual acts—even death—as tribal or community matters and so regulate their actions by unwritten "rules" that emerge through common experience. There exists not only shared understandings but shared belongings, communal marriages, and frequently the rule of "each for all." In this regard, he shows clearly that individualism is a *modern* not a primitive or premodern phenomenon. In brief, Kropotkin proceeds to present a social history that traverses the emergence, breakdown, and persistence of mutual aid practices, expanding his analysis to encompass the medieval period—which represents the flowering of his conception of community—and later the development of collective opposition to the evisceration of such institutions and structures in the form of popular revolts, strikes, and labor unions.

In many respects, Bookchin's *The Ecology of Freedom* is a modern-day equivalent to Kropotkin's classic work. Bookchin likewise explores the importance of beliefs and practices that are similar to or part of the genus, mutual aid, locating them within their respective social contexts. These include *usufruct* (the freedom in a community to appropriate items based on use, need, and function rather than ownership),[18] the *irreducible minimum* (assurance of the material means to subsistence despite one's contribution), and the *equality of unequals* (recognition of differences between humans along with compensation for inequalities). Bookchin finds each to be present—often unconsciously—in most organic cultures. Taken together, such practices help to maintain solidarity inside human society and to encourage consociation with the natural world. Bookchin points out that Kropotkin's mutualistic naturalism can also be applied morphologically, that is, among and within cellular forms. He cites William Trager on the importance of symbiosis and mutual cooperation between different types of organisms, finding Trager's biological judgment to contain an ethical component like Kropotkin's observations as a naturalist and an anarchist.

In this regard, symbionts are examples of forms of life that call into question our common belief in distinct organismic boundaries and independent functions by exhibiting tendencies toward mutual dependency, symbiosis, and cooperation. Lichen, for example, are composed of mutually supportive algal and fungal components, whereby the former provide photosynthesis and the latter organic structure and nutritive capacities. Thus, it is reasonable to ask whether a lichen is a discrete plant or a cooperative. Similarly, colonial organisms such as the Portuguese man-of-war consist of a community of individuals who perform distinct functions so as to maintain the whole.[19] Even humans partake of this biological ambiguity in that we exist in close cooperation and dependence upon intestinal bacteria and organelles that arrive in our cells as colonists and replicate independently. Having surveyed bacteria, insect, and sea life, biologist Lewis Thomas concludes in a like vein to Kropotkin and Bookchin that "most of the associations between the living things we know about are essentially cooperative ones, symbiotic in one degree or another; when they have the look of adversaries, it is usually a standoff relation, with one party issuing signals, warnings, flagging the other off."[20] On this point, Kropotkin's own words were prescient:

> And when a physiologist speaks now of the life of a plant or an animal, he sees an agglomeration, a colony of millions of separate individuals rather than a personality, one and invisible. He speaks of a federation of digestive, sensual, nervous organs, all very intimately connected with one another, each feeling the consequence of the well-being or indisposition of each. . . . Each organ, each part of an organ in its turn is composed of independent cellules which associate to struggle against conditions unfavorable to their existence. The individual is quite a world of federations.[21]

Bookchin claims, too, that mutualism is an "intrinsic good" because it promotes the development of natural variety, and he remarks on the possibility of extending it along lines suggested by Lynn Margulis, who found the practice to be central to the development of aerobic life.[22] In later work on the evolution of cities as ecocommunities, he presents the provocative idea that civic participation is a *social* analogue to biological mutualism just as civic history and citizenship are social counterparts to natural history and biotic involvement in an ecosystem, respectively.[23] However, while turning to *natural ecology* to inform and guide his *social ecology*, Bookchin disavows the idea that the latter can be reduced to (or crudely derived from) the former as, he asserts, the urban sociologists of the Chicago School attempted to do in the early twentieth century, or some sociobiologists have tried to do more recently. He maintains, rather, that social ecology *evolves* and emerges from natural ecology as a graded and mediated process involving differentiation, development, and transcendence. We will return to this matter when I consider the naturalistic ethics of Kropotkin and

Bookchin and the issue of whether they engage in the naturalistic fallacy. Before doing so, however, it is necessary to look first at their conceptions of nature and human nature, upon which ecological anarchism and social ecology are based.

Nature and Human Nature

Bookchin and Kropotkin each premise their ecoanarchist views on a conception of nature and human nature, as have most major political theorists. Kropotkin, who acknowledges that he studies human society from a biological perspective, claims that sociability and a need for mutual aid are intrinsic to human beings. In his view, we have been and always will be a social species, although he does not go so far as to assert explicitly that we are *zoon politikon* in the Aristotelian and later Arendtian senses.[24] In one important way, Kropotkin seeks to recover a conception of nature and humanity that is based on the idea that we (and many nonhumans) are inherently *social* animals, whereas Bookchin tries to recover the older, more fundamental insight that we are *political* animals whose life resides in public speech and action with other citizens of a community. In both cases, human community is thus entirely natural, whereas the state—which weakens or destroys tribal, group, and communal bonds—is not. According to Kropotkin, inequalities and egoism arise through a secondary drive of self-assertion that may lead one to seek out power over others. As noted earlier, Kropotkin challenges Hobbes's view of human nature for being unduly competitive and egoistic and his representation of the state of nature as a "war of all against all," but he also criticizes Rousseau's conception of a natural "love, peace and harmony" (as he puts it) since neither is an "impartial interpretation of nature."[25] He admits to the reality of competition, but he gives primacy to cooperation, especially as a normative goal.

Similarly, for Bookchin, human nature is real, biologically grounded, and formed through an organic process that involves consociation. From this perspective, love, cooperation, and mutual aid are as much natural attributes as they are cultural ones, and human personality and individuality are constituted through socializing, group work, and participation in a common culture. Nature remains present throughout this entire development as culture elaborates cooperative or associative tendencies in the natural world.[26] In his later work, Bookchin distinguishes further between *first* or biological nature (the natural world), *second* nature (culture and human communities), and their transcendent synthesis into a *free* nature and ecological society, which presumably would allow human nature to be realized fully and permit human culture to exist in a balanced relationship with the organic world.[27]

Concerning the merit of these underpinnings and approaches, there is much controversy. Arendt, for example, has argued that attempts to define human nature mistakenly treat a *who* as though it were a *what* and usually end in the flight to supernaturalism, theology, or deification:

> The problem of human nature . . . seems unanswerable in both its individual psychological sense and its general philosophical sense. It is highly unlikely that we, who can know, determine, and define the natural essences of all things surrounding us, which we are not, should ever be able to do the same for ourselves—this would be like jumping over our own shadows. Moreover, nothing entitles us to assume that man has a nature or essence in the same sense as other things. In other words, if we have a nature or essence, then surely only a god could know and define it.[28]

To avoid this problem, she proffers the notion of *human condition* (which includes as well those self-made things that enter into, alter, and condition our existence) in contradistinction to *human nature*. In this regard, neither Kropotkin nor Bookchin appeal to a deity or a transcendental realm, though Bookchin's analysis often takes on messianic qualities.

A related criticism is that an ecoanarchist position that presents humans as cooperative by nature (though affected adversely by artificial, cultural hierarchy and competition) conflates our *essential* nature with our *potential* (better) nature, constructing model institutions or practices upon an unrealized (or even unrealizable) ideal. In so doing, the argument goes, it falls prey to an essentialism that cannot account fully for or accommodate actions that do not fit the assumptions of solidarity or cooperation (e.g., antisocial behavior).[29] The briefest sketch of possible responses by an ecoanarchist to this line of criticism might be to defend claims and positions such as (1) our essential nature is in fact our *unrealized* potential nature; (2) we have not yet achieved a free society in which our true nature can emerge fully and be displayed; (3) in small, voluntary communities such as those formed or envisioned by anarchists, harmony, mutual aid, and cooperation are (would be) the norm; (4) social revolution could usher in a society in which avarice and strife would exist only in easily manageable forms; and (5) public opinion, social censure, and civic morality would be able to handle most cases of "antisocial" or egoistic behavior.

More generally, many anarchists have found in *nature*—not only in *human nature*—guiding inspiration for human sociation, and so a very "natural" connection between anarchism and ecology has emerged. For example, it has been argued that Taoism has a close affinity with both anarchist and ecological sensibilities since it locates humans completely within nature, and it is fundamentally antiauthoritarian, holistic, and radically critical of existing society.[30] About other historical examples, we find Bookchin arguing that, "the concept of living close to nature lent Spanish Anarchism some of its most unique features—vegetarian diets, often favoring uncooked foods; ecological horticulture; simplicity of dress; a passion for the countryside; even nudism."[31] It is no surprise, then, to find Kropotkin lauding the virtues of becoming nature-literate and of being "familiarised with the forces of Nature which some day [one] will have to utilise."[32] Woodcock has remarked of this association that anarchists have often looked to peasants as revolutionary figures since they are thought to be close to the earth and nature, and thus to be more "anarchic" in their responses. This turn toward "the natural, the spontaneous, the individual, sets [the anarchist] against the whole highly organized structure of modern industrial and statist society."[33]

The easy embrace of ecology by anarchists along with the early and continued influence of anarchism in the politicization of ecology thereby appear more understandable. In other words, it was neither accident nor opportunism that Kropotkin developed some of the first theoretical and practical links between biological field studies and a libertarian social vision or that Bookchin was able to discern the revolutionary potential of a nonmechanistic ecological science and to forge some of the initial ties with emerging social movements. The perspective of the oft-cited and influential ecological "Blueprint for Survival" (1972)[34] has also been compared with Kropotkin's much earlier work in terms of their stress on the need for small ecological communities, for example, even if differences exist between the two outlooks (e.g., the former relies more on governmental action).[35] Finally, it has been observed that Kropotkin

> laid the conceptual foundation for a radical theory of human ecology. He viewed nature and people in nature as organic, interrelated wholes—the actions of any one part affecting all other parts. Imbalances which exist in nature thus reflect imbalances which exist in human relationships.[36]

As we shall see next, Kropotkin and Bookchin appeal as well to nature (and a particular conception of it) as a ground for their ethics.

Ethical Naturalism

Kropotkin's *Ethics* can reasonably be read as an extension of *Mutual Aid* and an application of the latter book's ideas to the field of morality, which, he argues, is based upon solidarity and sociality. In this work, he attempts to ground ethics on a naturalistic basis, to show that nature retains a distinct moral dimension, and to locate mutual aid, solidarity, and justice as elements innate to human nature. Like Bookchin, Kropotkin argues that we can learn from a "sound philosophy of Nature" and, furthermore, derive our ideals from it. In the process, he points us back to the ancient Greeks (again like Bookchin) for guidance. He finds a new, widened conception of life emerging that allows us to conceive of matter as alive and undergoing the same cycles of growth and decay as living beings, a view which harks back to pre-Socratic *hylozoism*.[37] Kropotkin argues that a study of nature can provide us with an *explanation* of the sources of morality, including their rational origin, but he urges that we proceed further to find a *justification* for such a naturalistic ethic that takes as its end something *in life* rather than something outside it—that is, transcendental. "Ethics must demonstrate how moral conceptions were able to develop from the sociality inherent in higher animals and primitive savages, to highly idealistic moral teachings."[38] Regarding this issue, he believes it is possible to conceive of human history as the *evolution* of an ethical factor and an inherent tendency on our part to organize social life on the basis of mutual aid.

In *Ethics,* Kropotkin continues his critique of the view that nature is "red in tooth and claw." He recapitulates his argument for mutual aid *within* species as the predominant factor in progressive evolution and the preservation or welfare of species, calling this principle a "permanent instinct" always at work. Mutual sympathy contains the rudiments of moral conscience, permitting feelings of benevolence, group identification, and eventually a sense of justice or equity to emerge. The moral sense, moreover, is described as a natural faculty like our senses of sight, smell, or touch. Kropotkin, in fact, goes so far as to assert that *"the very ideas of bad and good,* and man's abstraction concerning 'the supreme good' have been borrowed from Nature."[39] They are reflections in the human mind of what has been observed of animal life.[40] Nature, he claims, must be seen as "the *first ethical teacher of man."*[41] On this point, Kropotkin devotes considerable attention to showing how primitive cultures exhibited a deep sense of kinship with, respect for, and understanding of the natural world. Nonhuman creatures in particular imparted feelings of belonging, consociation, and communication, along with "lessons" like "carnivorous beasts . . . never kill one another" and "the strongest beasts are bound to combine."[42]

Thus, unlike many writers on ethics who begin with the postulate that self-preservation is the strongest instinct or who proceed to derive their philosophies from what lies outside nature, Kropotkin (like Bookchin) begins *within* nature. When the former path is taken, "the triumph of moral principles . . . represent[s] a triumph of man over nature."[43] Although his *Ethics* makes little mention of anarchism, it is implied in related essays that a true ethics—that is, an evolutionist ethics—would likely be a kind of "anarchist morality" because, first, solidarity arises from "equality in mutual relations" (an anarchist ideal)[44] and, second, his conception of morality maintains a deep relation to a radical political vision. Kropotkin's maxim, "Without equity there is no justice, and without justice, there is no morality,"[45] can be compared favorably not only with Bookchin's discussion of the interrelationships between equality, justice, and freedom,[46] but also with the orientation of the pre-Socratics, who often grounded their cosmologies in a radical *isonomia* (equality) and appealed to naturalistic (rather than divine) conceptions of justice.[47]

At different times, Kropotkin lends greater or lesser emphasis to solidarity, sympathy, sociality, and instinct as constituents or bases of morality. However, it is not always clear what weighted role each plays and how they stand with respect to one another. Regarding instincts, he distinguishes three—social, parental, and comradely—of which the first is the earliest and strongest. When individual and social instincts conflict, the former yields to the latter. For three reasons, Bookchin distances his own perspective from such theories of instinct which seek to confirm mutualism. First, they often offer vague concepts in the place of serious arguments and explanations. Second, they can devolve easily into sociobiology. Third, they are frequently based on a selective survey of animals, ignoring in Kropotkin's case, for example, many solitary animals or advanced mammals, and confusing animal groupings such as herds, packs, troops, or communities with societies.[48]

Unfortunately, Kropotkin's other contributions to ethics—especially his essay "Anarchist Morality"—are sprinkled with spurious claims such as animals have conceptions of good and evil of the same kind as humans, odd invocations of utilitarian-like criteria (which consider what is useful to the preservation of a race as "good"), and remarks like the morality of animals can be summed up in a golden rule. His *Ethics* frequently refers to "laws of nature" that supposedly validate morality, and it exhibits a certain pretense to a "scientific ethics." Despite these weaknesses, Kropotkin develops an early case for a naturalistic ethics that is grounded in the organic world and that must be viewed in relation to a more encompassing radical social critique and vision.

One can observe similarities between Kropotkin's views and the work of Ernst Haeckel (1834–1919). Haeckel was a contemporary of the Russian anarchist and a figure who is given wide credit for coining the term "ecology" and for developing the field as a holistic nonmechanistic science. For Haeckel, nature implied a spirit of freedom. He viewed the natural order as progressive and characterized nature as benevolent. Like Kropotkin, he tried to derive human ethics from animal instincts. From his observations of sympathy and altruism in the natural world, Haeckel concluded that cooperation is the norm for both nonhuman and human animals.[49] Among anarchists, rough analogies to Kropotkin's notion of mutual aid can also be found in William Godwin's idea of "universal benevolence" and Tolstoy's use of love.

Bookchin opens *The Ecology of Freedom* with a quote from Kropotkin's *Ethics* that likewise stresses a double tendency in nature and human history toward the increasing intensity of life and a corresponding development of sociality, phenomena which Bookchin interprets together as "symbiotic naturalism." Thus, Bookchin signals his debt to Kropotkin and his desire to extend critically the Russian revolutionary's work. However, while both writers advance a form of naturalistic ethics, there are significant differences in what they accent, appeal to and conclude. Bookchin distinguishes, for example, more clearly than Kropotkin between the human and the nonhuman and their respective capacities (or lack thereof) for forming societies and making ethical decisions. Kropotkin maintained that intellectual differences between animals and humans were ones of degree (rather than of kind) so that one might admit the collective intelligence of an ant's nest or a beehive; acknowledge that horses, dogs, and other animals who live closely with humans have moral conceptions; and recognize that humans cannot judge the "morality" of worker-bees when they kill drones. Bookchin sees this kind of extensionism and projection as ill-founded and conceptually muddled. He claims that insects such as ants and bees are genetically programmed, that institutions and societies (as opposed to communities) are unique to human beings,[50] and that human animals (unlike nonhumans) possess the capacity for rational, abstract, and ethical reflection (although Bookchin appeals at times to an animistic sensibility that might belie some of these distinctions).[51]

Like Kropotkin, Bookchin advances a case for an objective ecological ethics[52]—one rooted in the natural order of things—within an evolutionary perspective, extending and deepening some of Kropotkin's claims or advancing on their simplicity in an attempt to develop a form of *dialectical naturalism*. Dialectical naturalism "ecologizes" the dialectic. It attempts to understand in fluid and processual terms the phenomena of evolution within the framework of a rational interpretation.[53] Dialectical reason, in turn, unlike instrumental and analytic reason, views reality as developmental rather than as static, acknowledging that Being is forever Becoming and that entities pass beyond themselves into a stage of otherness that retains former levels.[54] This general perspective is, of course, much earlier and more fully articulated in Hegel's idealism and Marx and Engels's materialism. Bookchin, however, interprets and develops Hegel's dialectic along naturalistic and organic lines. He rejects a cosmic *Geist* (spirit) and the culmination of the dialectic in an "Absolute." He modifies the role given to strife (by emphasizing as well reconciliation) and attends more to nature's "existential details" than had Hegel. In the same way, Bookchin criticizes Engels's appeal to rigid "natural laws," which are based in an outmoded physics and mechanical science, and he offers an alternative to Marx's images of matter and labor (the latter is the counterpart of the former), which are divested of or divorced from ethical substance and subjectivity.

According to Bookchin, dialectical naturalism provides an objective basis for ethical judgments in that "what-is" (an actuality or Hegel's *Realität*) can be measured against "what-it-should-be" (a fulfilled potentiality or Hegel's *Wirklichkeit*) in terms of the criteria of rationality and morality. In more Aristotelian language, dialectical naturalism is anchored in a natural ontology and causality that sees matter as informed with a latent potentiality and a *nisus* that allows it to strive toward an actualized form, which can be guided by intrinsic or extrinsic forces toward the realization of an immanent, though not predetermined, *telos*. This process contains a meaning that, in Bookchin's interpretation, is not only directive and purposive but rife with ethical content in its commitment to wholeness, completion, and fulfillment.

Kropotkin, it must be noted, rejected both teleological explanation and the dialectical method as it was applied to natural science. On these matters, a significant difference exists in the approaches of the two anarchists. Kropotkin argued that the only viable method was the scientific one, which proceeds through the use of induction and deduction. He described dialectics as antiquated, unable to produce a new nineteenth-century discovery in physics, chemistry, biology, psychology, anthropology, or astronomy. Unlike Bookchin, he does not find Hegel's work to be of value to the naturalist. Kropotkin, too, denies the merit of teleological factors and metaphysical frameworks. He agrees with Darwin that nature evinces no certain signs of evolution governed according to "preconceived aims" or as operating by a "guiding power." In this regard, Kropotkin's conception of anarchism is more of a mechanical—specifically, kinetic—interpretation of natural phenomena than a processual one.[55]

Kropotkin, however, spoke of a harmony in nature that is discoverable by the human mind[56] and maintained that when the feeling of solidarity surfaces that "it is the whole evolution of the animal kingdom speaking in us."[57] Despite his differences with Kropotkin on the value of dialectical and teleological thought, Bookchin recalls and develops older ideas of this kind in an attempt to recover reason as a self-organizing attribute of nature that allows us to "know nature within nature."[58] Such an understanding is made possible because humans are (at least potentially) nature become self-conscious, the "knowingness" in effect of the organic world. We are the very "embodiment of nature's evolution into intellect, mind and self-reflexivity."[59] Like Kropotkin, he claims that evolution exists not only around and about us but also within us.[60] Therefore, we *participate in* and can contribute *creatively* and *consciously* to its movement toward increasing complexity and variety (instead of destructively hindering its development and fostering simplicity and uniformity).[61] Our interventions into the natural world are thus "inherent and inevitable" and imply minimally human stewardship of the earth, ideas for which he has been challenged by some other ecological thinkers and activists, especially those associated with deep ecology.[62]

For Bookchin, as for many of the ancient Greeks (especially the pre-Socratic philosophers), the cosmos is laden with ethical meaning; it shows an ethical "grain" of sorts that can provide a "common ethical voice" for humanity and the natural world. Nature itself is not an ethics, though it can act as a *matrix* or *ground* for one, providing us with an "ethical ontology." "Nature," rather than humans, "is writing its own nature philosophy and ethics," argues Bookchin, advancing a claim that he has since qualified by the words "metaphorically speaking."[63] By this position, he means that reason, freedom, and selfhood (as well as human culture) should be seen as emergent from and intimately bound with (not opposed to) a first nature that exhibits a fecundity and latent, self-elaborating intentionality.

In articulating his ontological ethics, Bookchin counterposes it sharply to utilitarian,[64] relativistic,[65] and Kantian[66] positions that he critiques as inadequate. He is especially concerned to show the manner in which such a "libertarian ethics" (as he calls it at one point) is (1) tied to a conception of *techne* that recognizes limits of form and is rooted in natural evolution, (2) expressed in an economy that is rational and moral, and (3) embedded in a community and society that are nonhierarchical, radically democratic, and ecological. He is aware that one should be cautious in selecting nature as an ethical ground because of past attempts that provided ideological support for slavery, hierarchy, or oligarchy. Nevertheless, the issue as to whether such an ethic participates in the naturalistic fallacy—in which one mistakenly derives a valuational or prescriptive "ought" from a factual or descriptive "is"—is complex but needs to be mentioned.[67]

Social ecology's first reply to such a charge might be to deny the applicability of the fallacy to its version of naturalistic ethics, which conceives of reasoning as entailing a dialectical logic (A equals both A and not-A), not an analytic logic (A equals A). In dialectical logic, there are no "brute facts" without a history and a future.[68] What *is* must be seen in relation to what *was* and what is *becoming*, a process that implicitly delivers a criterion of judgment in terms of the fulfilled or unfulfilled potentiality. Second, social ecology grounds its ethic in a view of evolution that, though supported by biological and cultural evidence, is an assumption of sorts as well. To judge—even critically confront and assail—the current, dominant expressions of "second nature" (hierarchy, the state, capitalism, class, private property, etc.), Bookchin admits, makes sense "*only* if we *assume* that there is potentiality and self-directiveness in organic evolution toward greater subjectivity, consciousness, self-reflexivity."[69] Such an assumption appears to be warranted for Bookchin not only because of the weight of support it finds in evolutionary biology and natural ecology (the arguments and evidence in these fields are extremely contentious),[70] but also because of the drawbacks of nonnaturalistic ethics and denatured conceptions of reason, the failings of liberalism and socialism to create a rational ecological society and, finally, the utopian promise of social ecology's own vision. Since an image of nature offers us as well an image of society, it would behoove humanity to "choose" one that is not only compatible with but also contributive to social goals such as equality, freedom, or natural and cultural diversity.[71]

Third, just as social ecology is arguably not reducible to natural ecology nor second nature analyzable into a prior first nature, neither is a distinctly human ethics simply "plucked" out of or built upon blind, amoral natural laws and facts. Moreover, nature does not teach us easy "lessons" (to use Kropotkin's language), and what might be deemed "natural" is not ipso facto universally "good." Instead, an ethics is *educed*—rather than empirically *induced* or formally *deduced*—from the natural world in the sense that potentialities and possibilities are articulated and made manifest to the human mind and culture as what *should be*.[72] Whereas the naturalistic fallacy expresses itself in forms involving either the *definition* of value predicates in factual and empirical claims or the *deduction* of value judgments from a set of facts, the ecological ethics of social ecology neither appeals ultimately to brute, frozen facts and immediate, perceivable reality nor uses a hypothetico-deductive method (e.g., syllogistic reasoning) to derive an "ought" from an "is." Rather, such an ethics locates its sources and problems within a historical process from which it emerges, where boundaries are fluid and developmental transitions are significant.[73]

From State to Community: Anarcho-communism

In order to better understand the views of Kropotkin and Bookchin on ethics, human nature, and mutualism in natural and social ecology, we must examine their more encompassing commitments to and defenses of a form of anarcho-communism, which they distinguish from traditional socialism. As part of a shared vision, Bookchin follows Kropotkin in offering (1) a deep critique of the state and a critical recovery of an older, radically democratic conception of community; (2) an appeal to a particular conception of revolution in order to realize a free, nonauthoritarian society; (3) the advocacy of decentralist principles and local or regional forms of social and political organization; and (4) a stress on a new form of radical agriculture and its integration with small-scale industry (subjects considered in turn in the following sections). Like the areas discussed earlier in this essay, Bookchin often departs from Kropotkin's ideas, provides needed revisions of anthropological and biological findings, or makes advances on simplistic, scientific, or technophilic formulations. But one is also struck by the degree to which Bookchin's perspective remains part of a theoretical lineage that is continually informed or inspired by earlier work on (r)evolutionary anarchism.[74]

Anarcho-communism denotes for Bookchin a term that by definition denies the merit of any claim to domination and one that preserves the libertarian element in a viable revolutionary project.[75] Kropotkin's similar theoretical perspective on the subject is developed in *The Conquest of Bread* (1892) and numerous pamphlets.[76] In short, anarcho-communism is based on notions of voluntary stateless cooperation, collectivization of the means of production, abolition of the wage system, and free distribution, a concept and practice with roots older than anarchism itself. The tradition of anarcho-communism to which Kropotkin and Bookchin belong must be contrasted with individualist (e.g., Stirner and Tucker), syndicalist (e.g., Sorel and trade unions), mutualist (e.g., Proudhon), pacifist (e.g., Tolstoy), and collectivist (e.g., Bakunin) forms. In Kropotkin's first discussions of the idea, he emphasized the need for such a revolution to be carried out through and based upon local communes. He points out that capitalism leads not to *overproduction* but to *underconsumption,* a phenomenon that would be eliminated in a free society that encourages meaningful work and a new scheme of distribution. Kropotkin endeavored specifically to find a scientific basis for anarcho-communism, which he described as a nongovernmental system of socialism, a nonauthoritarian school of free communism, and a synthesis of economic and political freedom. For him, anarchism was in line with and an ultimate expression of the philosophy of evolution.[77]

With respect to a critique of the state, Kropotkin argued that it is not only the destroyer of creativity and initiative, but that for three hundred years the state systematically eliminated groups and institutions such as tribes, villages, fraternities, and guilds in which mutual aid practices flourished. It increasingly absorbed social functions and encouraged the rise of a narrow individualism. "In proportion as the obligations towards the State grew in numbers the citizens were evidently relieved from their obligations towards each other."[78] In a broad conclusion that later finds expression in the urban and technological narratives of Lewis Mumford and Bookchin, Kroptokin observed that throughout history there have been two opposed tendencies locked in conflict, which he names variously as the imperialist and federalist traditions, the Roman and popular traditions, or the authoritarian and libertarian traditions. In his view, the state, with its centralizing, antidemocratic orientation,[79] belongs clearly to the former of these poles.

Contrary to the emphasis of some anarchists historically, Bookchin argues that forging a free and ecological society is not just a matter of overthrowing the state. It also requires the creation or recovery of liberatory institutions (like those Kropotkin celebrated) and the reconstitution of human relationships on communal bases. This further step is necessary because the state has colonized and absorbed social life, just as it has bureaucratized and politicized the economy. The state is not simply a complex of political, military, or bureaucratic institutions; it also has a psychological history and fosters a distinct epistemology or a *state of mind* that derives from its own bureaucratic or militaristic form of organization. Its emergence—a slow evolution rather than a sudden eruption or revolution—was predicated, as Kropotkin had argued, upon the reworking of organic or traditional cultures and customs into forms that allowed for social domination. Its appearance was prefigured by the rise of warrior societies, priestly corporations, and political professionals. Bookchin, however, attempts to distinguish his own perspective slightly from that of Kropotkin (and Bakunin) whom, he asserts, saw the state either as "historically necessary" or as an "unavoidable evil." But there does not appear to be much substantiation for this claim in Kropotkin's case.[80]

Both Kropotkin and Bookchin differentiate between *society* and *state* and then further (at least clearly in Bookchin's case) between *community* and *society*, even if the two realms have been fused together in recent history.[81] Whereas Kropotkin looked to the medieval commune (or city) for his normative model, Bookchin turns more to the Hellenic polis.[82] In both instances, they seek to revive and revitalize the community as civic and cooperative sphere.[83] For Kropotkin, the medieval city emerges from the village community, which was a "universal phase of evolution" that had encouraged common fishing, hunting, fruit culture, and agriculture as an outgrowth of the earlier *gens* or clan. The medieval city, however, was not a state in modern terms because there existed neither centralization of functions nor territorial centralization, even though it did have the right to form alliances, to wage war, and to declare peace. Instead, the city was divided into numerous quarters or sections which often corresponded to a particular profession or trade. The city was akin to a "double federation" consisting of households belonging to small, local unions such as the section or parish, and of individuals united to guilds. As Kropotkin argued, the commune, too, was a voluntary association that brought together common interests rather than an extension of local government. In uniting with other communes, it formed a sphere of cooperation that was radically different from the centralized state. The availability of goods in such a sphere was based upon *need* rather than *contribution,* a feature that distinguishes Kropotkin's outlook from that of some other anarchists.

The guilds were federated with small village communities to form the city (the guild of guilds), and they possessed their own military forces. They were also marked by brotherly feelings and self-jurisdiction; each person referred to others as "brother" and "sister," and all were seen as equals. According to Kropotkin, the guild answered to a need inherent in human nature, and it was characterized by all those features that the state later appropriated for its police and bureaucracy. Comprehended as a whole, the commune was thus *mutui adjutorii conjuratio,* an oath of mutual aid, and a wholly natural organism. As Kropotkin puts it:

> The mediaeval city . . . was not simply a political organization for the protection of certain political liberties. It was an attempt at organizing, on a much grander scale than in a village community, a close union for mutual aid and support, for consumption and production, and for social life altogether, without imposing upon men the fetters of the State, but of giving full liberty of expression to the creative genius of each separate group of individuals in art, crafts, science, commerce, and political organization.[84]

The commune thereby exercised a great moral authority upon the people. Both nobles and ecclesiastics had to submit to the folkmote (people's general assembly), a principle embodied in the saying, "Who enjoys here the right of water and pasture must obey." Like Bookchin, Kropotkin often stresses the importance of the *ethical* and *utopian* dimensions to political and social practices.[85] For example, he notes that the ethical significance of communal possessions was greater than their economic value, and indicates that they functioned to maintain traditions and practices of mutual aid that served as checks upon the unbridled growth of avarice and individualism. He points out further that laborers enjoyed both prosperity and respect, observing that "not only many aspirations of our modern radicals were already realized in the middle ages, but much of what is described now as Utopian was accepted then as a matter of fact."[86]

According to Kropotkin, the dissolution and decline of community life began with the control exercised increasingly by lay or clerical lords over the villages. Medieval citizens made the error of not reaching out to newcomers and of basing their wealth upon industry and trade, resulting in the neglect of agriculture. Another cause of decline was a change in the chief ideas and principles that governed city life, including federalism, self-reliance, and sovereignty of all groups. Social and political salvation was sought with greater frequency in a centralized state, which was eventually thought to have near-divine authority. Apart from these major changes and the appropriation of communal lands by central authorities, some mutual aid practices nevertheless were maintained over time, and Kropotkin chronicles the attempts to reconstruct community on surviving forms of solidarity such as communist fraternities or the common use of the soil. Most importantly for present-day problems, Kropotkin concludes from his historical analysis that mutual aid should not exist solely in small associations. Rather, such groups must expand their ideas and practices to surrounding areas, or risk the possibility of being absorbed from outside.

Despite the many merits of his accounts, first, as balancing correctives to the views of Huxley, Hobbes, and other social historians and, second, as providing a plausible normative model for social change, Kropotkin did not seem to recognize very fully the frequent oppressiveness of custom and tradition in closely knit primitive societies, ancient towns, and medieval cities. He appears either to tacitly condone or, more likely, to naïvely overlook the possibility of forms of moral and cultural authoritarianism, however much they might have been mitigated by democratic and egalitarian practices. In his wide-ranging anthropology of anarchism, Harold Barclay has called attention to some tendencies within the medieval guild system toward a class orientation, rule by wealthy patricians, exploitation of apprentices (of whom strict obedience was often demanded), and the increasing development of a population of wage-earning proletarians.[87] Bookchin also alerts us to the fact that customs can be "highly opaque emotional sanctions" whereby the governing "morality" mystifies and hides a formerly coherent egalitarian orientation.[88] In this respect, Kropotkin's "ecocommunitarianism" is vulnerable to some of the charges that beset the visions of certain of the "new communitarians."[89]

Bookchin, too, charts a parallel but more complex history of political, social, and private institutions and practices, placing greater emphasis than Kropotkin on the changing roles of hierarchy and reason, the near hegemony of capitalism and a market economy, the development of modern industrial technics and science, and the failings of Marxism, liberalism, and representative government. As part of his plan for an ecological society, he argues for the importance of the polis as a guiding community model (as well as such experiments as Swiss communes and New England towns) and explores the possibility of creating *ecocommunities* and a new libertarian municipalism. The polis especially serves as a guide for reconstructing society along communitarian lines because of its organic emergence from prior forms like the family and village, its sense of aesthetic and ethical balance (intellect and body, town and countryside, art and craft), and its distinct commitment to both direct democracy (each citizen had a rule in turn in the *Ecclesia* or popular assembly) and a common nonprofessional, nonbureaucratic political life. As Hannah Arendt has observed, it embodied a unique sphere of civic freedom in which relationships were conceived as equal and public decisions were made in an agonistic *and* cooperative space.[90]

Moreover, as Bookchin argues, the *poleis* were characterized—at least ideally—by the well-roundedness of their citizens, who received a moral training through direct political participation. A further significant feature of the polis as a model for ecological communities is the fact that it encouraged *autarkeia* (self-rule and self-sufficiency) for the individual while limiting its own tendencies toward growth and expansion, toward passing beyond human scale into a *mega-polis* or *cosmopolis* in effect. "So long as the *polis* grows without losing its sense of unity, let it grow—but no further," wrote Plato.[91] Aristotle lent his support to this view as well, remarking that the citizens of a community must be familiar with each other's characters so that "the best limit of the population of a *polis*" is that which "can be taken in at a single view."[92]

The concept of ecological communities draws upon and extends the outlook of natural ecology where, for example, soil and forest communities can be identified and located within a broader environment. In Bookchin's view, ecocommunities would cultivate and retain the most desirable features of the Athenian polis, the medieval commune, or seventeenth-century New England townships. As he puts it in an early characterization that recalls Kropotkin, they would be further supported by a rational ecological technology that is scaled to local dimensions and help as well to restore an equilibrium between land and town, agriculture and industry.[93] An ecocommunity "must describe a decentralized community that allows for direct popular administration, the efficient return of wastes to the countryside, the maximum use of local resources—and yet it must be large enough to foster cultural diversity and psychological uniqueness."[94]

Revolution and Evolution

Closely bound with Kropotkin's and Bookchin's antiauthoritarianism (especially an opposition to statism) and communitarianism (more exactly, civic republicanism in Bookchin's case) is a notion of radical social change. The concept of revolution has been bound since its entrance into political theory with natural philosophy and particularly with astronomy, a natural science that studies the changing movements and cycles of the heavens and their seeming return to a point of origin.[95] Nineteenth-century theorists, moreover, often characterized nature in terms of an *evolutionary* process, and some like Kropotkin went a step further to present humans as part of a larger organic development that can progress by fits and bounds and not just slow growth.[96] Both Kropotkin and Bookchin link major shifts and advances in science with analogous and related upheavals in society.[97] In particular, evolution and revolution are interwoven for Kropotkin, while ecology and revolution are deeply imbricated for Bookchin.

Contrary to Bakunin's conception of revolution as apocalypse (an image Kropotkin invoked in the early 1880s), Kropotkin spoke later of revolutions as natural processes, as "periods of accelerated rapid *evolution*" that "belong to the unity of nature as do the times when evolution takes place more slowly."[98] An evocative and a provocative link between Kropotkin's evolutionary and revolutionary views is discernible in the similar language he applies to both the biological and political realms, particularly the notion of "direct action." Direct action—"propaganda by deed" as it has sometimes been called—is an idea and practice that is associated closely with anarchists. It is one that underscores the autonomy of individuals, illustrates the efficacy of collective activity, and reveals or even reverses the deeper relations of power. Kropotkin, however, also increasingly described "the direct action of the environment" as a major factor in evolution, in producing, for example, the adaptions in the outer forms and inner structures of plants.[99] In so doing, he questioned the more popularized overemphasis on natural selection, struggle, and chance variations (or unknown causes). Instead, he imparted a greater degree of influence to climate, temperature, air composition, food, sun level, and surrounding physical conditions—in short, one might say, to the earth itself—than had been previously acknowledged. When such an idea is understood in or translated into a social and political context, it is possible to see the manner in which the weight of authority, tradition, hierarchy, or fatalistic belief could be challenged by new developments in science. In brief, revolution might be a natural extension of evolution, facilitatible in part by direct action.

In Kropotkin's view, though, individual revolutionaries do not create revolutions; instead, they might guide, unite, and relate the efforts that begin with the people themselves. The revolution is a concrete event in which the participants achieve heightened consciousness of their own actions. Revolution should not end in a "revolutionary government" (which is self-defeating) but progress toward the goal of complete equality. Here one can note the similarity to Bookchin's view that "revolutionaries have the responsibility of helping others become revolutionaries, not of 'making' revolutions."[100] This activity, too, is only possible when the revolutionary herself or himself undergoes a process of radical self-liberation since deep changes in personal life must necessarily accompany revolutionary action. In fact, Bookchin suggests that revolution might be defined as the most developed kind of self-activity, "as direct action raised to a level where the land, the factories, indeed the very streets, are directly taken over by the autonomous people."[101]

In arguing for the necessity of radical social change, Bookchin emphasizes the value of revolutionary *ethical* resistance (to support a social and political stance, create a moral economy, and combat opportunism), direct action (as a sensibility not simply a tactic), affinity groups (rather than vanguards or mass actions), as well as creativity and spontaneity in group actions. In his view, revolution must be cultural as well as social and must challenge hierarchy and domination in *all* areas of life (social, political, personal, and economic) and in *all* its forms: humans over the environment, men over women, the old over the young, and so on. To date, however, there has been relatively less emphasis in Bookchin's work upon racial questions, matters of gender, international problems, or issues related to environmental justice than other theoretical or substantive areas. The utopian dimension to change is especially important to Bookchin, and he comments repeatedly upon and warns against the dangers of cooptation, conflicts internal to radical movements, and manipulation by "managerial radicals." Most recently, he has decried and decelebrated the avoidance of social revolution (or its reduction) by many anarchists, who have opted mainly for forms of "lifestyle" revolt or "metaphysical rebellion," to use Camus's phrase.[102] At the same time, he has challenged anarchists and the Left to embrace and articulate coherent ecological, feminist, and municipalist projects, especially those with a "libertarian ambiance."[103]

As he matured, Kropotkin emphasized the evolutionary and peaceful aspects of social resistance and reconstruction, as opposed to *traditionally understood* revolutionary dimensions. Increasingly, he tried to develop anarchist thought in an ethical direction rather than into a social or political program. Upon this change, Kropotkin reflected: "I gradually began to realize that anarchism represents more than a mere mode of action and a mere conception of a free society; that it is part of a philosophy, natural and social, which must be developed in a quite different way from the metaphysical or dialectical methods which have been employed in sciences dealing with men."[104] Woodcock argues that this shift in focus was in line with Kropotkin's temperament, his renewed interest in science, and his contact with the socialist movement in England, which retained libertarian impulses. In formulating his view of revolution, Kropotkin succumbed to the lure of productionism (a bête noire for many greens today), claiming that social revolution must first increase production and then permit increased consumption. However, in advancing his ideas, Kropotkin also helped to wed a conception of revolution with both ethical concerns and a vision of a nonhierarchical society.[105]

Decentralization and (Bio)regionalism

In developing their ecoanarchist perspectives, Kropotkin and Bookchin also turn to forms of decentralist and local or regional organization. In *Fields, Factories, and Workshops,* Kropotkin argued for the abolition of the sharp distinction between city and country through a creative combination of small-scale industry and agriculture. He advocated as well a "synthesis of human activities" that would overcome the separation of society into intellectual and manual labor. Kropotkin stressed the need for integration and balance—for the individual, the economy, and society—and underscored the importance of regionalism and decentralization (territorial and functional) in industry. His new definition of political economy as "a science devoted to the study of the needs of men and of the means of satisfying them with the least possible waste of energy"[106] is implicitly attentive to the ecological dimensions of the *oikos.* To this end, Kropotkin criticized neglect and waste of the land; advocated a new, radical agriculture; and envisaged farms along a horticultural and garden model that would be labor-intensive and support more people in the process.

In particular, Kropotkin rejected *division* of labor in favor of the ideal of *integration,* just as he sought a whole, balanced personality through progressive education[107] and a healthy, stable, and complex biological and social environment. The intellectual, agricultural, and industrial aspects of social life were to be harmoniously combined:

> A society where each individual is a producer of both manual and intellectual work; where each able-bodied human being is a worker, and where each worker works both in the field and the industrial workshop; where every aggregation of individuals, large enough to dispose of a certain variety of natural resources . . . produces and itself consumes most of its own agricultural and manufactured produce.[108]

In Kropotkin's view, we must find the best means to combine agriculture with manufacturing, but we must do so on a local and regional level so as to promote self-sufficiency. In this sense, he is critical of a global market economy that encourages dependencies of knowledge, skill, and goods and allows for ruthless competition, low wages, and myopic specialization. His message, like that of Bookchin, is both (*bio*)*regional* and *radically egalitarian.* He seeks to return power to localities. "Progress must be looked for in another direction. *It is in producing for home use.*"[109] Kropotkin prophesied that the paramount problem for Europe in coming years would be to return to an arrangement where food is grown and goods are manufactured for the use of those who produce them. In other words, each region is to become both its own producer and its own consumer.[110]

Bookchin, too, lauds the constructive possibilities of decentralization, finding in the small or human scale not just evidence of the beautiful (as Schumacher observed) but also of the *emancipatory.* He argues that the best case for decentralization occurs in Kropotkin's work because it views the human community as a kind of modern polis that admits of a great degree of "ecological integration of town and countryside, a highly flexible technology and communications system, a revival of artisanship as a productive form of 'aesthetic enjoyment,' and direct local democracy freed of the social ills, notably slavery, patriarchalism, and class conflict."[111] Bookchin points out, however, that appeals to decentralization (or even democracy) are meaningless unless accompanied by radical, nonhierarchical social changes as well as communal living and working relations:

> To demand "decentralization" without self-management in which every person freely participates in decision-making processes in every aspect of

life and all the material means of life are communally owned, produced, and shared according to need is pure obscurantism. . . . To leave questions like "who owns what" and "who runs what" unanswered while celebrating the virtues or beauties of "smallness" verges on demagoguery.[112]

He cautions us that decentralization can be divorced quietly by planners from both specific ecological technologies (e.g., wind or solar energies) upon which the notion is materially based, and separated further by technocrats from a broader, integrating radical social theory of which it is part. Decentralization is not merely a stratagem or "logistical alternative" to centralization and gigantism but a necessary element of direct democracy.[113]

In an attempt to clarify the practice and idea of decentralization, Langdon Winner has elaborated upon some of Bookchin's concerns and caveats, arguing that one must look at a variety of elements in order to assess its advantages or drawbacks. Such issues include the number of centers in question, their location, their relative power, and their diversity or vitality. He concludes that decentralization no longer necessarily implies a radical demand for social reconstruction, as it did in Kropotkin's time. Rather, it is more likely to mean something like "let us place greater faith in people's ability to make plans, shape policies, and manage their own public affairs."[114]

Robyn Eckersley has argued that the ecoanarchist case for local autonomy and decentralization is problematic, first, because progressive social and environmental changes have often emerged from central state governments or international agreements. Thus, a "supraregional perspective" involving multilateral state action is also necessary. She maintains, second, that strong arguments exist against *complete* decentralization because urban environments accommodate large populations better than the countryside and provide a needed counterpoise to "rural romanticism" among many greens.[115] In response to these challenges, ecoanarchists and social ecologists might agree to *some* extent with the thrust of the second point because many of them—and especially Bookchin—explicitly include cities within a framework for ecological communities, libertarian municipalism, and a new relationship for town and countryside. Bookchin, however, also urges that cities themselves be decentralized (and ecologized), remarking that "this is no longer utopistic fantasy but a visible necessity."[116] Regarding the first point, it might be replied that Eckersley ignores the degree to which the state (in both socialist and capitalist forms) and its military, industrial, and technological complex have contributed directly and deeply to the ecological crisis through warfare, pollution, simplification of ecosystems, and the like. Moreover, it is still arguable that real, fundamental, and lasting change (as opposed to short-term reform) is only possible through a recovery of community, local self-management, and decentralization, as Bookchin has long maintained.[117]

On a related note, Bookchin and other ecoanarchists such as Kirkpatrick Sale have spoken approvingly of *bioregions* as placial frameworks for situating and relating human communities within the encompassing natural world.[118] Bioregions are a geographical and biological concept with political import. Bioregionalism has an intellectual lineage in Kropotkin's radical geography and "dissident" biology as well as in Mumford's ideas on regional planning. Kropotkin, for example, notes that most people "know nothing about whence the bread comes which we eat" and counsels us generally to familiarize ourselves with and to support the regions and communities where we dwell.[119]

Finally, the apparent tensions within the writings of Bookchin and Kropotkin—and within ecoanarchist thought more generally—between *decentering* (of authority or decision making) and *integration* (of city and country or agriculture and industry), between *devolution* (of power) and *evolution* (of complexity), and between *community/cooperation/equality* and *autonomy/freedom/liberty* find a potential resolution in an underlying philosophical, ecological, and organic *holism*. Kropotkin constantly underscores the value of completion, synthesis, and adaptive integration, remarking, for example, that "we must inquire what is done with the territory taken *as a whole*."[120] Bookchin follows Hegel's dictum that "The True is the whole"[121] (not just a mystical oneness) and insists upon theoretical coherence, unity-in-diversity, and rounded individuals and communities, while challenging reductionism and dualism in its intellectual and institutional forms. Holism denotes for him a conscious attempt to determine the way in which a community is arranged qualitatively and how its structure has a history and logic that makes it more than the mere sum of it constituent parts, and more than an undifferentiated totality or universal. Holism involves differentiated development, actualization of potentiality (to use Aristotelian terms), and eventuation in *completeness*.

Radical Agriculture

Another major area of shared concern for the two thinkers is a new vision for our use of cultivated land. Kropotkin argued that the soil of Western Europe (and Great Britain in particular) could provide for more than the needs of its population through a more labor-intensive agriculture. He thereby sounded a theme that was to become a clarion call of the movement for ecology seventy years later, even if his estimates were overly optimistic. He pointed out how agriculture was neglected and noted that "land is going out of culture at a perilous rate,"[122] implying the increased severance of *ager* from *cultura,* field from farming community, and nature from human society. He named as *social* and *political* causes of such problems the concentration of land ownership in the hands of the few, high profit margins, the development of reserves for sport hunting, and the absence of institutions to spread practical knowledge about agriculture. In their stead, he argued for the necessity of intensive market-gardening, fruit culture, and the use of greenhouses, thus anticipating and encouraging these developments. His approach to agriculture is modeled in large part on a view of horticulture that seeks to develop farms along the lines of vegetable gardens. Associated labor, in turn, is the solution to growing more and working less, and it is to be supported by the use of machines and the application of experimental techniques in soil fertility and plant and animal breeding. Changes in agriculture must be accompanied by a new relation to industry and urban life. Agriculture is in great need of assistance from those who live in cities, observed Kropotkin, refusing to side with either an isolated romantic pastoralism or a global, urban cosmopolitanism.

As Colin Ward has noted, Kropotkin still adhered to Veblenite assumptions about the production of food, and he did not understand fully the economics of agriculture under capitalism.[123] He was wrong in his prediction that industrial nations would return to labor-intensive agriculture, since capital-intensive corporate agribusiness has been the rule in the twentieth century and many hundreds of thousands of small farmers have been driven out of their livelihood.[124] Ward adds that Kropotkin

> had a mechanistic nineteenth-century attitude to the land, was as cavalier about crop rotation systems as the modern British cereal farmer, and would probably have regarded contemporary factory-farming methods as just another indication of the indefinite expansion of production on a given area of land. Those who are worried about questions of conservation and pollution in the countryside will have little difficulty in pinpointing the farmer as the most serious polluter, through the use of herbicides, fungicides and pesticides and through the discharge of untreated effluents, and would ask whether the modern extension of Kropotkin's ideas would depend upon the exploitation and exhaustion of the soil.[125]

Nevertheless, Kropotkin's goal of creating "garden cities"—an idea further developed and popularized by Ebenezer Howard[126]—and of finding "what can and ought to be obtained from the land under a proper and intelligent treatment" still remains with us.[127]

Bookchin employs the Kropotkinian language of integration and communitarianism in advancing his comparable case for a new form of agriculture.[128] Radical agriculture would be an extension of early farming practices with their animistic sensibilities and sacred rituals, and it would belong to a "moral economy" that views the land as an *oikos* of living bacteria, insects, plants, and animals. This perspective stands in contrasts to industrial *agribusiness,* which belongs to a "market economy" that treats the earth as resource to be exploited.[129] Voicing themes that have been hallmarks of his social ecology, Bookchin celebrates radical agricultural practices for recognizing biotic communities, helping to restore human communities, and promoting complexity, stability, and natural and social variety. Such an orientation is not premised simply upon new ecological techniques. Instead, it is grounded in a "new non-Promethean sensibility" that views agriculture as a kind of culture, encouraging us to reinhabit the land rather than seeking escapist agrarian refuges upon it.

Bookchin, it should be noted, first considered the dangers of chemicals in food in the early 1950s and explored problems related to the environment, agriculture, and health in the 1960s, commencing with *Our Synthetic Environment.* This work shortly preceded Rachel Carson's better known *Silent Spring* and also exceeded it in terms of its deeper social and political analysis. At the same time, it helped to give rise to Bookchin's radical social ecology.[130] In this writing, he identified problems related to soil deterioration; spoke of the dangerous use of pesticides, hormones, and antibiotics in agriculture; and found links between environmental degradation and cancer. The concluding chapters, too, sketch urban regionalism and decentralized human scale communities as necessary elements in the search for plausible social and political directions.

Further Similarities and Differences

In terms of further similarities between Bookchin and Kropotkin,[131] one could note briefly, first, their unified opposition to capitalism in all its forms, whether state, free-market, or some combination thereof. For Bookchin, capitalism is a "social cancer"—"the greatest disease society ever suffered"[132]—one that is "inherently anti-ecological,"[133] responsible for undoing natural evolution, simplifying the organic and social worlds, promoting unrestrained technological growth, and destroying institutions and practices involving mutual aid and civic participation. Furthermore, the most significant contradictions of capitalism lie not within the system itself but between it and the natural world.[134] In a like vein, Kropotkin argues that the state is linked closely with capitalism, an association that promotes exploitation of labor, social injustice, atomistic individualism and political inequality.

Second, they both oppose mysticism, presenting it as an incoherent and incommunicable form of knowledge about the world.[135] Bookchin especially finds mysticism an ever-present danger in the ecology movement—particularly among deep ecologists—because it is often based upon a facile and fatuous opposition to dialectical reason, a rejection of important social concerns, and an appeal to an undifferentiated "oneness" that can lead easily into authoritarian, quietistic, or even ecofascistic positions.[136]

Third, both are extremely critical not only of Hobbesian characterizations of nature and human nature but of Malthusian perspectives on social issues. Kropotkin criticizes Malthus for providing a pseudo-scientific justification of inequality and notes the way sociology and biology have often been intertwined to the detriment of society. Bookchin, in turn, directs his polemic against neo-Malthusians, who fail to acknowledge the social origins of global population, hunger, and ecological problems, and who promulgate a dangerous antihumanism that sometimes border on misanthropy.[137]

Fourth, Kropotkin hints at and suggests what in Bookchin's early work becomes thematized as a "postscarcity" (i.e., freedom from scarcity) social perspective. Kropotkin holds that political economy adheres falsely to the belief in the impossibility of expanding productivity and satisfying wants, expressing his own productionist faith in the process. He argues that "political economy never rises above the hypothesis of a limited and insufficient supply of the necessities of life."[138] Kropotkin claims that for the first time in history humans are at a point where the means for satisfying needs exceed the needs themselves and that well-being can be achieved by all without placing the burden of oppressive or degrading work on anyone,[139] thereby exhibiting a technological optimism as had Bookchin in early essays like "Toward a Liberatory Technology."[140] He asks further:

> Are the means now in use for satisfying human needs, under the present system of permanent division of functions and production for profits, really *economical*? . . . Or, are they not mere wasteful survivals from a past that was plunged into darkness, ignorance and oppression, and never took into consideration the economical and social value of the human being?[141]

Kropotkin claims that—contrary to most other perspectives—the anarchist distinguishes between *real* wants and social and political forces (wars, conquest, ignorance) that inhibit or prevent the satisfaction of these needs.[142]

Bookchin's position on scarcity, postscarcity, and biological and social needs is developed in *Post-Scarcity Anarchism*, clarified in *Toward an Ecological Society*, and then defined in *The Ecology of Freedom*, where he argues that scarcity can be induced socially and occur even with material abundance. He points out that *wants* are often no longer related to *needs*, which themselves have been fetishized. He believes that in a free society needs would be formed through rational choice, autonomy, ethical deliberation, and consciousness (not through productive forces). A postscarcity society would reject "false" or "dehumanizing" needs because society itself would be largely free of need, having enough material resources and social freedom to engage in willful self-limitation.

In terms of differences of consequence between Bookchin and Kropotkin beyond those previously noted, Bookchin criticizes the traditional anarchist concept of contract and exchange which, he claims, is found in Kropotkin's work.[143] According to Bookchin, it is based upon equivalence and eventuates in "bourgeois conceptions" of right.[144] Bookchin finds this notion of equity to be foreign to a deeper and older notion of freedom, and he challenges the restricted concept of freedom implied in contractual language. "The higher conception of 'no revenge for wrongs,' of freely giving more than one expects to receive from his neighbors, is proclaimed as being the real principle of morality—a principle superior to mere equivalence, equity, or justice, and more conducive to happiness."[145] In "Anarchist Morality," however, Kropotkin opposed the model of mere equity in trade relationships, arguing that "something grander, more lovely, more vigorous than mere equity must perpetually find a place in life." He implied further that it would be "greater than justice."[146] Thus, to some extent, they both gesture toward an idea of *supererogation* in ethics—that is, one that goes beyond duty, need, or obligation. Bookchin also remarks that Kropotkin was often a technological determinist who was enamored with economic progress.[147] On this point, there is more textual support since Kropotkin's technological and scientific optimism is apparent throughout his work. He believed, for example, that "mankind [was] entering upon a new era of progress," and like others of his era, he often celebrates uncritially the "wonders of industrial technique."[148]

Conclusion: The Ecoanarchist Tradition

Although their views diverge on selected ideas and issues, it should be clear that Bookchin shares much with Kropotkin in terms of advancing a radical social critique of existing authoritarian and state institutions, advocating a philosophical and ethical naturalism grounded in evolutionary theory, developing local and decentralist strategies for social change, and supporting communitarian institutions and ideals for rural and urban life. We are, no doubt, *very* far as a society from their respective and collective visions of integrated work, an ethical economy, human-scale technics, and ecological communities. Since Kropotkin's time, decentralization has been an exception rather than a norm with the rise of corporate industry and agribusiness. Capitalism has revealed an unimagined capacity to adapt to or coopt libertarian challenges and critique. The integrity of the land, waters, and air are constantly threatened by petrochemicals, monoculture, clear cutting, species extinction, and human overpopulation. Cities and communities are assaulted more and more by an unbridled urban sprawl. High technology seems to recognize few (if any) ethical limitations, and even natural ecology and evolution are confronted with unprecedented losses of complexity, variety, and stability.

In his assessment of Kropotkin's efforts, George Woodcock perhaps goes a bit too far when he writes that Kropotkin's "irrepressible optimism, his exaggerated respect for the nineteenth-century cult of evolution, his irrational faith in the men of people, deprived him of true scientific objectivity," suggesting that "his real contribution was rather the humanizing of anarchism."[149] As Ward implied by appending a final word to the title of *Fields, Factories, and Workshops,* Kropotkin's ideas were projected into a *tomorrow.* Unfortunately, this day has not yet arrived.

As to the influence of Kropotkin's work, Lewis Mumford's magisterial and informative social histories of urban life, technics, and utopian thought can be viewed in certain respects as constructive and critical extensions of some of Kropotkin's most suggestive ideas on face-to-face community relations, integrated education, regionalism, decentralization, and "green" or "garden" cities. They also serve as a prelude to and inspiration for some of Bookchin's work on alternative technology, municipalism, and social ecology. Like Kropotkin, Mumford explores the medieval commune and cloister for the triumph of *communitas* over *dominium,* the wide freedom granted to cities, and the measure of democratic participation, autonomy, and corporate equality that they achieved. And like Bookchin, Mumford celebrates the Greek polis for its ideals of citizenship, formative integration with village and countryside, democratic principles, public culture, and connection with the *agora* (marketplace). Of Kropotkin, Mumford observed:

> Almost half a century in advance of contemporary economic and technical opinion, he grasped the fact that the flexibility and adaptability of electric communication and electric power, along with the possibilities of

intensive, biodynamic farming, had laid the foundations for a more decentralized urban development in small units, responsive to direct human contact, and enjoying both urban and rural advantages. . . . With the small unit as a basis, he saw the opportunity for a more responsible and responsive local life, with greater scope for the human agents.[150]

In fact, given Mumford's orientation and concerns, it is arguable that he belongs as well to the ecoanarchist tradition as a libertarian historian and a "forgotten environmentalist."[151]

It should be mentioned as well that groups like the Institute for Local Self-Reliance and New Alchemy have been influenced directly by the ideas of Kropotkin and Bookchin, particularly in their emphasis on ecological technologies, decentralization, and self-governing communities. Kropotkin's studies and theories, too, have found a second life in the ecological "Blueprint for Survival"; continually inspired grassroots, anarchist, and environmental movements; and provided a healthy countervailing addition to evolutionary theory. Bookchin, in turn, has developed a theoretically sophisticated critique of the idea of the domination of nature (finding its origins in the domination that humans exercise over each other) and has done much to underscore or advance the critical and reconstructive potential of ecological thought and practices for political theory and engaged social activists.

In the preceding sections, I have highlighted some of the main areas and perspectives upon which Bookchin and Kropotkin share a general agreement. The thought of both figures is rooted in a tradition of ecological anarchism that provides a historical framework for situating and understanding, advancing and applying, or critiquing and superseding their work.[152] From the foregoing discussion, it should be evident, then, that social ecology as a form of ecological anarchism is much older and richer than Bookchin's writings alone. Without question, Bookchin has articulated and improved upon many of Kropotkin's ideas or formulations. As Bookchin himself has warned, "If anarchist theory and practice cannot keep pace with—let alone go *beyond*—historic changes that have altered the entire social, cultural, and moral landscape and effaced a good part of the world in which traditional anarchism was developed, the entire movement will indeed become what Theodor Adorno called it—'a ghost.' "[153] While acknowledging the truth of this conditional, it is still imperative to see that Kropotkin's contributions on mutualism and naturalism, regionalism and decentralism, evolution and revolution remain important to critical social theory and integral to developing a new politics of nature.[154] In order to address critically and to respond to our ecological and political challenges, we must recognize the extent to which the ecological anarchist tradition—which finds many of its earliest inspirations and insights in Kropotkin's wide-ranging body of work and its most recent expression in Bookchin's representation of social ecology—is still vital and relevant. Despite its drawbacks, this living, evolving tradition continues to provide a reservoir of revolutionary alternatives to present-day problems.

Acknowledgment

I wish to thank Andrew Light for suggestions that helped to improve the organization of this essay.

Notes

1. Kirkpatrick Sale, "Anarchy and Ecology: A Review Essay," *Social Anarchism,* Vol. 5, No. 2, 1985, p. 23.

2. John Clark, "What Is Anarchism?," in J. Roland Pennock and John W. Chapman, eds., *Anarchism: Nomos XIX* (New York: New York University Press, 1978). Bookchin identifies four similar principles of traditional anarchism in terms of an opposition to statism; advocacy of confederated, decentralized municipalities; commitment to direct democracy; and the goal of attaining a libertarian communist society. See Murray Bookchin, *Social Anarchism or Lifestyle Anarchism: An Unbridgeable Chasm* (San Francisco: AK Press, 1995), p. 60.

3. Although anarchism and ecoanarchism are usually associated with the political Left, Paul Goodman has remarked somewhat provocatively, "It is only the anarchists who are really conservative, for they want to conserve sun and space, animal nature, primary community, experimenting inquiry," employing *conservative* in a deliberately illicit manner. As to the degree to which anarchist theory is utopian, there is some dispute. Daniel Guerin, for example, argues that it is decidedly constructive rather than utopian; see Guerin, *Anarchism* (New York: Monthly Review Press, 1970), pp. 41–42. Kropotkin himself qualifies the tag of utopian in his *Encyclopedia Britannica* article, "Anarchism" (1905), by noting that the anarchist perspective is not built upon an a priori method but rather upon an examination of tendencies that are already at work.

4. Murray Bookchin, *Post-Scarcity Anarchism* (Palo Alto, CA: Ramparts Press, 1971), pp. 77–78.

5. Bookchin remarks that Kropotkin's contribution is unique because of "his emphasis on the need for a reconciliation of humanity with nature, the role of mutual aid in natural and social evolution, his hatred of hierarchy, and his vision of a new technics based on decentralization and human scale" (*The Ecology of Freedom* [Palo Alto, CA: Chesire Books, 1982], acknowledgments

to the book). Elsewhere, Bookchin has noted that "Kropotkin was perhaps one of the most far-seeing of the theorists I encountered in the libertarian tradition." ("Deep Ecology, Anarcho-syndicalism, and the Future of Anarchist Thought," in *Deep Ecology and Anarchism* [London: Freedom Press, 1993], p. 55).

6. On Kropotkin's life, see George Woodcock and Ivan Avakumovic, *The Anarchist Prince* (New York: Schocken Books, 1971), and Martin A. Miller, *Kropotkin* (Chicago: University of Chicago Press, 1976), as well as Kropotkin's own *Memoirs of a Revolutionist* (New York: Horizon Press, 1968).

7. Thomas Huxley, "The Struggle for Existence," in Appendix to Peter Kropotkin, *Mutual Aid* (Boston: Extending Horizon Books, n.d.). Originally published in 1902.

8. "Those communities which included the greatest number of the most sympathetic members would flourish best, and rear the greatest number of offspring," wrote Darwin in *The Descent of Man and Selection in Relation to Sex* (New York: Appleton, 1896).

9. For an overview of the debate about social Darwinism, see Richard Hofstadter, *Social Darwinism in American Thought* (Boston: Beacon Press, 1955). For perspectives on sociobiology, see Arthur L. Caplan, ed., *The Sociobiology Debate* (New York: Harper & Row, 1978). For Bookchin's view, see *Which Way for the Ecology Movement?* (San Francisco: AK Press, 1994), pp. 49–75.

10. In terms of intellectual influences, Kropotkin acknowledged a debt to a Professor Kessler, who in 1880 propounded a law of mutual aid that was said to exist along with a law of mutual struggle. He quotes Kessler in words that are similar to his own: "In the evolution of the organic world—in the progressive modification of organic beings—mutual support among individuals plays a much more important part than their mutual struggle" (*Mutual Aid*, p. 8). Kropotkin also noted that Goethe had speculated about such a general law, though he never carried out a specific study on the subject.

11. Kropotkin, *Mutual Aid*, p. 5. One should take note here of Kropotkin's use of distinctly human social terms such as "warfare" and "extermination" that are misleadingly read back into the natural world. To his credit, Bookchin warns repeatedly against this practice, arguing that in nature there is predation but not cruelty, physical pain but not suffering, and needs but not scarcity.

12. Ibid., p. 6.

13. For more recent support of the role of cooperation in evolution or among animals, see W. C. Allee, *The Social Life of Animals* (New York: W. W. Norton, 1938); and *Animal Aggregations* (Chicago: University of Chicago Press, 1931); as well as G. G. Simpson, *The Meaning of Evolution* (New Haven, CT: Yale University Press, 1951).

14. Again, Bookchin would disagree with using such terms, arguing that society "is *institutionalized* community, structured around mutable organizational forms that may range from totalitarian despotism to libertarian municipalism. As such, society is specific to human beings; indeed, an expression like 'social insects' is, from my standpoint, nonsensical and oxymoronic, conflating

a fixed, genetically programmed aggregation of animals with the developmentally structured consociation of humans" (*The Philosophy of Social Ecology* [Montreal: Black Rose Books, 1995], p. xii). For a characterization and defense of the social life of higher animals, see Adolf Portmann, *Animals as Social Beings* (New York: Harper & Row, 1961).

15. Kropotkin, *Mutual Aid,* p. 58.

16. Ibid., p. 75.

17. The turn to Hobbesian assumptions, arguments, and conclusions is prevalent today in writings on political ecology. See, for example, William Ophuls, *Ecology and the Politics of Scarcity* (San Francisco: W. H. Freeman, 1977). For a critique of this use of Hobbes, see Frank Coleman, "Nature as Artifact: Thomas Hobbes, the Bible and Modernity," in David Macauley, ed., *Minding Nature: The Philosophers of Ecology* (New York: the Guilford Press, 1996), and David Macauley, "Greening Philosophy and Democratizing Ecology," also in *Minding Nature: The Philosophers of Ecology.*

18. Bookchin distinguishes usufruct from mutual aid because mutual aid, like reciprocity and exchange, is "trapped within history's demeaning account books with their 'just' rations and their 'honest' balance sheets" (*Ecology of Freedom,* p. 51).

19. Neil Evernden, *The Natural Alien* (Toronto: University of Toronto Press, 1985), p. 38.

20. Lewis Thomas, *The Lives of a Cell: Notes of a Biology Watcher* (New York: Bantam, 1974), p. 6.

21. Kropotkin, "Anarchism: Its Philosophy and Ideal," in *Kropotkin's Revolutionary Pamphlets,* ed. Roger N. Baldwin (New York: Dover, 1970), pp. 118–119. In addition, Kropotkin suggested something akin to symbiosis in his observations that disparate animals aid one another in hunting or hiding.

22. Margulis's work, in turn, has been developed into the Gaia hypothesis, which speculates that the Earth functions as one organism that actively maintains itself. See Lynn Margulis, *Symbiosis in Cell Evolution* (San Francisco: W. H. Freeman, 1981); James Lovelock, *Gaia: A New Look at Life on Earth* (Oxford: Oxford University Press, 1979); and William Irwin Thompson, ed., *Gaia: A Way of Knowing: Political Implications* (Great Barrington, MA: Lindisfarne Press, 1987).

23. Murray Bookchin, *The Rise of Urbanization and the Decline of Citizenship* (San Francisco: Sierra Club Books, 1987), p. x.

24. It is important to note with Arendt that the Latin rendering of this definition into *animal socialis* wrongly removes the distinctly political understanding that the Greeks imparted to humans, replacing it with the "social," a term of Roman origin. Aristotle, of course, adds a second definition of humans as *zoon logon ekhon* (a living being capable of speech), which also loses much in the translation to *animal rationale.*

25. Kropotkin, *Mutual Aid,* p. 5.

26. Earlier in the same work in which this characterization appears, Bookchin writes that human nature also "seems to consist of proclivities and potentialities that become increasingly defined by the instillation of social needs" (*Ecology of Freedom,* p. 114).

27. See Bookchin, "Thinking Ecologically: A Dialectical Approach," in *Philosophy of Social Ecology*. The distinction is first briefly introduced in *Ecology of Freedom*, p. 279.

28. Hannah Arendt, *The Human Condition* (Chicago: University of Chicago Press, 1958), p. 10.

29. See Robyn Eckersley, *Environmentalism and Political Theory* (Albany: State University of New York Press, 1992), pp. 170–171, for one formulation of this objection.

30. E.g., see John Clark, *The Anarchist Moment* (Montreal: Black Rose Books, 1984), Chapter 7, and the special issue of *Journal of Chinese Philosophy*, Vol. 10, No.1, March, 1983. Bookchin appears to be critical of (perhaps even opposed to) the social, rather than the distinctly ecological, dimensions to this interpretation of Taoism; see *Philosophy of Social Ecology*, p. 104.

31. Murray Bookchin, *The Spanish Anarchists: The Heroic Years, 1868–1936* (New York: Harper & Row, 1977), p. 5.

32. Peter Kropotkin, *Fields, Factories, and Workshops Tomorrow*, ed. Colin Ward (New York: Harper & Row, 1974), p. 195. A new edition of this work, edited with commentary by George Woodcock, has recently been published by Black Rose Books.

33. George Woodcock, *Anarchism: A History of Libertarian Ideas and Movements* (New York: Meridian, 1962), p. 26. See also Eric Hobsbawm, *Primitive Rebels* (New York: Norton Library, 1959).

34. Edward Goldsmith et al., *Blueprint for Survival* (Reprint, Boston: Houghton Mifflin, 1972).

35. See Colin Ward, *Anarchy in Action* (London: Freedom Press, 1982), pp. 140–141, and David Pepper, *The Roots of Modern Environmentalism* (London: Croom Helm, 1984), pp. 188–191.

36. Myrna M. Breitbart, "Peter Kropotkin, the Anarchist Geographer," in D. R. Stoddart, ed., *Geography, Ideology, and Social Concern* (Totowa, NJ: Barnes and Noble, 1981), p. 139.

37. See John Ely, "Animism and Anarchism," for a discussion of the way in which anarchist and animist sensibilities dovetail, in John Clark, ed., *Renewing the Earth: The Promise of Social Ecology* (London: Greenprint, 1990), pp. 49–65.

38. Peter Kropotkin, *Ethics: Origin and Development* (New York: Dial Press, 1924), p. 67.

39. Ibid., p. 16; emphasis in original.

40. Specifically it was from animals that humans learned their first lessons in the "valorous defence of fellow-creatures, self-sacrifice for the welfare of the group, unlimited parental love, and the advantages of sociality in general" (*Ethics*, p. 21).

41. Ibid., p. 45; emphasis in original.

42. Ibid., p. 52.

43. Ibid., p. 43.

44. Kropotkin sometimes defines anarchism as the "morality of equality"; see "Anarchist Morality," in *Revolutionary Pamphlets*, p. 105.

45. Elsewhere he writes: "*Mutual aid—Justice—Morality* are thus the

consecutive steps of an ascending series, revealed to us by the study of the animal world and man. They constitute an *organic necessity* which carries in itself its own justification, confirmed by the whole of the evolution of the animal kingdom . . . and gradually rising to our civilized human communities" (*Ethics,* p. 31).

46. See Bookchin, *Ecology of Freedom,* esp. Chapters 6 and 7.

47. On the relation of the pre-Socratics and Empedocles in particular to nature philosophy, see David Macauley, "Be-wildering Order: Toward an Ecology of the Elements in Ancient Greek Philosophy and Beyond" (unpublished Ph.D. Diss., State University of New York at Stony Brook, 1998), ch. 2.

48. Murray Bookchin, "Deep Ecology, Anarcho-syndicalism, and the Future of Anarchist Thought," in *Deep Ecology and Anarchism,* p. 56.

49. E.g., see Ernst Haeckel, *The Riddle of the Universe* (New York: Harper & Row, 1900) and *The Wonders of Life* (London, 1905).

50. Animals "do not have consciously formed ways of ordering their communities that are continually subject to historical change" (*Rise of Urbanization,* p. 226).

51. Regarding Bookchin's remarks on animism, see *Ecology of Freedom,* pp. 98–101.

52. Bookchin distinguishes *morality* (standards of behavior not yet subjected to rational analysis) from *ethics* (which is subject to rational claims).

53. Bookchin, *Philosophy of Social Ecology,* p. 15.

54. "Dialectic is *development,* not only change; it is *derivation,* not only motion; it is *mediation,* not only process; and it is *cumulative,* not only continuous" (Bookchin, *Philosophy of Social Ecology,* p. 125).

55. See Peter Kropotkin, "Place of Anarchism in Modern Science" and "The Theory of Evolution and Mutual Aid," in *Evolution and Environment* (Montreal: Black Rose Books, 1995), pp. 51–55 and pp. 117–138.

56. Peter Kropotkin, "Anarchism: Its Philosophy and Ideal," in *Revolutionary Pamphlets,* p. 120.

57. Peter Kropotkin, "Anarchist Morality," in *Revolutionary Pamphlets,* p. 98.

58. Bookchin, *Ecology of Freedom,* p. 39. Kropotkin and Bookchin share a belief in the Enlightenment assumption that reality is orderly, rational, and knowable.

59. Ibid., p. 38.

60. Ibid., p. 23.

61. At one point, Bookchin says that humanity is *presently* a "curse" on natural evolution rather than its fulfillment.

62. Extending this controversial note, Bookchin adds: "Nature without an active human presence would be as unnatural as a tropical rainforest that lacked monkeys and ants" (*Philosophy of Social Ecology,* pp. 131–132).

63. Bookchin, *Ecology of Freedom,* p. 355, and *Which Way for the Ecology Movement?,* p. 63.

64. See Bookchin, *Ecology of Freedom,* pp. 164–165.

65. Murray Bookchin, "History, Civilization, and Progress: Outline of a Criticism of Modern Relativism," in *Philosophy of Social Ecology.*

66. See Murray Bookchin, "Toward a Philosophy of Nature—The Bases for an Ecological Ethics," in Michael Tobias, ed., *Deep Ecology* (San Diego, CA: Avant Books, 1985).

67. For the first elaboration of the naturalistic fallacy (which is adumbrated in Hume's *Treatise of Human Nature*) and a defense of a nonnaturalistic theory of ethical judgments, see G. E. Moore, *Principia Ethica* (Cambridge, UK: Cambridge University Press, 1903). For an overview of the debate, see W. D. Hudson, ed., *The Is/Ought Question: A Collection of Papers on the Central Problem in Moral Philosophy* (New York: St. Martin's Press, 1969).

68. Here one might recall as well Nietzsche's comment that the only things that can be defined have no history, or his claim that there are no facts proper, only interpretations of them. Nietzsche, however, was no friend of the dialectic, or objective ethics for that matter.

69. Bookchin, *Philosophy of Social Ecology*, p. 32; second emphasis added. This caveat is also expressed in *Ecology of Freedom*, where he writes: "If we assume that the thrust of natural evolution has been toward increasing complexity . . . a prudent rescaling of man's hubris should call for caution in disturbing natural processes" (p. 24).

70. A consideration of this matter requires a separate paper. For a range of positions, consult Ernst Mayr, *Toward a New Philosophy of Biology* (Cambridge, MA: Harvard University Press, 1988), esp. pp. 38–66 on the multiple senses of teleology; Ronald Munson, ed., *Man and Nature: Philosophical Issues in Biology* (New York: Dell, 1971); Richard Levins and Richard Lowontin, *The Dialectical Biologist* (Cambridge, MA: Harvard University Press, 1985); and C.H. Waddington, *The Ethical Animal* (Chicago: University of Chicago Press, 1967)—to name just a few works.

71. A dilemma arises, of course, in reconciling the claims and demands of objectivity with the "choice" of assumptions, even if it can be shown that some conceptions of nature are more liberatory than others.

72. See Murray Bookchin, "Ecologizing the Dialectic," in John Clark, ed., *Renewing the Earth*, p. 205.

73. For a critique of Bookchin's view of evolutionary ethics, see Robyn Eckersley, "Divining Evolution: The Ecological Ethics of Murray Bookchin," reprinted as the first part of Chapter 2, this volume. For Bookchin's response, see Murray Bookchin, "Recovering Evolution: A Reply to Eckersley and Fox," *Environmental Ethics*, Vol. 12, 1990, pp. 253–274. For a relevant discussion of Kropotkin, see Stephen Jay Gould's essay, "Kropotkin was no Crackpot," in *Bully for Brontosaurus* (New York: Norton, 1991).

74. It would be wrong, however, to conclude with Graham Purchase that "Bookchin has done *little more* than update these [Kropotkin's, Fourier's, and Reclus's] ideas and present them in modern form" (emphasis added) or agree entirely with Michael E. Zimmerman's judgment that "his work is not strikingly original." See Graham Purchase, "Social Ecology, Anarchism and Trades Unionism," in David Goodway, ed., *For Anarchism: History, Theory, and Practice* (New York: Routledge, 1989), and Michael E. Zimmerman, *Contesting Earth's Future: Radical Ecology and Postmodernity* (Berkeley and Los Angeles: University of California Press, 1994), p. 151. At the very least, Bookchin's

work is philosophically original in the way he transforms the Hegelian dialectic, integrates ecology into radical social theory, and develops a new historical account of the relation between human hierarchy and the goal of subduing the natural world. What makes his work of social and political merit is the way he presents this understanding and framework with a high degree of coherence, orienting it toward relevance for present-day problems.

75. Murray Bookchin, *Toward an Ecological Society* (Montreal: Black Rose Books, 1980), p. 224. Bookchin claims his commitment to anarchism is "an invisible moral boundary that has kept [him] from oozing over to neo-Marxism, academicism and ultimately reformism" (Ibid., p. 22).

76. See *The Conquest of Bread* (New York: Vanguard Press, 1926) and the articles collected in *Kropotkin's Revolutionary Pamphlets*.

77. Peter Kropotkin, "Anarchist Communism: Its Basis and Principles," in *Revolutionary Pamphlets*.

78. Kropotkin, *Mutual Aid*, p. 227.

79. See also Kropotkin's work, *The State: Its Historic Role* (London: Freedom Press, 1903).

80. Bookchin, *Ecology of Freedom*, p. 325. Kropotkin's remarks on the "historical necessity" of the state are made primarily in the context of an elaboration of Bakunin's views. For example, see "Modern Science and Anarchism" in *Revolutionary Pamphlets*, p. 165.

81. Exhibiting the Aristotelian tendency toward *distinguo,* Bookchin remarks upon further significant differences that he finds between, for example, urbanization and cities, politics and statecraft, and representation and sovereignty.

82. In actuality, Kropotkin and Bookchin extol the merits of *both* the commune and the polis, though there is clearly a large difference in where they find their primary focus. See Bookchin's early work, *The Limits of the City* (New York: Harper & Row, 1974), esp. pp. 37–51, where he writes that "like the *polis,* these [medieval] towns formed a complete and rounded totality." See Kropotkin's *Mutual Aid*, where he remarks that the two greatest historical periods were the eras of the ancient Greek city and of the medieval city.

83. Theodore Roszak claims: "Communitarianism is the political expression of ecological intelligence. Anarchism has always been, uniquely, a politics swayed by organic sensibility; it is born of a concern for the health of cellular structure in society and a confidence in spontaneous self-regulation" (Roszak, *Where the Wasteland Ends* [Garden City, NY: Anchor Books, 1973], p. 389). On the close connection between anarchism and community, see also Michael Taylor, *Community, Anarchy, and Liberty* (Cambridge: Cambridge University Press, 1982).

84. Kropotkin, *Mutual Aid,* p. 186.

85. Bookchin claims contrarily that Kropotkin's thought—like that of other nineteenth-century socialists and anarchists—was often dystopian.

86. Kropotkin, *Mutual Aid,* pp. 194–195.

87. Harold Barclay, *People without Government* (London: Kahn and Averill, 1982), p. 93. Barclay concludes that Kropotkin was not justified in treating the medieval commune as an example worthy of early anarchy.

88. Bookchin, *Ecology of Freedom,* p. 116. Nor should we forget that the life of the polis was built around the exclusion of women and the use of slave labor.

89. See Amy Gutmann, "Communitarian Critics of Liberalism," *Philosophy and Public Affairs,* Vol. 14, 1985, pp. 308–322, and Marilyn Friedman, "Feminism and Modern Friendship: Dislocating the Community," *Ethics,* January, 1989, pp. 275–290. For an overview of recent debate, see Markate Daly, *Communitarianism: A New Public Ethics* (Belmont, CA: Wadsworth, 1994).

90. On Arendt's relevance for ecological politics, see David Macauley, "Hannah Arendt and the Politics of Place: From Earth Alienation to *Oikos*" in Macauley, ed., *Minding Nature: The Philosophers of Ecology.*

91. Plato, *Republic* (423b), in Edith Hamilton and Huntington Cairns, *The Collected Dialogues of Plato* (Princeton, NJ: Princeton University Press, 1989); translation altered.

92. Aristotle, *Politics,* Book 7.4 (1326b25), in Richard McKeon, *The Basic Works of Aristotle* (New York: Random House, 1941).

93. See Bookchin, *Limits of the City,* pp. 137–139.

94. Murray Bookchin, "The Concept of Ecotechnologies and Ecocommunities," in *Toward an Ecological Society,* p. 110.

95. For relevant discussions, see Karl Griewank, *Der Neuzeitliche Revolutionsbegriff* (Weimer, Germany: Hermann Böhlhaus Nachfolger, 1955), pp. 171–182, and Hannah Arendt, *On Revolution* (New York: Viking Press, 1965).

96. Among anarchists, Proudhon and Godwin were especially fond of arguing for the ability of society to progress. For example, see Proudhon, *Philosophie du Progres* (Brussels, 1853).

97. See Bookchin, *Ecology of Freedom,* p. 20, and "Ecology and Revolutionary Thought," in *Post-Scarcity Anarchism.*

98. Cited by Woodcock, in *Anarchism,* p. 19. As Stephen Jay Gould would argue many years later, Kropotkin observed that evolution does not proceed as slowly and or as evenly as commonly claimed.

99. Kropotkin, "The Direct Action of Environment on Plants," *Evolution and Environment,* pp. 139–158.

100. Bookchin, *Toward an Ecological Society,* p. 262.

101. Bookchin, "Beyond Neo-Marxism," *Telos,* Vol. 36, 1978, p. 21

102. See Bookchin, *Social Anarchism or Lifestyle Anarchism.*

103. See Murray Bookchin, "New Social Movements: The Anarchic Dimension," in David Goodway, ed., *For Anarchism: History, Theory, and Practice* (New York: Routledge, 1989).

104. Kropotkin, quoted by Roger Baldwin, in "The Story of Kropotkin's Life," *Revolutionary Pamphlets,* p. 17.

105. Kropotkin, "Anarchist Communism," in *Revolutionary Pamphlets,* p. 77. Bookchin's view of revolution has also changed slightly over the years in terms of the emphasis he places on spontaneity, organization, the revolutionary subject, municipalism, and the like. See, for example, Bookchin, "Were We Wrong?," *Telos,* Vol. 65, Fall, 1985, pp. 59–74, for his discussion of the com-

plexities of capitalism and revolution. For a recent essay on the subject, see "Defining the Revolutionary Project" in Murray Bookchin, *Remaking Society: Pathways to a Green Future* (Boston: South End Press, 1990), pp. 127–158.

106. Kropotkin, *Fields, Factories, and Workshops Tomorrow,* p. 17.

107. See Michael Smith, "Kropotkin and Technical Education," in David Goodway, ed., *For Anarchism,* pp. 217–234.

108. Ibid., p. 26.

109. Ibid., p. 39.

110. Ibid., p. 40.

111. Ibid., p. 104.

112. Bookchin, *Toward an Ecological Society,* pp. 52–53.

113. Specifically, it is "a function of individuation, *of a comprehensible public sphere that fosters authentic personal autonomy and empowerment*" (Bookchin, "Finding the Subject: Notes on Whitebook and 'Habermas Ltd.,' " *Telos,* Vol. 52, 1982, pp. 78–98).

114. Langdon Winner, *The Whale and the Reactor* (Chicago: University of Chicago Press, 1986), p. 96.

115. Robyn Eckersley, *Environmentalism and Political Theory: Toward an Ecocentric Approach* (Albany: State University of New York Press, 1992), pp. 173 ff.

116. Murray Bookchin, "Radical Agriculture," in Richard Merrill, ed., *Radical Agriculture* (New York: New York University Press, 1976), p. 12.

117. See Paul Goodman, *People or Personnel: Decentralizing and Mixed Systems* (New York: Vintage Books, 1968), esp. pp. 3–27, for an anarchist defense of decentralization.

118. Bookchin, *Ecology of Freedom,* p. 33, and "A Letter of Support," in *North American Bioregional Congress Proceedings,* May 21–25, 1984, pp. 77–78. Bookchin speaks of "a systematic recolonization of the land along ecological lines." See "Radical Agriculture," in Merill, ed., *Radical Agriculture,* p. 12.

119. Kropotkin, *Fields, Factories, and Workshops Tomorrow,* p. 106. For a constructive critique of bioregionalism as mystifying the concept of region, see Donald Alexander, "Bioregionalism: Science or Sensibility?," *Environmental Ethics,* Vol. 12, Summer, 1990, pp. 161–173.

120. Ibid., p. 57; emphasis added.

121. Bookchin adds that one can also maintain the reverse, that "the whole is the True," since "the true lies in the self-consummation of a *process* through its development, in the flowering of its latent particularities into their fullness or wholeness" (*Ecology of Freedom,* p. 32).

122. Ibid., p. 52.

123. See Colin Ward's very helpful appendices in *Fields, Factories, and Workshops Tomorrow.*

124. Kropotkin's nineteenth-century productionism—which frequently does not recognize the value of natural entities in themselves—is evident in remarks such as "a meadow remains a meadow, much inferior in productivity to a cornfield." Kropotkin fails to see how the former is a complex ecocommunity and the latter a monoculture. He also writes that "the dearest of all varie-

ties of our stable food is meat," but proceeds oddly enough to note how much land is required for a single head of cattle and what could be produced otherwise in grain—a much more efficient, economical, and ethical converter of protein. For a consideration of issues related to the domestication of plants and nonhuman animals, see David Macauley, "Be-wildering Order: On Finding a Home for Domestication and the Domesticated Other," in *The Ecological Community: Environmental Challenges for Philosophy, Politics, and Morality,* ed. Roger Gottlieb (New York: Routledge, 1997); Macauley, "Political Animals: A Study of the Emerging Animal Rights Movement in the United States," *Between the Species: A Journal of Ethics,* Vol. 3, Nos. 2 to 4, no. 2 (serialized in five parts from Spring 1987 to Spring 1988); Macauley, "Consuming Passions: An Intercourse on Animals and Food," *Lomakatsi,* Vol. 3, Spring, 1988.

125. Ibid., pp. 116–117.

126. Ebenezer Howard, *Garden Cities of Tomorrow* (London, 1902). Howard reintroduces the Greek notion of *natural limits* to expansion of human organizations and urban life, arguing for a more organic kind of city. In order to integrate countryside with city and to achieve more internal unity in the latter, he proposed surrounding the city with an agricultural "greenbelt." These greenbelts would be comparable to the walls that enclosed the polis.

127. Kropotkin, *Fields, Factories, and Workshops Tomorrow,* p. 55.

128. See esp. Bookchin, "Radical Agriculture," in Richard Merrill, ed., *Radical Agriculture,* and *Our Synthetic Environment,* rev. ed. (New York: Harper and Row, 1974).

129. On the distinction between market and moral economy, see Murray Bookchin, *The Modern Crisis* (Montreal: Black Rose Books, 1987), pp. 77–98.

130. Of *Our Synthetic Environment,* Yaakov Garb writes: "It is hard to imagine what an America that took this almost unknown work to heart to the extent it did Carson's would have been like, and what it might be like today as a result"; see Garb, "Change and Continuity in Environmental World-View: The Politics of Nature in Rachel Carson's *Silent Spring,*" in David Macauley, ed., *Minding Nature: The Philosophers of Ecology.*

131. Space does not permit a more complete discussion of the similarities and differences of perspective between Kropotkin and Bookchin. In addition to the subjects discussed briefly in this section, one should take note, first, of their disagreements with or challenges to Marxism. Kropotkin was not extremely familiar with Marx but believed his "scientific socialism" to be unscientific in a naturalistic sense. He also opposed the centralizing, parliamentarian tendencies of Marxists. Bookchin's view is too complex to encapsulate here. See especially "Listen Marxist!," in *Post-Scarcity Anarchism,* and "Marxism as Bourgeois Sociology," in *Toward an Ecological Society.* Second, they both critique syndicalism. Kropotkin came to think it should be a subsidiary, not a primary, element in social change. Bookchin sees the commune or muncipality rather than the factory as the proper locus of attention, as did Kropotkin. Third, both emphasize the dimensions of *social* revolution that involve ethical integrity, spontaneity, and direct action rather than expediency, violence, and organizational control.

132. Bookchin, Interview *Kick It Over,* Winter, 1985–1986, p. 9.

133. Bookchin, *Post-Scarcity Anarchism*, p. 16. In *Social Anarchism or Lifestyle Anarchism*, he states this position even more boldly, remarking that capitalism *produced* the modern environmental crisis (p. 35).

134. Bookchin, "Deep Ecology, Anarcho-syndicalism, and the Future of Anarchist Thought," p. 57.

135. E.g., see Kropotkin, *Ethics*, p. 10, and Bookchin, "The Future of the Ecology Movement," in *Which Way for the Ecology Movement?*

136. On the issue of mysticism and mystification in ecological thought, see also Stephan Elkins, "The Politics of Mystical Ecology," *Telos*, Vol. 82, Winter, 1989–1990, pp. 52–70, and Tim Luke, "The Dreams of Deep Ecology," *Telos*, Vol. 76, Summer, 1988, pp. 65–92.

137. The general historical parallels could be developed further because, like Kropotkin, who challenged social Darwinists such as Herbert Spencer, Bookchin criticizes modern sociobiologists such as E. O. Wilson and Richard Dawkins. For Bookchin's writings on these matters, see especially "The Population Myth" and "Sociobiology or Social Ecology?," in *Which Way for the Ecology Movement?* Kropotkin critiques social Darwinism in *Ethics* and considers Malthus in *Fields, Factories, and Workshops Tomorrow*, p. 77ff.

138. Kropotkin, *Fields, Factories, and Workshops Tomorrow*, p. 78.

139. Kropotkin, *Ethics*, p. 2.

140. See Bookchin's *Post-Scarcity Anarchism*, where he writes that "for the first time in history, technology has reached an open end. The potential for technological development, for providing machines as substitutes for labor is virtually unlimited" (p. 95).

141. Kropotkin, *Fields, Factories, and Workshops Tomorrow*, p. 193.

142. Kropotkin, "Anarchist Communism," in *Revolutionary Pamphlets*, p. 47.

143. See Robert Graham, "The Role of Contract in Anarchist Ideology," in David Goodway, ed., *For Anarchism*, pp. 150–175, for a consideration of anarchist views on contractarianism. Graham argues that the contract originally served a double role in anarchist thought: to attain economic justice and to safeguard individual liberty. He points out that anarcho-communists have generally agreed that there can be no real freedom of contract between capitalists and workers. Kropotkin questioned the moral ground of contractarianism because it was rooted in "a shopkeeper's mentality."

144. Bookchin, *Ecology of Freedom*, p. 320.

145. Ibid., pp. 299–300.

146. Kropotkin, "Anarchist Morality," *Revolutionary Pamphlets*, p. 107. It is not clear what this "something" would be however.

147. Bookchin, *Ecology of Freedom*, p. 325.

148. Kropotkin, *Ethics*, pp. 1–2. See Frank Harrison, "Science and Anarchism: From Bakunin to Bookchin," in Dimitrious Roussopoulos, ed., *The Anarchist Papers 3* (Montreal: Black Rose Books, 1990), pp. 72–84, for a relevant discussion of the role of science in anarchist thought.

149. Woodcock, *Anarchism*, pp. 220–221.

150. Lewis Mumford, *The City in History* (New York: Harcourt, Brace and World, 1961), pp. 514–515.

151. On Mumford's ecological orientation, see Ramachandra Guha, "Lewis Mumford: The Forgotten American Environmentalist: An Essay in Rehabilitation," in David Macauley, ed., *Minding Nature: The Philosophers of Ecology,* and Mark Luccarelli, *Lewis Mumford and the Ecological Region: The Politics of Planning* (New York: Guilford Press, 1996).

152. It should be observed briefly that other figures belong (at least peripherally) to a broadly defined tradition of ecoanarchism, even if their respective works often differ greatly in approach from Kropotkin and Bookchin. In addition to Mumford, such individuals include Gustav Landauer, William Morris, Charles Fourier, Elisée Reclus, Henry David Thoreau, Ebenezer Howard, and Patrick Geddes. More recently, the writings of Theodore Roszak, Kirkpatrick Sale, John Clark, and John Ely are generally continuous with this tradition. It has been suggested recently that postmodern theory shares affinities with green politics—and anarchistic views in particular—in its questioning of the Enlightenment project, its critique of instrumental rationality, its embrace of "otherness," its appreciation of beauty, and so on. See David Pepper, *Eco-Socialism: From Deep Ecology to Social Justice* (London and New York: Routledge, 1993), pp. 55–58. Bookchin, of course, would strongly oppose this association. For some of my own very early reflections on ecoanarchism, see Macauley, "Animals, Ecology, and Anarchism," *Lomakatsi,* Vol. 1–2 (serialized in two parts from Spring 1987 to Summer 1987).

153. Bookchin, "Deep Ecology, Anarcho-syndicalism, and the Future of Anarchist Thought," p. 58. Bookchin elaborates on this position in "New Social Movements: The Anarchic Dimension," in David Goodway, ed., *For Anarchism,* where he calls for a "rigorous and conscious re-examination of anarchism itself" (p. 273). In *Social Anarchism or Lifestyle Anarchism,* he challenges the withdrawal of many anarchists from the social realm and their retreat into individualistic, aesthetic, primitivistic, antirational, and metaphysical forms of revolt or "lifestyle" insurrection.

154. In one very real sense, Kropotkin's ideas have never been given the serious attention that is due to them within either critical social theory or evolutionary biology. For example, of Kropotkin's most well-known work, *Mutual Aid,* Ashley Montagu has written that it is "now a classic . . . [but] few people read it. . . . Yet no book in the whole realm of evolutionary theory is more readable or more important"; Montagu, *Darwin: Competition and Cooperation* (Westport, CT: Greenwood Press, 1973), p. 42.

Chapter 11

Reconsidering Bookchin and Marcuse as Environmental Materialists
Toward an Evolving Social Ecology

ANDREW LIGHT

Introduction

For some time those interested in the intersection between radical political theory and environmental issues have been either participating in or listening attentively to a debate between two strong voices, deep ecology and social ecology.[1] Both sides have attempted to provide an integrated strategy for articulating answers to a wide range of questions concerning the relationship between humans and the nonhuman natural world, and to offer insights into the peculiarities of human social life.[2]

Even though some authors have tried to write the brief history of the development of what is known today as "political ecology" as a struggle between these groups, these two positions do not exhaust the possible range of answers to political–environmental questions. Other programs such as socialist ecology and ecofeminism stand as viable alternatives.[3] Many of these positions however do share certain fundamental assumptions that I will argue justify maintaining a general distinction between "ontologists" and "materialists" in political ecology.

In this chapter I will explore this distinction by looking at two examples of environmental materialism in political ecology, one espoused by Murray Bookchin and the other promulgated by Herbert Marcuse. My reason for choosing these two authors is that while their positions are arguably compatible and mutually supporting, at least one of these writers is skeptical of this connection. Even though he has occasionally recognized the importance of Marcuse's theory to his own thought, Bookchin has criticized Marcuse's writings as too contradictory and imbued with neo-Marxist categories of analysis to be of much use in a liberating social critique of either human social relations or the impact of humans on nature. And in one of his more recent books, *The Philosophy of Social Ecology,* Bookchin has continued this critique of other members of the Frankfurt School, focusing in particular on Horkheimer and Adorno's *Dialectic of Enlightenment.* Bookchin claims that while these thinkers provide powerful critiques of positivism, we must not look to them for an ecological social theory:

> Despite some recent nonsense to the effect that the Frankfurt School e - connoitered a nonhierarchical and ecological view of society's future, in no sense were its ablest thinkers, Max Horkheimer and Theodor Adorno, resolutely critical of hierarchy and domination. . . . Attempts to make them into proto-social ecologists, much less precursors of bioregionalism, involve a gross misreading of their ideas, or worse, a failure to read their works at all.[4]

Certainly, one cannot read a critique of Horkheimer and Adorno as also a critique of Marcuse, especially given the critical differences that developed between these thinkers, particularly during the post-World War II period. Neither could it be said that a full-fledged political ecology can be gleaned from the works of any member of the Frankfurt School. But surely it is not true that such figures are as beyond the pale for the further development of political ecology as Bookchin's comment suggests. The underlying structure of the critical theory developed by the Frankfurt School, and in particular by Marcuse, does share something with Bookchin's social ecology. The further development of social ecology should proceed at least in part through a serious study of the works of the Frankfurt School. Rather than concluding that this body of work cannot be counted as a form of social ecology, as Bookchin has maintained, a more charitable reading of the Frankfurt School could develop the constructive lesson that critical theory has a contribution to make in the continuing elaboration of social ecology.

Developing this suggestion, that critical theory is important for the further evolution of social ecology, is the main goal of this chapter. To make that argument, though, I need to revisit and expand on the point of my original comparison between Bookchin and Marcuse. I need to show that a reading of the works of Bookchin and Marcuse casts some doubt on Bookchin's mitigated skepticism of the utility of critical theory for his own work and instead help to place Bookchin and Marcuse together as advocating a "thin" environmental materialism. Such an approach, which is concerned primarily with the material basis of relationships between humans, technology, and the environment, can be contrasted with a framework that proceeds from a more ontological bent. If Bookchin's social ecology and Marcuse's critical theory are found to be compatible as forms of environmental materialism, even though they disagree on a wide range of particular issues, then a mutually beneficial dialogue between their positions may yield a stronger critique of technological and hierarchical society and its effects on environmental problems than either school of thought alone has been able to muster. I expect that it will be up to the inheritors of these two traditions to carry on this dialogue.

I will begin this discussion by looking briefly at Bookchin's social ecology, comparing this theory with Marcuse's early statements on technology and nature, and providing some background to the debates each has engaged in on a similar range of issues. From such a vantage point we can see how both Marcuse and Bookchin can be fairly described as environmental materialists. In the context of this description, I will turn to more substantial comparisons of these two thinkers. Next, I will return to Bookchin's specific critique of Marcuse, answer it, and show how a materialist analysis of these authors demonstrates how much more important their metatheoretical similarities are than their differences. In doing so I will briefly discuss what I think to be Marcuse's most surprising discussion of the role of nature in its relation to the human struggle with technology, which demonstrates that environmental materialism is not completely incompatible with a reserved form of environmental ontology. If further interactions between social ecologists and critical theorists take place, this last point concerning the potential compatibility of the materialist and ontologist positions may prove important for a reconsideration by social ecologists of their past debates in political ecology.

Materialists and Ontologists

The basic premises of environmental materialism can be seen at work in Bookchin's opposition to any political system that sees nature only through the lens of demands for unlimited economic growth. Bookchin attacks both market-driven capitalist economies and statist-planned political economies for their view of nature as an object only to be used to garner more resources. Bookchin argues that any economy structured around the maxim "Grow or Die" will necessarily pit itself against the natural world and inevitably lead to ecological ruin.[5] In the press for economic growth, Bookchin says, the state and its people replace the organic with the inorganic, soil with concrete, living forest with barren earth, and the diversity of life forms with simplified ecosystems. For him, this is a "turning back of the evolutionary clock . . . to a world incapable of supporting complex life-forms of any kind, including the human species."[6] Crucial for understanding Bookchin's views, however, is the fact that his political position rests first on a critique of hierarchy among humans and an advocacy of its dissolution. From here Bookchin makes a social analysis of the relationship between humans and the natural world. For Bookchin, "The imbalances man has produced in the natural world are caused by the imbalances he has produced in the social world."[7]

In liberal society, as Bookchin reads it, power relationships are structured around a hierarchically organized social status quo that serves the interests of property owners (among others). Therefore, the interests of the non-property-owning public will always be sacrificed to the power interests of the landlord, the developer, and the exploiter of natural resources.[8] The best that liberal environmentalists can do, because they are committed to the preservation of the free-market system (or more accurately, they are unwilling to oppose it), is to garner a few compromises and trade-offs with business interests within the market, which renders their approach ineffectual as an authentic challenge to the property-owning interests of the market system. Bookchin concludes that the willingness of liberal environmentalists to accept such piecemeal advances must rest on a presupposition that certain fundamental institutions in the liberal, capitalist state are here to stay: "All of these 'compromises' and 'trade-offs' rest on the paralyzing belief that a market society, privately owned property, and the present-day bureaucratic nation-state cannot be changed in any sense."[9] Instead of this piecemeal environmentalism, Bookchin argues, we need an ecology, specifically, a social ecology. Bookchin maintains that the power of the term "ecology" has to be preserved, as opposed to the merely reformist "environmentalism." He takes issue, for example, with Arne Naess's distinction between "shallow" and "deep" ecology by suggesting that "ecology" is not "applicable to *everything* that invokes environmental issues." Nonetheless, Bookchin's description of liberal environmentalism as short-sighted and reformist is very similar to Naess's description of shallow ecology as a form of environmentalism that resists a deeper questioning that addresses the roots of environmental problems.[10]

But for my purposes, more important than the comparison with Naess, Bookchin's critique of capitalist society's paralysis on dissent resonates with Herbert Marcuse's evaluation of advanced industrial society in *One-Dimensional Man*. There, Marcuse argues that opposition to the economic and political systems of advanced industrial societies has been socially eliminated for those classes that had traditionally acted as opponents to the system.[11] Certainly, Marcuse's analysis of the causes of capitalist hegemony is not exactly the same as Bookchin's, but the two views are nonetheless compatible. For both theorists it is the perception of an ahistorical permanence and rigidity that empowers the capitalist system, along with the social, economic, and political structures that ensure that permanence. Wound up in this static conception of the status quo is the idea that technology, and more specifically technological forms of production, help to maintain this order—not just technological systems themselves, but the uses of technology to "harness" nature.[12] In an earlier essay, Marcuse had argued that technologically based systems in market economies favor large-scale industry, which must be owned by vast corporate enterprises.[13] It is most likely the case that the same big business interests that are exploiting people and damaging nature in Bookchin's account are also the only ones able to control large-scale industry and technological processes in Marcuse's account. In both cases, there is overlap between the exploiter of humans and the preserver of the exploitative system, which in turn has harmful consequences for nature.

A crucial difference, however, between Bookchin and Marcuse, at least in the literature I have so far discussed, is their relative degree of social hope. Bookchin has argued in response to the comparison with Marcuse that the two of them hold contrary assessments of the potential for social resistance.[14] Therefore, even if I am right in claiming that both figures share a critique of the relative quietism of what passes for resistance to a technologically controlled capitalist system, Marcuse's relative pessimism regarding the possibility for revolt (expressed in *One-Dimensional Man*) may belie a fundamental difference with Bookchin in his assessment of the structure of capitalist hegemony.[15] But we can surely understand Marcuse's skepticism concerning the formulation of a viable resistance to capitalism in the late 1950s and early 1960s when *One-Dimensional Man* was being written. Although this work was published just as the New Left was emerging, at the time Marcuse saw only the faintest glimmer of hope for a rejection of one-dimensional society, and that hope resided in the early civil rights movement. Charitable readers of Marcuse recognize the extent to which his views changed over time and how he later articulated new forms of liberatory potential in the growing radical community (a point that I will discuss below). Thus, both Marcuse and Bookchin are significantly similar in their evaluation of the social realm of resistance and the relation of that resistance to the material structure of society. Both find it crucial to express hopes for the creation of new states of material organization in their analysis of the proper political response to human exploitation and environmental problems.

We can characterize the similarities between Bookchin and Marcuse pointed out so far—and others to be discussed below—as constitutive of a perspective I call "environmental materialism."[16] For both Bookchin and Marcuse, the appropriate human response to environmental problems must primarily involve an analysis of the causes of those problems in the organization of human society through the material conditions of capitalist (or state capitalist) economies, and the social and political systems that sustain those societies. Material conditions, such as who owns and controls the technological processes that are used to stimulate economic growth, expand markets, and consume natural resources, are, for these thinkers, the starting points for unpacking the complex web of environmental problems. From such an analysis, Bookchin and Marcuse both conclude that the solutions to environmental problems should be based on an evaluation of human social, political, and economic systems. A radical political ecology would push the boundaries of these systems and champion a change in the material organization of society as a whole.[17] Certainly, such a view could not embrace what Bookchin calls a "crass" instrumentalism, where human concern for nature rests only on self-interest.[18] The characterization of Bookchin as an environmental materialist captures suggestions Bookchin himself has made in the past while comparing his own views with those of deep ecologists and other theorists. Bookchin argues that laying blame for environmental destruction on humans in general is mistaken because it "masks the fact that our ecological problems are fundamentally social problems requiring fundamental social change."[19]

By way of a contrast, those theorists who see more potential in diagnosing environmental problems as primarily involving the idea of the human self in relation to nature, rather than the idea of social, political, or material systems in relation to nature, are constitutive of the position I call "environmental ontology." For environmental ontologists, social, political, and material problems are the symptom of a crisis involving the relation of the self with nature, not the root cause. The primary cause of environmental problems is our ontological disconnection from nature. For ontologists, the principle location of solutions to environmental problems is to be found in changing the "consciousness" (for lack of a better term) of individual humans in relation to the nonhuman natural world. Theorists embracing this broadly identifiable view argue that humans should be identified with nature not as a separable organism or set of organisms, but as an integrated part of a larger life/world system. Environmental ontologists focus their critique of liberal environmentalism on its general lack of analysis of individual human self-perceptions with respect to the nonhuman natural world. The focus for political reform for ontologists is on a rethinking of the self as expressed in individual identity, rather than the materialist focus on society and its institutions as the primary mechanism for environmental renewal. Perhaps the clearest example of an ontological theory can be seen in the development of deep ecology, originally by Arne Naess, and later by his American and Australian disciples.

Deep ecology, if we formulate it through Naess's original statements on the theory, primarily embraces an ontological and not a materialist criticism of human interaction with the nonhuman natural world. Naess takes great pains to distinguish his theory from an environmental ethic. Naess has said: "I am not interested in ethics or morals. I'm interested in how we experience the world. . . . If deep ecology is deep it must relate to our foundational beliefs, not just to ethics. Ethics follow from how we experience the world. If you articulate experience then it can be a philosophy *or* a religion."[20] More recently, Naess has explicitly identified his approach as consistent with this distinction: "I am for what I call a focus on environmental ontology, how you see the world, how you *see* it, how you can bring people to *see* things differently."[21]

I emphasized the "or" in the passage before this last one to highlight a source of tension between materialists and ontologists, particularly the two exemplary representatives of Bookchin's social ecology and Naess's deep ecology. The openness by Naess to articulate the experiential basis of deep ecology as a philosophy or a religion is at the root of Bookchin's worries about deep ecology as a theory and as a practice. But Bookchin often groups many forms of environmental ontology into one collection characterized only by its most outrageous representatives and their most counterintuitive claims.[22]

The focus of Bookchin's argument against deep ecology is that due to its spiritual dimensions (and I would suggest environmental ontological commitments), it is notable for its "absence of reference to social theory," which makes it incompatible with his social ecology (and hence with the priorities for reform in his environmental materialism). When Bookchin moves from an analysis of the philosophy of deep ecology to its proponents in the deep ecology movement (a distinction we should note that Bookchin rarely, if ever, acknowledges) a clear struggle emerges for social ecologists:

> In America, the rapidly forming Green movement is beset by a macho cowboy tendency [the reference is to deep ecologists of the Dave Foreman/Earth First! variety] that has adopted Malthusianism with its racist

implications as a dogma, an anti-humanism that among some of the wilderness oriented "campfire" boys has become a brutalized form of misanthropy, and a "spiritualist" tendency that tends to extol irrationalism and view ecology more as a religion than a form of health naturalism. It has become primarily the task of American ecoanarchists to develop a sustained resistance to these primitivistic, misanthropic, and quasi-religious tendencies.[23]

The philosophical and practical terrain that gets divided here is momentous on Bookchin's scheme, and can reveal important differences between the deep ecology and social ecology camps that cannot simply be reduced to sloganeering.

Take, for example, Bookchin's critique of deep ecologists for diminishing distinctions between humans and nonhumans as part of their claim that the proper view of the human self is one of the self in relation to nature. True to this charge, Warwick Fox comments that implicit in the deep ecology movement "is the idea that we can make no firm ontological divide in the field of existence: That there is no bifurcation in reality between the human and the non-human realms. . . . To the extent that we perceive boundaries, we fall short of deep ecological consciousness."[24] Bookchin argues that failing to distinguish between humans and nonhumans ignores important, peculiarly human, social dimensions of environmental problems. To reduce humans merely to one species among many is to diminish the importance of human social distinctions, such as class distinctions, that are not found in nonhuman species.[25] When deep ecologists do not distinguish between human social classes, by lumping human and nonhuman animals together, the result is a broad distribution of the responsibility for environmental problems among humans. If no social distinctions are recognized between classes, then all humans are equally to blame for the destructive impact of humans on the environment. Without an acknowledgment of the specific sources of environmental deterioration, it becomes possible to "comfortably forget that much of the poverty and hunger that afflicts the world has its origin in the corporate exploitation of human beings and nature—in agribusiness and social oppression."[26] One way of generalizing this critique is to say that some forms of deep ecology lack an environmental materialist base.

But the elimination of class distinctions is a comparatively less damning problem with deep ecology than others unearthed by Bookchin. Much more famous are Bookchin's cluster of claims that deep ecologists hold a body of views "that openly welcomes the AIDS epidemic, Ethiopian famines and restrictions on the immigration of fleeing Central Americans in the name of 'population control' and that, to add insult to injury, views the presence of Latin immigrants as a sinister danger to our 'northern European' culture (read: Aryan?)."[27] While charges like these can be justified, for it is certainly true that a reading of the early Earth First! literature reveals that such ridiculous claims were actually made by self-professed deep ecologists, Bookchin's overall attitude toward deep ecology usually prohibits any careful discussion of its philosophical foundations in social ecology circles. After John Clark circulated a version of his contribution to this volume (Chapter 5) at the world gathering of the international network for social ecology at Danoon, Scotland, in August 1995, in part arguing for a rethinking of the deep ecology–social ecology debate, Bookchin responded with an extended critique chastising Clark for blurring differences between deep and social ecology "at a time when it is of essential importance to sharply distinguish them," and characterized Clark's views as "mystical" and "reactionary."[28] But Bookchin's approach, characterized by his wholesale condemnation of all deep ecology, is exemplary of the uncharitable reading of each other's views that makes much of environmental political theory difficult to translate into practice. One must wonder why, after the initial salvos in the deep ecology–social ecology debate ten years ago, it is still imperative to sharply distinguish the two forms of political ecology. Are the differences between the two theories still that unclear? What exactly is going to be gained by this continual refusal to try to form a unified ecological front, rejecting the worst extremities and excesses of both sides of this debate? This problem is made more salient when we consider how formulating effective environmental practices may require large-scale coalitions across a broad array of philosophical commitments.[29] But it is not only at the theoretical level that such divisions are drawn. It is clear that Bookchin really does see the practical struggle for ecoanarchists to be directed as much against deep ecologists as it is against liberal environmentalists and growth-oriented polluters of the earth.

In trying to negotiate the deep ecology–social ecology debate, the materialist–ontological distinction serves a useful purpose: it more generally describes the ground upon which these two schools of thought disagree. But certainly, the materialist–ontologist distinction is not as clean-cut as the range of distinctions implicit in Bookchin's critique of deep ecology. Further, if Bookchin's argument with deep ecology can be situated within the more general materialist–ontologist distinction, then we may be able to get a better perspective on the territory covered by social ecology and deep ecology. And if it were the case that other materialists, with whom Bookchin shares some ground, were in fact closer to forms of environmental ontology, then perhaps social ecology, as a form of environmental materialism, is closer to deep ecology than

Bookchin is willing to admit. Or more modestly, perhaps the philosophical underpinnings of these two views are not so far apart as to warrant the necessary exclusion of the concerns of deep ecology (such as those Clark finds important) from the program of social ecology. Are there, then, materialists with coherent ontological commitments?

For one thing, there is a clear ontological dimension to Marcuse's work, as can be seen in *Counter-revolution and Revolt*. There, as I will suggest later, Marcuse may be arguing that the transformation of technology from its repressive forms in advanced industrial capitalism occurs within a framework that includes an ontological dimension, in a similar sense to the use of ontology by Naess. Such a combined materialist–ontological framework is necessary for Marcuse to articulate a viable radical alternative to the existing technological society.[30] But even with such environmental ontologist intuitions, Marcuse does have certain materialist priorities. Ultimately, in his critical theory, changes in the material conditions of society cause further changes in individual consciousness, which then may consequently lead to more changes in the material conditions of society. For example, as we will see below, even though Marcuse suggests in *One-Dimensional Man* that material social change may be achieved through individual participation in a radical aesthetic, the causes of alienation that motivate such artistic creations are the debilitating material conditions of society.

Nonetheless, Marcuse's environmental materialism has an ontological dimension that cannot be ignored. Presumably, then, if it can be shown that Bookchin and Marcuse are both environmental materialists, and it is the case that Marcuse's views can sustain some ontological commitments, then perhaps one can embrace a robust social ecology informed by Bookchin's work without necessarily rejecting the relevance of all ontological dimensions to the description of environmental problems. The materialist–ontologist distinction therefore need not mark a mutually exclusive ground, but rather the poles of a continuum.

On the other side of the environmental spectrum, Arne Naess includes in his discussions of an ontologically based ecology a dimension that is clearly materialist. In Naess's *Ecology, Community, and Lifestyle,* he devotes several chapters to discussing the social and political order, the types of economic organization that are best suited to a "no growth" society, and references to the types of policy changes that are needed to enact a deep ecology program.[31] More recently, Naess has explicitly stated that one of the positive effects of the deep ecology–social ecology debate has been to push deep ecologists "more in the direction of political thinking." Naess has argued that deep ecologists "must point to what we mean politically, particularly pointing to the green political theorists and the green economists who we side with." Deep ecology theorists, he maintains, engage too much in "spiritual thinking."[32] Additionally, Naess has explicitly stated that capitalism itself, and contemporary industrial society, "is out."[33] In Naess's home country of Norway, he has stopped supporting the Green Party (which he describes as too "fundamentalist") and instead supports the Sosialistik Venstre Parti, which he describes as a "radical socialist party" advocating "green socialism."[34] Naess even describes himself as "supportive" of social ecology, even though he does not "agree with everything they say."[35] While I still want to categorize Naess as an environmental ontologist, because of his strong commitment to the priority of ontological changes in the transformation to and preservation of any lasting shift in the human relationship with nature, we should acknowledge that he also has a strong commitment to the materialist dimension of political ecology in both theory and practice.

Bookchin has objected to this materialist–ontologist distinction, though in doing so has arguably not attempted a thorough assessment of the justification I have previously provided. I will try to separate and summarize his objections in three points. First, Bookchin claims that I have overlooked the ethical dimension of his own thought which problematizes the claim that he is an environmental materialist.[36] Bookchin is correct in this claim. There is an ontological dimension to his work that he has referred to as an "ethical ontology," partly developed through his expansion of Hans Jonas's argument that a theory of ontology can serve as an objective ground for an ethical conception of nature and freedom.[37] But by arguing that his work contains an ethical dimension, he is only affirming my distinction by highlighting the contrast between his work and the work of those deep ecologists who argue that their commitment to ontology is transethical. Because deep ecology is transethical, and experientially based (rather than necessarily rational), such a position, as Naess notes, can serve as the basis for a religious viewpoint. Such a claim emphasizes the connection between Naess's environmental ontology and that of a more explicitly religious form of environmental ontology, such as that found in the work of Martin Buber.[38] Bookchin himself has criticized the religious stance of deep ecologists, so he must acknowledge some metatheoretical contrast between his position and theirs. My way of putting this difference is that materialists and ontologists not only disagree on substantive positions on ethical questions, but that an environmental ontology represents an entirely different approach to ethical dilemmas in general. Certainly materialists, in the restricted use to which I am putting the term here, have an ethical dimension to their thought. Nonetheless, their ontology will not eclipse their ethics. The materialist–ontologist distinc-

tion then, again, only helps us to understand the difference in the starting points of various environmental theories.

Second, Bookchin suggests that I have characterized him and Marcuse as "pure materialists," which of course would entail that they are
ontologists of a particular sort.[39] But as I tried to take pains to clarify
early on (in this chapter and in my previous treatment of this subject), I
only argue that at least Bookchin and Marcuse are "thin" materialists,
specifically with respect to their political analysis of environmental
problems. The term, "environmental materialism" should not be confused with the traditional philosophical usage of "materialism" (e.g., in
Marx's work), which as a form of metaphysics entails a form of ontology. Environmental materialists are materialists only insofar as they pri-
oritize analysis of social formations, especially social structures, institutions, and material culture, in their description of the causes of
environmental problems and in their prescription of what should be
done to resolve these problems. This label does not neuter other possible descriptions, or self-descriptions, of a theorist's work (as Bookchin
seems to worry), whether he or she be socialist, anarchist, syndicalist,
municipalist, or whatever.

Third, Bookchin characterizes the distinction as "awkward at
best" since he "never accepted existentialist uses of the word *ontology*
to refer to psychological, subjectivistic, and largely personalistic concerns." Bookchin emphasizes that the "antonym of *materialism*" in his
view "is still *idealism.*"[40] Here, I can only say that we have a difference
in terminology which, while expected, does not in itself count as an objection to the rationale for my distinction. Certainly, Bookchin is correct when he notes that I am using the terms "materialist" and "ontologist" in nontraditional forms (though I should note that it is not
unprecedented to use the term "materialism" to refer to different kinds
of positions in different subfields of philosophy). But in some sense, at
least with my use of "materialism," I am invoking an ordinary language use of the term. For the environmental materialist, the most important environmental questions concern the material organization of
nature through social institutions. The distinction is only awkward
given a rather ambitious assumption about the work that one expects
from philosophical distinctions in general: if one expects them to provide a complete account of the substance of the theories they describe,
then most distinctions will be rather awkward. This distinction certainly does not provide a full characterization of the theories it seeks to
categorize. The point of the distinction is much more akin to the convention in the history of philosophy of seeing enough relevant similarities in Fichte, Schelling, and Hegel to call them "idealists" while recognizing that finer distinctions can be employed between them. No

serious user of the term "idealist" is much disappointed at the prospect that, say, Fichte and Schelling upon inspection are found to hold substantively different positions on some particular question. Nor are they confused at using "idealism" to refer to some proposition not heretofore articulated by some particular idealist, or to use "idealism" in opposition to general modes of analysis other than its traditional antipodes. In Bookchin's original criticism of this distinction, he repeatedly points out that he and Marcuse hold different positions on some particular issue as if this refutes my argument that they are both, in my sense, (thin) environmental materialists. Such differences do not change the propriety of the application of my distinction in this case, as the distinction is gauged as a general description of the overall approaches of the two theorists.

The materialist–ontologist distinction will help us to predict the propensity for certain serious points of contention between critical ecologists, such as the earlier example of our relation to nonhuman animals, but it need not be the case that such disagreements are necessary for the distinction to work. The distinction identifies where one's priorities fall on general political questions of environmental reform. But even assuming that split, we can imagine materialists and ontologists agreeing on a particular question of policy. Given, for example, the choice of whether a particular site should be preserved as a national park, both groups might agree that the area should be set aside even if they differed on the role of parks in the ultimate transformation to a new social order (or to the correct individual attitudes toward nature). From the theorists presented so far, it is noteworthy that Bookchin has even reached out to members of the deep ecology camp to at least talk about what they can potentially agree upon.[41]

Attempts at mixing these positions demonstrate the usefulness of developing both types of theories, materialist and ontologist, in the pursuit of answers to questions pertaining to both. In the development of a materialist conception of political ecology, certain ontological lessons will be learned along the way that can ultimately be useful to a strict environmental ontologist. The same is true in the other direction. It is too early in the development of environmental thought to try to fully discount most theories as wholly untenable, even though there are plenty of fair criticisms to be levied against theorists throughout the spectrum of political ecology. Whatever the ultimate solutions prove to be for environmental problems and the heretofore troubling relationship of humans to nature, such solutions are sure to contain both ontological and material dimensions.

If it is legitimate to make this materialist–ontologist distinction, and if Bookchin and Marcuse can both be identified as materialists, then even if Bookchin and Marcuse differ on their formulation concerning the connection between environmental problems and capitalist or statist economies, as well as on the kind of political and economic structures they envision as alternatives, they can be read together in a project of further developing a materialist political ecology. More specifically, a careful study of their similar positions, as developed through very different ideological commitments, should ultimately aid in the further development of social ecology, perhaps resulting in a stronger materialist position, or at least decreasing the ideological isolation of social ecology. But certainly a more thorough account of these two theorists is needed to strengthen this intuition. In the next two sections I will provide a more detailed comparison of parts of their respective positions.

Social Evolution, the Self-Negation of Technology, and Aesthetics

In his critique of deep ecology, we saw that Bookchin argued that the proper focus of a political ecology should first be directed at divisions within human society itself, and not at conflicts between humans and nature per se. Human impacts on the environment alert us to the fact that we must take a closer look at our own institutions and forms of social organization. For Bookchin, the category of domination is something that is only properly applied to relations within human society and not to the human treatment of nature. Bookchin argues that we must focus on the "deep-seated conflicts between humans and humans that are often obscured by our broad use of the word 'humanity.' "[42] This part of Bookchin's reaction to the perceived antimaterialist stance of deep ecology points to an important part of his philosophy that may carry some implications for a more detailed comparison with Marcuse. The specific argument that is important to this connection is found in Bookchin's conception of social evolution.

Throughout his work, Bookchin argues that social evolution is derived from natural evolution.[43] The evolution of life is termed "first nature" and the evolution of society is "second nature." Humans are products of nature and the "natural order of the world." Based on this fact, Bookchin argues that human social relations are also a product of the natural world. He writes: "In the most intimate of our human attributes, we are no less products of natural evolution than we are of social evolution. As human beings we incorporate within ourselves eons of organic differentiation and elaboration."[44] So far this claim is straightforward. A fair interpretation of the point is that social evolution is a form of natural evolution because human cognitive capabilities, and the products those capabilities produce, are generated by the evolution of nature. But Bookchin does not stop here. Rather than resting on this notion of a connection between social and natural evolution (which from a traditional Darwinian standpoint is defensible), Bookchin goes on to argue that the potential, or tendency, for diversity in social evolution, as derived from natural evolution, is a source of freedom "not only in terms of new choices . . . but also in terms of the richer social background that diversity and complexity create."[45] Natural evolution thus provides a basis for an "ecological ethics of freedom." There is no necessary direction to natural evolution toward freedom, but this potentially ideal (or at least better) state of nature has normative consequences for humans. Sometimes, in what appears to be stronger language, Bookchin refers to the "immanent striving of life-forms toward various degrees of freedom."[46] At previous times in human history we were closer to this beneficial potential of natural evolution toward freedom in our social evolution. Bookchin claims that anthropological data suggests that "participation, mutual aid, solidarity, and empathy were the social virtues early human groups emphasized within their communities."[47]

For Bookchin, even though every social evolution is a product of natural evolution, "we are substantially less than human today in view of our still unknown potential to be creative, caring and rational."[48] To be "fully" human is to live up to our evolutionary potential; thus evolution sets a normative standard for human society. Bookchin's evidence for our lapse includes the fact that humans are having an adverse effect on the environment. One premise of this argument must be that this natural human potential for creativity, care, and so on, involved a tendency toward a form of life incompatible (or at least less compatible) with environmental destruction. Behind this argument is Bookchin's belief that over time humanity was separated from its roots in nature—in other words, that the favorable potential of social evolution was derailed. If humans had continued to live naturally (e.g., according to the parameters of mutual aid), without resorting to current relationships of inequitable hierarchy and domination (racism, sexism, ageism, etc.), then they could have fully developed the potential for technical insight, culture, and self-reflective thought that is indicative of a more sustainable society.[49] But then one may wonder what forms of technical insight are consistent with the parameters of social evolution.

Bookchin's position on technology is that while current forms may be inconsistent with an enlightened form of social organization, technology would have some redeeming qualities as a product of a properly realigned social evolution. But as I argued earlier, this is not to say that just any technological product of a creative consciousness is a good product. Against some current technologies he writes: "The certainty that technology and science would improve the human condition is mocked by the proliferation of nuclear weapons, by massive hunger in the Third World, and by poverty in the First World."[50]

The potential benefits of technology arise in what Bookchin calls a "humanly scaled alternative technology," which provides an alternative to the high technology of capitalism and statism.[51] Technology and science may instead serve to "monumentalize natural evolution itself." For Bookchin, the products of our design provide evidence that we are an intelligent life form as opposed to a "mindless insect"; here he clearly contrasts his view with that of the deep ecologists.[52] By inference, one can see that given his account of the best development of the potential of human social evolution, the production of a technology that is debilitative or alienating is the product of a nonnatural social evolution, or is at least inconsistent with the preferred (though not necessary) direction of social evolution. A new technology, consistent with the tenets of social ecology, would enrich individuals with a "new sense of self-assurance."[53] Finally, Bookchin envisions a time when humans, now in tune with the full potential of social evolution, or the capacities of a "conscious, ecologically oriented life-form," would have the inclination, for example, to aid in the preservation of other species.

In the original version of this essay, I took issue in a footnote with Bookchin's description of the relationship between human social evolution and natural evolution. My brief, and awkward, claim at the time was that what I perceived as Bookchin's argument for a direction in social evolution was at odds with the dominant skeptical interpretation of the possibility of directionality in nature, such as that found in Stephen Jay Gould's work. As such, I suggested that Bookchin could not claim a legitimate ground for ethics in evolution. Bookchin correctly replied that, first, he does not advocate the claim of necessity in the direction of evolution, and second, that I had overlooked his criticism of Gould.[54] But even if I concede that he is correct when he says that he does not argue for a necessary direction in evolution, but only for a potentiality (or perhaps more accurately, a "developmental unfolding of potentiality"), it does not follow that this potentiality has ethical, and certainly not political, implications, as Bookchin maintains. I do not think that Bookchin can easily claim that a happier end of the evolutionary story (as a potential or as a necessity) is either good or bad.

When Bookchin argues that contemporary social formations are faulty, he seems to be asserting that they have gone awry from some previous state of affairs that is implicitly given normative weight. What are these states of affairs? Perhaps they are the earlier human social formations that he maintains were evidenced by the instantiation of a tendency toward greater freedom (and its benefits like mutual aid). Indeed, Bookchin suggests that we are "less human" when we act against our evolutionary potential for freedom realized through nonhierarchical social relationships. Cooperation with each other is good and domination is bad not just because of some judgment about the inequalities that they may produce, but because somehow domination is inconsistent with our evolutionary potential as exemplified in a preexisting state of affairs. "The central problem we face today," according to Bookchin, "is that the social evolution of 'second nature' has taken a wrong turn."[55] But such a conclusion only serves as a reason to change existing social institutions if this description of the "wrong turn" is based in a substantive normative claim. This normative claim is one many of us would no doubt like to be entitled to: there is some secure sense in which social evolution has a right direction, and by implication, so does natural evolution. It is not just that evolution has *a* potential, but that it has *a preferred* potential.

What would be wrong with such an argument? Well, for one thing, to argue that there is a preferred direction to evolution that we may equate with a normative description of justice, rightness, or more generally, the good, risks the naturalistic fallacy. As G. E. Moore argued, to move from the "is" of attribution to the "is" of identity of the good in any description of a property of the world is a problem.[56] For example, the utilitarian move from the claim that pleasure is good (the "is" of attribution) to the claim that pleasure is identified with the good, improperly assigns a normative property to an attribute of human beings, namely, that they contingently find those sensations we call "pleasures" good things. We could of course follow other ethicists in rejecting the naturalistic fallacy, thus eliminating this problem for Bookchin. But if we do not, we can at least apply a version of it to Bookchin's "ecological ethics of freedom," which would then complicate his picture and weigh in on the side of contingency theorists in evolution like Gould. Thus, even if Bookchin is correct in arguing that nature has no necessary direction, he is assuming a substantial argumentative burden. If evolution has no necessary direction, but if it does have a preferred outcome, then that outcome must be contingent. But if the preferred tendency in evolution is contingent, then it cannot be identified, at least on a view like Moore's, with the good. Thus evolution does not easily serve as a basis for ethics. Certainly Bookchin does not assume that his argument is without problems, but this specific worry still may give us pause. Bookchin's claim that he doesn't advocate a direction to evolution does not eliminate all problems with his socially normative appropriation of evolution that I had raised before.[57]

Setting aside my objections to Bookchin's account of the normative role of social evolution in his critique of contemporary social formations, there is an interesting connection to be made between the structure of Bookchin's account of social evolution as a basis for his critique of modern technology and Marcuse's account of the "self-negation of technology" in *One-Dimensional Man*. Technology, argues Marcuse, does not have to enslave us; it can instead be our most liberating tool. Advanced industrial society may contain the qualitative change that will liberate all humans from their technological bondage.[58] For Marcuse, technology, broadly construed is self-negating, meaning that technology contains within itself the potential for transformation to more liberating forms in the form of a new aesthetic dimension.[59] Reason, which has been defined through technological systems as instrumental reason, must be liberated in part through an aesthetic framework that will challenge coercive, hierarchically organized social formations.[60] This framework exists as a potential within technology.

Compared to Bookchin's relatively clear thesis of social evolution, Marcuse's account of the aesthetic transformation of society is very difficult to follow. But while we can be sure that the forces of social transformation for both authors are different, there is a notable similarity between the structures of their two theories. Both authors ground their critique of the development of modern technologies specifically, and material culture in general, (1) in an assumption of a strong normative basis for their critiques ground in an assessment of the best that human material culture could be; and (2) in a dialectical argument that the resources for a transformation of material culture exist within the social formations that both find objectionable. Both are motivated to such analysis in part through a recognition of human social and material inequalities, and the adverse effects of humans on nature.

1. For Bookchin, the normative ground for his critique of modern society is his account of social evolution. This normative ground is evidenced in prehierarchical social formations. There is, according to Bookchin, an identifiable history of material culture that informs us of the wrong turn that has been taken in social evolution. We can look at the ways in which some societies organized themselves, including their technologies of social organization, to see what a society would look like that is more consistent with the preferred potential of social evolution. Our questions on resolving environmental problems should focus then on the issue of how this transformation to an antinatural material culture emerged.[61] The answer to these questions is going to be found for Bookchin, as it is for Marcuse, in a transformation of the *human* material conditions that support this warped state of affairs. Unlike many environmental ontologists and mainstream environmental ethicists, Bookchin argues that it is a reform of human relationships, and

the structures that maintain them, that must necessarily precede consideration of environmental issues. The same is true for the Marcuse of *One-Dimensional Man*: the normative ground for the transformation of society lies in part in a human aesthetic dimension that contains the traces of a preferred realm of autonomy. This aesthetic dimension is no less natural than Bookchin's social evolution even though its lineage is not as clearly identifiable. It is natural because it represents an innate (though not necessary) property of humanity that can ground a resistance to the dominant commodity-influenced aesthetic. A Brecht play, on Marcuse's account, contains an inherent element of resistance to advanced industrial society.[62] Our potential to resonate to the play reflects a capacity that we all have. Even though there is an ontological dimension involved in this appeal to aesthetic experience, the foreseen transformation of society is founded in changes in the material basis of social formations that produce the aesthetics of material culture. The aesthetic transformation of technology results in new material forms that help to redefine, among other things, social relationships between humans, and eventually, by implication, material relations with the nonhuman natural world.

2. Additionally, both Bookchin and Marcuse argue that the potential for a corrected alignment of material culture is embedded in the same realm in which they locate their critiques. For Bookchin, even though social evolution has gone awry, it dialectically contains the resources for the needed transformation of society. For Marcuse, the aesthetic dimension contains the correct forms of reason that organizing structures in the world should reflect, even though it is in the commodity aesthetics of contemporary culture that he locates part of the basis of his critique of modern society.

But we must not gloss over the critical differences between these two authors. Clearly, the role of aesthetic experience is much more important for Marcuse than for Bookchin, and Marcuse offers us nothing like Bookchin's sophisticated account of social evolution. But differences like these are a virtue for the purposes of my overall argument in this chapter. Remember, I maintain that the further development of social ecology will be improved by an investigation into and, as appropriate, incorporation of the unique developments of critical theory. We can assume that the work of Marcuse will be more important for this project if it is arguably the case that both Bookchin and Marcuse are members of the same broad category of political ecologists, namely, environmental materialists. We can be even more certain of the utility of looking to Marcuse's work given the similarity in the structure of their respective critiques of the material basis of modern society just discussed.

Looking forward to the further evolution of social ecology, it is interesting to note that aesthetic experience bridges the distinction that Bookchin makes between the social and the political realms. For Bookchin, everyday lifestyle practices, including the consumption of cultural artifacts, is part of the social sphere of life, separate from the public, or political sphere. The social sphere for Bookchin, which we can reasonably assume would include many (if not all) aesthetic issues, is off limits for his critical apparatus: "It is the business of no one to sit in judgment of what consenting adults freely engage in sexually, or the hobbies they prefer, or the kinds of friends they adopt, or the mystical practices they may choose to perform."[63] Bookchin uses this distinction in his criticism of John Clark's attempt to transform social ecology (for Bookchin it is a corruption of the theory) in such a way that it goes beyond the political sphere and includes issues that Bookchin considers part of the social realm. Without getting into the debate concerning the relative virtues of Bookchin's distinction, his dispute with Clark, or an assessment of the accuracy of his interpretation of Clark, one can readily see that aesthetic issues could easily cross the boundary between Bookchin's social and political spheres. Especially if, as Bookchin maintains, we are to identify politics in the "Hellenic" sense of the term, there is a long history of an aesthetic dimension to the political sphere at the very least as regards the ethical and political role of drama in Greek life.[64] Thus, social ecology will need a strong account of the aesthetic dimensions of political life if it maintains a classical conception of the political sphere. If Bookchin responds with the claim that any aesthetic dimension to the Hellenic conception of politics cannot include individual aesthetic preferences (or, more probably, that I am taking his Hellenic description of the political sphere too literally), then I will need to go back to his dispute with Clark. For now, I will only say that there is a lively debate to be continued concerning the role of the possible reform of the cultural sphere in the further development of social ecology. If it turns out that there is a place for a politicization of the cultural sphere beyond what Bookchin envisions, then there will be a clear place for Marcuse's aesthetic dimension to enter into the discussion—if only as a prelude to looking at other political discussions of aesthetics.

Materialism and False Needs

Another important similarity between these two figures concerns their respective accounts of the production of human needs by the material organization of society. Here we have a similarity between the two views that is not just structural, but also substantive. It is the expansion of human needs that leads to the overextraction of natural resources. As materialists, both thinkers believe that eliminating the structures of production of such "false" needs is a necessary prerequisite to the creation of an ecologically sustainable society. On this point, an even closer connection can be made between these two figures than was found in the last section.

Early on Bookchin characterized his vision of the future as a "postscarcity" society. This society would reject all the false, dehumanizing needs that are stimulated by technologically based market and statist systems. The recognition of these false needs, according to Bookchin, will enable us to adopt a simpler way of material life, provide a greater degree of social freedom, and encourage a better conception of social relations.[65] The contrast of a postscarcity society is found in the present-day organization of growth-oriented societies for the production of enough goods to fulfill all the possible needs of their citizens. By Bookchin's reasoning, the fulfillment of such needs "is more likely to perpetuate unfreedom," in contrast to a more traditional society that may lack the needs commonly found in a technologically sophisticated society, but that promotes more freedom. Which needs do traditional societies lack? Needs for "sophisticated energy sources, dwellings, vehicles, entertainment and a steady diet of food that Euro-Americans take for granted."[66] But Bookchin is no primitivist. He does not romanticize traditional cultures, nor does he advocate a self-deceived adoption of aboriginal practices. Bookchin is a stalwart critic of such trends. The postscarcity society that Bookchin foresees is "modern." It is able to reject "false, dehumanizing needs precisely because it can be substantially free of need itself."[67] To "be free of need" means that a society can set voluntary limits on need. Such self-imposed limits allow a culture to slow its growth and change its patterns of material life. Additionally, Bookchin argues that the "realm of necessity," the cultural drive to accept false needs that go beyond what is minimally required, is a "historical phenomenon," not a "natural" form.[68] Thus, there is no natural drive for the pursuit of economic freedom.

Such a position is very close to the one taken by Marcuse concerning false needs in his *One-Dimensional Man*. Marcuse begins his discussion of false needs by suggesting that the "most effective and enduring form of warfare against liberation is the implanting of material and intellectual needs that perpetuate obsolete forms of the struggle for existence."[69] As with Bookchin, the locus of true needs is at the material level while all others are "preconditioned." Those needs created in advanced industrial society serve the "prevailing societal institutions and interests."[70] False needs are defined as those that are "superimposed upon the individual by particular social interests in his repression. . . . Most of the prevailing needs to relax, to have fun, to behave and consume in accordance with the advertisements, to love and hate what others love and hate, belong to the category of false needs."[71] For Marcuse, these needs must be overcome by a transformation of society. And like Bookchin, Marcuse thinks that while the ability to fulfill false needs may appear to be indicative of a society that has expanded its citizens' freedom, the imposition of these needs is actually a sign of repression. The only needs that "have an unqualified claim for satisfaction are the vital ones—nourishment, clothing, lodging at the attainable level of culture."[72] To supplant false needs with true ones is termed by Marcuse the "abandonment of repressive satisfaction."[73] Finally, like Bookchin, Marcuse situates false needs in a historical context, as is the determination of their objectivity.[74]

For both writers, technological organization systems (embedded in forms of political hierarchy and domination) order relations of production and create false needs upon which society gorges itself. To reject these false needs, which for both authors is a necessary step toward achieving environmental sustainability, the political and economic systems that perpetuate them must be transformed into new forms that do not encourage the consumption of cheap material goods. Some personal transformation is necessary on both theorists' accounts to reveal the danger of a reliance on material wealth as a gauge of individual welfare. For both Marcuse and Bookchin a healthy society is not measured by its ability to provide its members with easily reproduced material goods but by its provision of a greater degree of "real" freedom and autonomy for all humans. But before this new society can develop, its future members must be able to recognize that their economic output can be used to provide the material conditions for greater freedom and a fair standard of living for everyone, rather than luxury goods only for a few.

Bookchin suggests that there is a disjunction between the economic output of a society and the willingness of the members of that society to realize the alternative uses of their wealth: "Gauged merely by our current agricultural and industrial output, North Americans and Europeans clearly have the material means for making such a judgment; gauged by our social relations, on the other hand, we lack the freedom, values and sensibility to do so."[75] Marcuse identifies the same problem in his discussion of false needs, and there provides reinforcement for his claim that advanced industrial society has a built-in social and psychological ability to preserve itself and suppress dissent. Both writers have therefore located the bulk of the origin of technological manipulation of nature, necessary to produce a plethora of unneeded goods, in the material conditions of society. More importantly, both authors predicate their vision of a transformation of society for the better on a change in material conditions, specifically a change in the use of already existing social institutions and forms of production. Add to these similarities the fact that both maintain that there is a distinction to be made between true and false, or authentic and inauthentic, needs, and it becomes clear that their positions are very close. Even if there are subtle differences in the formulation of this last distinction, the idea of identifying false needs through a materialist critique of consumption patterns demonstrates a striking connection between these two authors.

In Bookchin's reply to my original comparison of his work with that of Marcuse, he stated that (1) it should be noted that the distinction between true and false needs was not a unique contribution by Marcuse, and (2) that the attribution of the true needs/false needs distinction to his own work was perhaps "off base, since unlike Marcuse's, [his] work usually gives centrality to 'choice' rather than 'needs.' "[76] As to (1), of course, Bookchin is correct. I presume that he thinks that pointing out this fact mitigates my comparison of his work with Marcuse's, but this is not the case. I have not argued that the true needs/false needs distinction is unique to Marcuse, only that he and Bookchin share it, including particular details of how assessing these different sets of needs works in a social analysis. This similarity helps to make my case that both theorists are environmental materialists with an interesting set of common views. As to (2), Bookchin's argument here is very interesting. Bookchin does not deny that he makes use of the true needs/false needs distinction, but only that his emphasis on choice makes the distinction relatively less important than the role it plays for Marcuse. Still, if both theorists employ the distinction in different contexts, then, as with the account given at the end of the last section, this difference will count as a virtue for a further development of either theory through a comparison with the other.

But when Bookchin substantiated argument (2) in his reply, he belied how similar his view actually is to Marcuse's. After the passage concerning the centrality of "choice" from his response cited above, Bookchin quotes himself, saying:

> I have always emphasized that everything that privileged people today enjoy must be made *available* to everyone so that they can rationally choose in a materially abundant society what they really want and need to *enjoy* life, indeed to "break the grip of the 'fetishization of needs,' to dispel it" and to recover the *freedom of choice*. (EF [*The Ecology of Freedom*], p. 69)[77]

But in this passage Bookchin has left out an important part of his original argument. In the full original citation from *The Ecology of Freedom*, Bookchin says: "To break the grip of the 'fetishization of needs,' to dispel it, *is* to recover the *freedom of choice*, a project that is tied to the freedom of the *self* to choose."[78] To understand how Bookchin uses the ideas of choice, needs, and the attainment of a free society, it is crucial that we unpack the "is" (the emphasis here is mine) in this last sentence, as well as its last clause.

Now, remember that Bookchin's original claim in his reply to me was that the true needs/false needs distinction did not reflect his views so much because of the centrality in his argument of choice rather than needs. So, the issue we need to address is whether Bookchin's rejection of false needs (here "to break the grip of the 'fetishization of needs'") involves simply (a) making a choice to do so, or (b) recovering the ability to freely engage in choice making. Bookchin's suggestion that rejecting false needs "*is* to recover the freedom of choice" more reasonably indicates a preference for option (b). If not, then Bookchin's argument that "everything that privileged people today enjoy must be made *available* to everyone so that they can rationally choose" could be interpreted as a claim that choice is made rational by the range of choices provided to the choice maker, not by the way in which choices are made. I take it that because Bookchin says in the full passage from *The Ecology of Freedom* that recovering freedom of choice is "a project that is tied to the freedom of the *self* to choose," he would reject the view that real choice is made possible only by increasing the range of choices.[79] Real choice will involve a reconsideration of what it is to choose, including the issue of the reassessment of the relationship between the freedom of the self and the articulation of needs. Recovery of the ability to make free choices will then include a rejection of false needs. But this means that the project of rejecting false needs will not solely be encapsulated in increasing the number of choices that can be made. So, rejecting false needs cannot be reduced solely to a rejection of the actual things that privileged people have, but also must involve a rejection of those needs which, according to Marcuse, mask privilege, or forms of social hierarchy (e.g., meeting the perceived "need" of TV shows that mask class inequality by appealing across class lines).[80] As I read it, then, Bookchin's argument is that rejecting false needs is necessary in part to help achieve a free society so that in turn everyone will recover the ability to make choices. But the ability to make choices, or more accurately, to be able to freely choose, is presumably not possible in a nonfree society, namely, one that is imbued with false needs. Therefore, the connection between choice and needs is much closer in Bookchin's work than he suggested in his response to this part of my

comparison between him and Marcuse. Even if this comparison between the two theorists is still not admitted, the foregoing discussion of
Bookchin's views on false needs and choices may reveal that the relationship between choices, needs, and freedom is so complex and imbued with chicken-and-egg puzzles in Bookchin's work that the further
evolution of this part of social ecology would benefit from looking at
other accounts that also employ the true needs/false needs distinction.

Marcuse and Bookchin on Individual versus Collective Action

Now, given such marked similarities in their positions, it is interesting
to see how Bookchin eventually criticizes Marcuse. It is certainly true
that Bookchin has found occasion to praise Marcuse's work, and even
to defend him against some critics.[81] But more important to me here is
one of Bookchin's central criticisms of Marcuse, namely, that the latter's work is inconsistent. Bookchin argues that Marcuse has anarchistic tendencies with reference to individuals, but that he easily slips into
a more familiar neo-Marxist mode when analyzing larger social groups.
Of course, again, given the structure of my comparison between Marcuse and Bookchin as environmental materialists, we should not be surprised that first, there will be significant differences between them, and
second, that they themselves would have pointed out these differences.[82] But the fact that such differences exists does not mean that
both figures cannot be placed in the same broad category of political
ecology. Therefore, I did not need to refute Bookchin's substantive criticisms of Marcuse in order to make the first fundamental point of this
essay that both figures are environmental materialists. But if I am to go
on to say that a further elaboration of social ecology would benefit
from a consideration of Marcuse's critical theory, and in particular his
common concern with Bookchin regarding the effects of technological
rationality and industrial society on human self-organization and interactions with nature, then I need to look more closely at Bookchin's
criticism of Marcuse. The reason is that Bookchin's criticism suggests
that Marcuse's work contains a fatal flaw that places it outside the
bounds of the politics of social ecology. On Bookchin's account, Marcuse's Marxism makes his work much less relevant to the further development of social ecology. I will argue that such a view is flawed, first,
because Bookchin provides a misleading interpretation of the implications of Marcuse's views, and second, because Bookchin's work arguably has the same problem he assigns to Marcuse. After discussing these
issues I will return, in my last section, to Bookchin's criticism of Mar-

cuse and suggest that part of Bookchin's critique discussed in this section leads to a surprising conclusion not foreseen by Bookchin in his assessment of critical theory.

In his essay "On Neo-Marxism, Bureaucracy, and the Body Poli-
tic," Bookchin alternately praises and criticizes Marcuse on a number
of fronts. He suggests that Marcuse makes a variety of mistakes, including sloppiness concerning his historical account of the Spanish Civil
War. But Bookchin's praise for Marcuse's investigations into "direct democracy, decentralization, representation, spontaneity and liberatory
structures" is mitigated by what appear to be comments suggesting that
Marcuse's work is the best of a bad tradition. Bookchin states that it
would be a grave error to "view my remarks on Marcuse as a critique
of Marcuse as an *individual* thinker. Inasmuch as his theoretics have
dealt more directly with social problems than that of any other neo-
Marxist body of theory, they more clearly reveal the limits of the neo-
Marxist project."[83] In short, Marcuse's work reveals the inherent incompatibility of critical theory with social ecology through Marcuse's
commendable, but evidently failed, efforts to address some of the most
important issues for social ecology. But the most important part of
Bookchin's critique is his suggestion that in Marcuse's wedding of
Freud with Marx, Marcuse muddles an account of the capabilities of
individuals with the capacities of social institutions and classes. Bookchin argues that the prior tendency, to see individuals as the primary
actors in their own liberation, represents a form of anarchism, and the
later move, to see social classes as the main vehicle for liberation, is
more akin to neo-Marxism. How, he asks, can these two approaches be
reconciled?[84] How can the autonomy of the individual, freed from the
slavery of the technocracy of advanced industrial society, be exercised
in a neo-Marxist framework? What would that framework be? Bookchin is not explicit about his worry here. We could charitably assume
that what Bookchin has in mind is the potential authoritarian collectiv-
ism of a Marxist state, even one tempered by neo-Marxism.

Bookchin makes no mention of technology here, but he does indicate that his criticisms are at least in part based on his reading of *One-
Dimensional Man*.[85] It is unclear, though, whether Bookchin's criticism
takes into consideration the ease with which individual and class interests can be merged in a materialist account of the corruption of technological systems. But more importantly for the comparisons I have offered so far, Bookchin offers no argument as to whether Marcuse's
aesthetic transformation of technology, largely an individual effort, is
incompatible with the individual freedom that will thrive after the overcoming of false needs. If the society that Marcuse assumes would be
necessary to sustain such freedom closely approximates Bookchin's

postscarcity society, then why should we presume that Marcuse would tolerate a society that would suppress the cultivation of true needs? Does Bookchin think that the author of *Soviet Marxism* is blind to the authoritarian potential of centralized control, or only indifferent to that possibility?

Bookchin's suggestion here of a necessary incompatibility between individual freedom and collective action apparently reduces the collective aspirations of all of neo-Marxism to a singular form. In other works, Bookchin makes similar reductions concerning other aspects of Marxism and generally does not acknowledge the serious differences between orthodox Marxism and Western forms of Marxism.[86] For Bookchin, all Marxist theories of the family, feminism, and ecology (no matter how nuanced or self-critical about their intellectual tradition) are doomed because "Marx negates the issues they raise, or worse, transmutes them into economic ones."[87] Bookchin finds a "reactionary aspect" to the Marxist and socialist projects, in that they retain "the concepts of hierarchy, authority and the state as part of humanities' post-revolutionary future."[88] But, of course, such claims represent massive generalizations that ignore important developments in Western Marxism from the Gramscian recognition of the importance of culture to the philosophical subtleties of analytic Marxism.[89] Further, in the first edition of *The Philosophy of Social Ecology,* where Bookchin greatly expands on his own theory of "dialectical naturalism," he makes a point of distinguishing it from "Hegel's empyrean, basically anti-naturalistic dialectical idealism and the modern, often scientific dialectical materialism of orthodox Marxists."[90] But what modern form of Marxism, orthodox or not, does he have in mind that could accurately be described as embracing a scientific dialectical materialism? If there is such a school of thought still in existence, it is such a relic of a dead age that to oppose oneself to it is not to take seriously a comparison with the best that Marxism has to offer. At least it is clear that the Frankfurt School's stand against instrumental reason would resist any characterization that hints of "scientific dialectical materialism." And this is not even to go into the rejection of the metaphysics of historical materialism that is integral to Marcuse's critical social theory.

An important part of Bookchin's misreading of the materialism of Marcuse, and the resulting contradictions he identifies in Marcuse's work, can be found in Bookchin's lack of acknowledgment of the shifts in Western Marxism on the potential of the proletariat as a revolutionary class. In one essay Bookchin suggests that

it is no longer possible to ignore the fact that the proletariat is not only less susceptible to revolutionary ideas than it was in the past; worse, the proletariat itself is dwindling in numbers and in economic power—that is, unless one wants to recast all the actors in the Marxist drama and make professionals, managers, technicians, and white-collar employees into a "new working class."[91]

Quite so. Given that this comment comes right after a short critique by Bookchin of Marx's idea of the "hegemonic working class," it is clear that Bookchin is criticizing the idea of a necessarily revolutionary working class. Certainly, such an idea is difficult to demonstrate. If Bookchin is correct in arguing that Marcuse mixes appeals to individual and class actors as the means for social transformation, then, according to Bookchin's interpretation of what counts as a Marxist conception of a class, Marcuse would certainly have a problem. But Bookchin never acknowledges that Marcuse and the rest of the Frankfurt School share his idea of the limitations of class action, right down to the details of how the proletariat has been supplanted in the workplace with a class of managers and technicians, thereby eliminating the possibility of the development of a proletarian revolutionary consciousness.

If Bookchin persists in overlooking similarities like this when pressing his interpretation of Marcuse (and other socialists interested in political ecology), then he may be opening himself up to criticisms along the same lines. Particularly given the similar structures of Bookchin's idea of social evolution and Marcuse's self-negation of technology, Bookchin would be hard-pressed to explain how his idea was not somehow also susceptible to the criticism he has leveled against Marcuse concerning individual versus class action. Bookchin's account of the relationship between social evolution and the motivation to transform material society also mixes the potentials of individuals as the particular repository of the gifts of social evolution with the institutions and practices needed to collectively set in motion the new ecological order. Even though this social order would be restricted to small communities, it would still necessitate the cooperation of individuals and smaller social groups to organize a postscarcity society. As much as with Marcuse, especially the Marcuse of *One-Dimensional Man*, the collections of individuals working for Bookchin's new social formations would be vehicles for social transformation. In the 1960s Bookchin wrote that the agents of social transformation would be collective, though not in the traditional Marxist sense of a class. Revolutionary potential would come out of the "emergence of an entirely new class, *whose very essence is that it is a non-class.*"[92] More recently Bookchin has written of the "reemergence of 'the People,'" in contrast to the decline of 'the Proletariat.' "[93] This transclass constituency, for Bookchin, has produced "entirely new issues, modes of struggle and forms of organization and calls for an entirely new approach to theory and praxis."[94] But class or not, this constituency is collective, rather than individual. Furthermore, it is the collection of these actors together that has produced new approaches to theory and practice, presumably different from what individuals alone would develop. If Bookchin persists in criticizing Marcuse for conflating class action and individual action, and if it is true that Marcuse does not have an orthodox Marxist conception of a class, then Bookchin's criticisms would either be false or would also apply to his own work. While Bookchin's notion of a "transclass" formation means that he is not susceptible to the claim that his own approach is neo-Marxist, the relative probability of the communal forms of transformation he embraces lapsing into authoritarianism should be as big a worry for him as for Marcuse. I would rate this probability as very low for both theories.

The views of Bookchin and Marcuse are compatible as materialist accounts of the critique of industrial society and the two theories are comparably comprehensible. A few more brief comments on Marcuse's later thoughts on nature should show that while Marcuse's analysis eventually diverges more sharply from Bookchin's, it is not mutually exclusive with an expanded social ecology. In some ways Marcuse's critical theory may help in the evolution of social ecology, at least for those proponents of the theory who wish to reconcile their views with environmental ontologists, potentially producing a form of radical ecology that includes the best ideas from the spectrum of political ecologies.

A Possible Route for Reconciliation?

By 1972, Marcuse had reconsidered some of his positions on technology and had incorporated a more explicit analysis of nature into his critical theory. In *Counter-revolution and Revolt* he argues that the liberation of nature from its "objectified" technological realm is the first step in the liberation of humans. After some consideration of the growth of the New Left, Marcuse suggests:

> What is happening is the discovery (or rather, rediscovery) of nature as an ally in the struggle against the exploitative societies in which the violation of nature aggravates the violation of man. The discovery of liberating forces of nature and their vital role in the construction of a free society becomes a new force in social change.[95]

Marcuse suggests that when we commercialize, pollute, or militarize the natural environment, human environments are also destroyed. This is a position that Bookchin could agree with, as long as it is not reducible to a crass instrumentalism. As I indicated above, in the development of his political ecology, Bookchin has always argued that the causes of environmental problems are to be found in material disruptions in the human social world. Whether Marcuse and Bookchin would completely agree with each other depends on the interpretation of what Marcuse means by the potential of "nature as an ally in the struggle against the exploitative societies." If Marcuse means that a description of environmental problems helps to reveal the roots of exploitation in human society, then he and Bookchin are arguing similar positions. They are even closer if by "violation" of nature, Marcuse does not mean something like "domination" of nature, and if the role of "liberating forces of nature" with their "vital role in the construction of a free society" includes evolutionary forces. But even if Marcuse and Bookchin wind up in complete disagreement about these particular issues, the most important point for my analysis is that Marcuse has more fully developed his environmental materialism in this work from the argument that we must reconstruct in *One-Dimensional Man*. Nature is important here for Marcuse because the implications of an analysis of its problems aids in the consideration of human social problems. Marcuse's clear object of concern is the construction of a "free society"; certainly the same is true for Bookchin.

Marcuse goes on to argue that if technology and mechanistic reason represent dangers to the way humans treat each other, then the liberation of humans goes hand in hand with a liberation of nature from the same oppressive social relations. And here, perhaps more closely than in the last section, we can see that Bookchin's earlier assessment of the structure of Marcuse's account of social transformation was correct, at least insofar as this account involves two distinct (though not necessarily incompatible) elements. For Marcuse, liberation from such technologically influenced thinking must encompass an individual and social struggle in order to achieve the goal of human and ecological liberation. The individual struggle will include a change in individual consciousness concerning nature. Marcuse's environmental materialism therefore requires an ontological dimension. Marcuse argues that while

> the historical concept of nature as a dimension of social change does not imply teleology and does not attribute a "plan" to nature, it does conceive of nature as subject-object: as a *cosmos* with its own potentialities, necessities and chances. And these potentialities can be, not only in the sense of their value-free function in theory and practice, but also as bearers of *objective values*. . . . Violation and suppression [of nature] then mean that human action against nature . . . offends against certain objec-

tive *qualities* of nature—qualities which are essential to the enhancement and fulfillment of life. And it is on such objective grounds that the liberation for man to his own humane faculties is linked to the liberation of nature—that "truth" is attributable to nature not only in a mathematical but also in an existential sense. The emancipation of man involves the recognition of such truth in things, in nature.[96]

This passage may be interpreted in many ways, indeed, probably too many ways. Generally, Marcuse goes beyond acknowledging the economic connections between the sustainability of nature and the survival of humans to a recognition of the link between humans and nature on another level. Nature exists, before we come to violate it, free to develop in the same way that humans are free to develop. Nature has, through the processes of evolutionary change, a myriad of possible options for growth and development other than those that are thrust upon it by one particular species. Humans have taken from nature in order to direct our economic growth toward a certain path, the path that best serves our particular needs at this stage in our history. But Marcuse's claim is not just that we should avoid environmental destruction because the Earth is our habitat; he argues that in doing so we restrict objective values in nature that exist as autonomous potentialities.[97] The question of why these objective values are important is possibly another issue. Marcuse could be saying that we must respect these values in a way similar to the respect we owe to the value of other persons. Such an interpretation would make Marcuse's work a neglected precursor of the more substantive theories of intrinsic value in nature in contemporary environmental ethics. But because Marcuse says that these values are objective qualities "which are essential to the enhancement and fulfillment of life" we are faced with a critical dilemma. On the one hand, if by "life" Marcuse means only human life, then Marcuse shares a view similar to Bookchin's, identifying the social importance of nature in relation to humans. (Of course, Bookchin would find objectionable the terms Marcuse uses to express this view, such as the anthropomorphic idea of the "liberation" of nature.[98]) If by "life" Marcuse means all life (which I think is most likely true), then he may very well be arguing beyond Bookchin that nature is a subject on par with humanity. As such, the full revolutionary potential of a political ecology on Marcuse's view would include acknowledgment of the subjectivity of nature, hence further revealing the environmental ontology of Marcuse's later work.

But even with these ambiguities in the interpretation of Marcuse's argument, it is again the case that the structure of Marcuse and Bookchin's theories are similar as forms of environmental materialism.

For both, there is a conceivable state of nature such that, if we are not connected to it, then we are not connected to our capacity for living in harmony with the nonhuman natural world. For Bookchin, the preferred state of nature is the positive developmental unfolding of a potentiality in natural evolution, and the proper relation of social evolution to it. If social evolution becomes corrupted, then human relationships suffer, and in turn humans produce imbalances in nature. For Marcuse, nature has potential objective value that one way or another requires a particular human alignment to those values. Both agree that there is some preferred state of humans and nonhumans (either a hypothetical or an actual state depending on one's reading of either theorist) that has been lost, or that simply can be imagined but does not exist (thanks in part to the distorting importance of false material goods), and that this state can only be restored by some cooperation between humans with implications for nature predicated on a change in the organization of material production. Certainly, given his remarks in *Counter-Revolution and Revolt,* Marcuse would agree with Bookchin's warning that society will be askew until it uses its "collective wisdom, cultural achievements, technological innovations, scientific knowledge and innate creativity for the benefit of the natural world."[99]

But Marcuse is more open than Bookchin in suggesting that such commitments may require a form of ontological realignment with nature, namely, an ontology that does not necessarily erode a thin materialist or social emphasis to a radical ecology, or digress into a form of spiritualism. In a later work, published posthumously, Marcuse more explicitly makes this claim concerning the importance of an ontological transformation of the relationship between humans and nature. There he defines radical change as a transformation not only in the basic institutions of society but also in the individual consciousness of members of that society. His point is that industrial society maintains the established political order in part by reproducing itself in the consciousnesses and unconsciousnesses of individuals in that society. Such control must be confronted on the individual as well as on the social front.[100] Again, then, Bookchin was right in arguing that both the individual and collective elements were at work in Marcuse's formulation of social transformation. It is striking to note, though, that the implications of this view are not that Marcuse's pull toward collective transformation would lead him down the road of repressive centralism, but that his form of individualism may lead the political ecology of critical theory down the path blazed by Naess's interpretation of deep ecology. In short, the "anarchical" tendencies in Marcuse's work of the 1960s that Bookchin admired—namely, Marcuse's early focus on the importance of individuals in social transformation—are actually closer to

Naess's work and have arguably evolved into a more serious challenge to Bookchin's political ecology than Marcuse's collectivist intuitions, which remained securely materialist in orientation.

But it would be a mistake to conclude from this last observation, that social ecologists should now avoid critical theory in order to further solidify their distance (at least as marked by Bookchin) from deep ecologists in particular, and environmental ontologists in general. Nonetheless, Bookchin's position on the utility of Marcuse's work seems clear. In his response to the original version of this essay he made a strong case that Marcuse's critical theory was not sufficient to warrant a serious consideration for a political ecology.[101] No doubt Bookchin is correct in many of his criticisms of Marcuse. But I certainly have not argued that Marcuse's work can stand as a political ecology on its own. There are parts of his thought that I have described as a form of environmental materialism which can aid in the development of a much more sophisticated theory. Anyone who wanted to rest on Marcuse's work alone as the basis for political ecology would have a very incomplete theory indeed. But a reading of Bookchin with Marcuse *could* help to inform social ecologists that a commitment to environmental materialism can also sustain some important ontological considerations. Marcuse's critical theory stands as an example of a credible, yet incomplete, materialist analysis of nature that is commensurable with a certain kind of claim that nature can possess a form of subjectivity which we have to consider in the identification of our selves. Additionally, Marcuse retains in this recognition a commitment to an environmental analysis that focuses on human social distinctions. Insofar as I have shown the compatibility of Marcuse's position with Bookchin's as a form of environmental materialism, social ecologists interested in exploring the potential of their school of thought should be able to recognize that not all kinds of environmental ontological commitments, possibly including a kind similar to that embraced by deep ecologists, should be disregarded in the formation of a radical ecological philosophy. Along these lines, Henry Blanke has argued that while Marcuse's position is indeed an amalgam of both the materialist and the ontologist positions, Marcuse's thought is especially relevant today because it offers a corrective to precisely those problems that Bookchin finds in deep ecology: misanthropy, primitivism, antirationalism, quietism, and so on.[102]

Such a theoretical expansion of social ecology could eventually lead to a practical realignment of the activities of groups on either side of the materialist–ontologist divide. As I noted in my first section, Naess acknowledges that his serious consideration of social ecology has improved his own form of deep ecology. I do not point out this improvement in Naess's work because, as Bookchin seems to think, I am a deep ecologist. I am not a deep ecology theorist or activist. I am not even arguing that social ecologists should engage in a reciprocal study and appreciation of deep ecology itself. I have instead taken the more restricted view that a careful consideration of critical theory may serve as a forum in which social ecologists can address the broader issue of the importance of an environmental ontology in the formation of a more rigorous and complete social ecology. Importantly, this forum is uncorrupted by the legacy of unfortunate pronouncements of xenophobia, racism, or antihumanism. A critical but open comparison of social ecology with other forms of political ecology may result in an abandonment of the excessiveness and dismissiveness common in past debates among greens, and get us on the road to developing a more complete political ecology.

Notes

1. This chapter is a completely revised version of my "Rereading Bookchin and Marcuse as Environmental Materialists," *Capitalism, Nature, Socialism*, Vol. 4, No. 1, March 1993, pp. 69–98. I have made several changes in light of Murray Bookchin's critique of the original article in his "Response to Andrew Light's 'Bookchin and Marcuse as Environmental Materialists,'" *Capitalism, Nature, Socialism*, Vol. 4, No. 2, pp. 101–113, including expansion on my original reply to Bookchin, "Which Side Are You On?: A Rejoinder to Murray Bookchin," *Capitalism, Nature, Socialism*, Vol. 4, No. 2, June 1993, pp. 113–120. At some points in the chapter I have explicitly returned to Bookchin's response and addressed his worries. In other places I have corrected my original presentation of Bookchin's views without explicitly acknowledging his criticism. Nonetheless, in this chapter I attempt to incorporate or answer all of the criticisms Bookchin made in his critique of my original article. As such, this chapter should replace "Rereading Bookchin and Marcuse" and "Which Side Are You On?"

I wish to thank Bill Chaloupka, John Clark, Andrew Feenberg, Joel Kovel, David Macauley, Ron Perrin, and Alan Rudy for their helpful comments. I also feel compelled to note that while I have undertaken a major revision of my original piece in order to clarify my views on these issues too awkwardly put before, my initial motivation for engaging in the comparison between Bookchin and Marcuse has diminished greatly in the last four years. I am no longer sure how important a further elaboration of utopian theories is for meeting the current challenges facing those interested in the relationship between politics and nature. Thus, I offer this revised chapter mainly to those who will continue working on these topics. I encourage them to do such work, but as for myself, I have nothing personally invested anymore in a social ecology tempered by critical theory. It is also important to note that in preparing this new version of my previous paper I have not attempted to incorporate discussions of any further material from Marcuse's work than I had previously. Certainly, a more thorough treatment of Marcuse on nature and ecology would have to include discussions of *The Aesthetic Dimension* (Boston: Beacon Press, 1978) and *Eros and Civilization* (Boston: Beacon Press, 1966). A more substantial analysis of Marcuse's work on ecology is available in Tim Luke's *Ecocritique* (Minneapolis: University of Minnesota Press, 1998), Chapter 7.

2. For an overview of the issues and stakes in this debate, see Murray Bookchin, "Social Ecology versus Deep Ecology," *Socialist Review,* Vol. 18, No. 3, July–September, 1988, pp. 9–29; Kirkpatrick Sale, "Deep Ecology and Its Critics," *The Nation,* May 14, 1988, pp. 670–675; and Murray Bookchin, "As If People Mattered," *The Nation,* October 10, 1988, p. 294.

3. For examples of treatments of political ecology that seem to focus only on the social ecology–deep ecology debate, see Murray Bookchin, "Introduction to Elkins," *Telos,* 82, Winter, 1989–1990, pp. 47–51, and Stephen Elkins, "The Politics of Mystical Ecology," *Telos,* 82, Winter, 1989–1990, pp. 52–70. An argument can also be made that Michael E. Zimmerman's *Contesting Earth's Future* (Berkeley and Los Angeles: University of California Press, 1994) suffers from the same problem. Zimmerman, however, is very thorough in his treatment of ecofeminism, even though there is no role in his account of the debates in political ecology for socialist ecology or ecological Marxism. I no longer count this as so much a flaw of Zimmerman's scholarship as it is a reflection of how the debates in political ecology have been shaped by influential scholars and activists like Bookchin.

4. Murray Bookchin, *The Philosophy of Social Ecology,* 2nd ed. (Montreal: Black Rose Books, 1996), p. 95, note 2. This edition is an updated version of a book originally published in 1990. In this version Bookchin makes a point of saying that he has undertaken revisions to excise "favorable references to the Frankfurt School and Theodor Adorno." Bookchin says that "like Leszek Kolakowki, I have come to regard much of Adorno's work as intellectually irresponsible, wayward, and poorly theorized, despite the brilliance of his style (at times) and his often insightful epigrams" (p. ix). For an alternative view on the Frankfurt School on nature see Steven Vogel, *Against Nature: The Concept of Nature in Critical Theory* (Albany: State University of New York Press, 1996).

5. Murray Bookchin, *Toward an Ecological Society* (Montreal: Black Rose Books, 1980), p. 15.

6. Murray Bookchin, *Remaking Society: Pathways to a Green Society* (Boston: South End Press, 1990), p. 20.

7. Murray Bookchin, "Ecology and Revolutionary Thought," in S. Blau and J. B. van Rodenbeck, eds., *The House We Live In* (New York: Macmillan, 1971), pp. 442–443.

8. Bookchin, *Remaking Society,* p. 15.

9. Ibid.

10. Bookchin, "Social Ecology versus Deep Ecology," p. 15. For Naess's shallow versus deep distinction, see "The Shallow and the Deep: Long-Range Ecology Movements: A Summary," *Inquiry,* Vol. 16, 1973, pp. 95–100.

11. Herbert Marcuse, *One-Dimensional Man* (Boston: Beacon Press, 1964), p. xv.

12. It should be noted that Bookchin, more than Marcuse, explores the possibilities of an alternative technology that is more compatible with natural systems. See Bookchin's essay, "Toward a Liberatory Technology," in *Post-Scarcity Anarchism,* 2nd ed. (Montreal: Black Rose Books, 1986 [1971]), pp. 105–161. This paper, however, does not represent a philosophical treatment of the relationship between technology and nature. For a more thorough treatment of the theory of technology in Bookchin's work, see the essays by David Watson and Eric Higgs in this volume. Alan Rudy and I have briefly surveyed the change in Bookchin's views on technology over the years, concluding that his "ecotopic visions have become increasingly low-technology affairs"; see Alan Rudy and Andrew Light, "Social Ecology and Social Labor: A Consideration and Critique of Murray Bookchin," *Capitalism, Nature, Socialism,* Vol. 6, No. 2, June 1995, pp. 84–85.

13. Herbert Marcuse, "Some Social Implications of Modern Technology," *Studies in Philosophy and Social Science,* Vol. 9, 1941, p. 141.

14. Bookchin, "Response to Andrew Light's 'Bookchin and Marcuse,' " p. 104.

15. Both John Clark and Alan Rudy have pointed out to me that the same could be said for Bookchin. He too tends to vacillate between pessimism and hope. The Introduction to Bookchin's early *Our Synthetic Environment* (New York: Harper & Row, 1962), for example, is a similarly pessimistic text.

16. By materialism, I mean a "thin" interpretation of the term. This point is clarified at the end of this section.

17. In another much more thorough treatment of the ecological implications of Marcuse's thought, Henry T. Blanke has agreed with my suggestion that Marcuse can be fairly categorized as an environmental materialist. See Blanke, "Domination and Utopia: Marcuse's Discourse on Nature, Psyche, and Culture," in *Minding Nature: The Philosophers of Ecology,* ed. David Macauley (New York: Guilford Press, 1996), pp. 186–208.

18. Bookchin, *Philosophy of Social Ecology,* p. 74.

19. Murray Bookchin and Dave Foreman, *Defending the Earth: A Dialogue between Murray Bookchin and Dave Foreman,* ed. Steve Chase (Boston: South End Press, 1991), p. 32.

20. Cited in David Rothenberg's Introduction to Naess's *Ecology, Community, and Lifestyle,* trans. David Rothenberg (Cambridge: Cambridge University Press, 1989), p. 20; emphasis added.

21. Andrew Light, "Deep Socialism?: An Interview with Arne Naess," *Capitalism, Nature, Socialism,* Vol. 8, No. 1, March 1997, p. 84; emphasis in original.

22. See the first chapter of *Remaking Society,* which indiscriminately lumps goddess worship with deep ecology. While some deep ecologists may invite this comparison, certainly not all warrant it—see, for example, David Rothenberg's remarks on the differences between various branches of theoretical deep ecology in his Introduction to *Ecology, Community, and Lifestyle.*

23. Murray Bookchin, "New Social Movements: The Anarchic Dimen-

sion," in *For Anarchism: History, Theory, and Practice,* ed. David Goodway (London: Routledge Press, 1989), p. 273.

24. Cited in Bill Devall and George Sessions, *Deep Ecology: Living as if Nature Mattered* (Salt Lake City, UT: Peregrine Smith Books, 1985), p. 16.

25. Bookchin, *Remaking Society,* p. 12.

26. Ibid., p. 10.

27. Bookchin, "As If People Mattered," p. 294.

28. Murray Bookchin, "Comments on the International Social Ecology Network Gathering and the 'Deep Social Ecology' of John Clark," *Democracy and Nature,* Vol. 3, No. 3, 1997, pp. 154–155.

29. Though he does not address the deep ecology–social ecology debate, Bryan Norton, in his *Toward Unity among Environmentalists* (Oxford: Oxford University Press, 1991), offers perhaps the most sustained argument that debates like this one should not distract us from unified action. More importantly, he argues that empirically there are few if any serious differences in the kinds of policies the different sides in the ecology debates would embrace, at least among their reasonable proponents. One can also point to the fact that groups like Earth First! have gone through substantial changes in the last few years with the exit of Foreman and others from the organization. In part, Foreman became uncomfortable with movements in the group toward social and economic justice issues of the sort that Bookchin endorses. See Bookchin and Foreman, *Defending the Earth,* p. 119.

30. For a more thorough discussion of this materialist/ontological tension in Marcuse's work, see Andrew Feenberg, "The Bias of Technology," in *Marcuse: Critical Theory and the Promise of Utopia,* ed. Robert Pippin, Andrew Feenberg, and Charles Webel (South Hadley, MA: Bergin and Garvey, 1988), pp. 225–256, and Blanke, "Domination and Utopia." In a direct comparison with my analysis, Blanke says that what I identify as Marcuse's combined materialist–ontologist perspective is the same as the tension he identifies between Marcuse's "Marxist and mystical elements." Describing Marcuse's ontology as "mystical" may even be a more powerful way of demonstrating its connection with views like that found in the deep ecology literature, though it also risks the same general critique that Bookchin levies against all forms of mysticism. An extension of such a critique to Marcuse would be mistaken for reasons pointed out in note 102, below.

31. Also see Arne Naess, "The Politics of the Deep Ecology Movement," reprinted in *Wisdom in the Open Air,* ed. Peter Reed and David Rothenberg (Minneapolis: University of Minnesota Press, 1993), pp. 82–99.

32. Light, "Deep Socialism?: An Interview with Arne Naess," p. 80.

33. Ibid.

34. Ibid.

35. Ibid.

36. Bookchin, "Response to Andrew Light's 'Bookchin and Marcuse,' " pp. 101–102.

37. Murray Bookchin, *The Ecology of Freedom* (Palo Alto, CA: Cheshire Books, 1982), p. 365.

38. I characterize this connection further in my "The Role of Technology

in Environmental Questions," *Research in Philosophy and Technology,* Vol. 12, 1992, pp. 83–104.

39. Bookchin, "Response to Andrew Light's 'Bookchin and Marcuse,' " p. 102.

40. Ibid., p. 101.

41. Bookchin, *Defending the Earth.* Even in this fairly conciliatory exchange the priority of Bookchin's environmental materialism shines though. See my analysis of the book in "Which Side Are You On?"

42. Bookchin, *Remaking Society,* p. 32.

43. Ibid., p. 23.

44. Ibid., p. 31.

45. Bookchin, *Philosophy of Social Ecology,* p. 92.

46. Ibid., p. 82.

47. Bookchin, *Remaking Society,* p. 23. For a full critique of this set of claims, see Alan Rudy's chapter in this volume. For a thorough appreciation of Bookchin's contribution to evolutionary theory, see Chapter 3 by Glenn Albrecht, this volume.

48. Bookchin, *Remaking Society,* p. 35.

49. Ibid., p. 27.

50. Ibid., p. 20.

51. Ibid., p. 18. Detailed examples of what Bookchin had in mind with such technology can be found as early in his writings as his "Toward a Liberatory Technology."

52. Ibid., pp. 35–36.

53. Ibid., p. 27.

54. Bookchin, "Response to Andrew Light's 'Bookchin and Marcuse,' " pp. 105–106. For the sake of argument I will concede that Bookchin does not argue that evolution of any type has a necessary direction. I do, however, still think that despite Bookchin's claims to the contrary, he does edge up to this position on a number of occasions. See Chapter 2 by Robyn Eckersley in this volume.

55. Bookchin, *Defending the Earth,* p. 33.

56. See G. E. Moore, *Principia Ethica* (Cambridge, UK: Cambridge University Press, 1994).

57. Bookchin has a footnote in *The Philosophy of Social Ecology* (note 23, p. 182) concerning what he calls the "epistemological fallacy." I do not know if he means by this the naturalistic fallacy. His explanation of the epistemological fallacy seems similar but not identical. But if he thinks his claim that dialectical naturalism avoids whatever problem he is answering because it is "structured around the *reality* of potentiality," he is not giving a sufficient answer to the naturalistic fallacy. A fact about the world, such as the fact that there is a potential for or tendency toward diversity, does not make diversity morally right. For the record, I do think there is something else that makes this diversity morally right, but it is not the fact that diversity is an evolutionary tendency with consequences I prefer. Remember that the naturalistic fallacy was formulated as part of a defense of moral realism.

58. Marcuse, *One-Dimensional Man,* p. xv.

59. This argument has many problems as it is presented in Marcuse's work. For the most thorough attempt to resolve these problems and update this view (to a new theory of the "subversive rationalization" of technology), see the following works by Andrew Feenberg: *Critical Theory of Technology* (Oxford, UK: Oxford University Press, 1992); *Alternative Modernity: The Technical Turn in Philosophy and Social Theory* (Berkeley and Los Angeles: University of California Press, 1995); and *Questioning Technology* (London: Routledge, forthcoming).

60. Marcuse, *One-Dimensional Man*, pp. 228–239.

61. Bookchin, *Remaking Society*, p. 38.

62. Marcuse, *One-Dimensional Man*, p. 70.

63. Bookchin, "Comments on the International Social Ecology Network Gathering," p. 158. Bookchin does say that "matters of human rights" in the social sphere can be politicized.

64. See, e.g., Martha Nussbaum, *The Fragility of Goodness: Luck and Ethics in Greek Tragedy and Philosophy* (Cambridge, UK: Cambridge University Press, 1986).

65. Bookchin, *Toward an Ecological Society*, p. 25. It should be noted that although the importance of the concept of a postscarcity society has diminished for Bookchin over the years, he still maintains that it is a "precondition" of "exorcising the hold of the economy over society." See Bookchin's new edition of *Post-Scarcity Anarchism*, p. 45.

66. Ibid., pp. 24–25.

67. Ibid., p. 25.

68. Ibid., p. 26.

69. Marcuse, *One-Dimensional Man*, p. 4.

70. Ibid.

71. Ibid., p. 5.

72. Ibid.

73. Ibid., p. 7.

74. Ibid., p. 6. I would like to point out that I do not endorse this highly controversial distinction between true needs and false needs. Both Bookchin and Marcuse can and have been criticized for employing the true needs/false needs distinction, which easily lends itself to a form of elitism. For an interesting discussion of some problems with this distinction, see William Leiss, *The Limits to Satisfaction: An Essay on the Problem of Needs and Commodities* (Toronto: University of Toronto Press, 1976). My own view is that while the early version of this distinction should be rejected, the general intuitions behind the utility of the distinction should be preserved. A complete radical social theory, including a theory of political ecology, must include a theory of the propriety and relative worth of various human needs. For a much more thorough and sophisticated account of human need than found in the social ecology or critical theory literature, see Len Doyal and Ian Gough, *A Theory of Human Need* (New York: Guilford Press, 1991).

75. Bookchin, *Toward an Ecological Society*, p. 183.

76. Bookchin, "Response to Andrew Light's 'Bookchin and Marcuse,'" pp. 103–104.

77. Ibid., p. 104; emphasis in original.

78. Bookchin, *Ecology of Freedom,* p. 69; emphasis added.

79. Further down on the same page in *Ecology of Freedom,* Bookchin says, "In a truly *free* society . . . needs would be formed by *consciousness* and by *choice*" (emphasis in original). I take it that by "free" here he means a more substantive sense of the term than simply a society with abundant goods with no prices.

80. Marcuse, *One-Dimensional Man,* p. 8.

81. Bookchin outlines his positive interaction with Marcuse in his "Response to Andrew Light's 'Bookchin and Marcuse,' " pp. 110–111.

82. Bookchin gives a very compelling account of differences between his work and Marcuse's which are not dealt with in this chapter in his response to my original article.

83. Bookchin, *Toward an Ecological Society,* p. 222; emphasis in original.

84. Ibid., p. 225.

85. By way of support for this supposition, Bookchin clearly points out his uneasiness with Marcuse's rationale for "centralized control" in *One-Dimensional Man* in his response to my original article. See Bookchin, "Response to Andrew Light's 'Rereading of Bookchin and Marcuse,' " p. 110.

86. Bookchin is not alone in equating all forms of Marxism. Kate Soper identifies this tendency among environmental theorists as the most important reason (albeit political rather than intellectual) for why Marxism is considered to be "no friend of ecology"; see Kate Soper, "Greening Prometheus: Marxism and Ecology," in *Socialism and the Limits of Liberalism,* ed. Peter Osborne (London: Verso, 1991), pp. 272–273.

87. Bookchin, *Toward an Ecological Society,* p. 209.

88. Bookchin, *Post-Scarcity Anarchism,* p. 20ff.

89. For a much more thorough exposition of the ecological potential of Marxism conjoined to a critique of Bookchin's social ecology for its wholesale refusal to entertain the possibility of retaining at least some of the subtleties of Marxist analysis, see Alan Rudy and Andrew Light, "Social Ecology and Social Labor," pp. 75–106.

90. Murray Bookchin, *The Philosophy of Social Ecology* (Montreal: Black Rose Books, 1990), p. 16.

91. Bookchin, "New Social Movements," pp. 272–273.

92. Bookchin, *Post-Scarcity Anarchism,* p. 185.

93. Murray Bookchin, *The Modern Crisis* (Philadelphia: New Society Publishers, 1986), p. 153.

94. Bookchin, *Post-Scarcity Anarchism,* p. 209.

95. Herbert Marcuse, *Counter-revolution and Revolt* (Boston: Beacon Press, 1972), p. 59.

96. Ibid., p. 69; emphasis in original.

97. I am still unsure if this is the same argument that Bookchin makes about the normatively positive potential of social evolution. It is hard to tell because Marcuse, unlike Bookchin, does not try to ground this claim in an argument that a past time in human history realized these potentialities in nature

(by which I mean Bookchin's second nature). I think that Marcuse may be saying that to restrict any potential—good, bad, nonnormative—is a problem. And that the reason why any particular potential is preferred is not ground in its roots as a potential in nature. This may or may not also incur the naturalistic fallacy. I would not be surprised if it did. Again, incurring this fallacy may not be a problem, though it does increase one's argumentative burdens.

98. Bookchin, "Response to Andrew Light's 'Rereading Bookchin and Marcuse,' " p. 105.

99. Bookchin, *Remaking Society,* p. 39.

100. Herbert Marcuse, "Ecology and the Critique of Modern Society," *Capitalism, Nature, Socialism,* Vol. 3, No. 3, September, 1992, p. 30. This essay, and the helpful commentaries that follow it, provide an excellent summary of Marcuse's final comments on radical ecology.

101. Bookchin, "Response to Andrew Light's 'Rereading Bookchin and Marcuse,' " pp. 108–113.

102. Blanke, "Domination and Utopia," p. 205. Also see Steven Vogel's *Against Nature* for a thorough account of the resources available for a political ecology (including a critical sociology of science) in Marcuse's work, as well as in the Frankfurt School more generally, especially Habermas.

Index

Contributors

Glenn A. Albrecht, PhD, is a lecturer in Environmental Studies at the University of Newcastle, Callaghan, New South Wales, Australia. He completed a PhD on "Organicism" in 1988 and has helped to create undergraduate and postgraduate degrees in environmental studies at the University of Newcastle. Glenn teaches subjects on environmental ethics, policy, and politics within these degrees. He has published essays on the philosophy of social ecology and organicism. He is active in local and regional conservation of habitat for biodiversity and has a particular interest in ornithology. Glenn has acted as an advisor to regional industry and local government on the policy of ecologically sustainable development.

John Clark is Professor of Philosophy and co-chair of Environmental Studies at Loyola University, New Orleans, LA. He has written or edited a dozen works in social and ecological philosophy, including *The Anarchist Moment: Reflections on Culture, Nature, and Power*; *Environmental Philosophy: From Animal Rights to Radical Ecology*; and the forthcoming *Liberty, Equality, Geography: The Social Thought of Elisée Reclus*. He is presently completing a major critique and reformulation of social ecology as a philosphy of dialectical holism. He has for many years been an activist in the green and bioregional movements and also works in ecological forestry.

Regina Cochrane is a postdoctoral fellow in the Department of Philosophy at the University of Toronto. She completed a PhD in Political Science at York University in Toronto where she defended her thesis entitled "Feminism, Ecology, and Negative Dialectics: Toward a Feminist Green Political Theory."

Robyn Eckersley is a senior lecturer in the Politics Department at Monash University, Clayton, Victoria, Australia. She holds a law de-

gree, a master's in philosophy, and a PhD, and has published widely in
the fields of environmental philosophy and green political theory.

Adoph G. Gundersen grew up in the politically glacial but environmentally unique Driftless Area of southwestern Wisconsin. He holds graduate degrees from both eastern and midwestern universities and was until recently Associate Professor of Political Science at Texas A&M University, where he taught democratic theory and environmental philosophy. He writes and lives in Madison, Wisconsin.

Eric Stowe Higgs, PhD, is Associate Professor, Department of Anthropology, University of Alberta, Edmonton, Alberta, Canada.

Joel Kovel, MD, Trained as a psychiatrist and psychoanalyst, Joel Kovel is currently Alger Hiss Professor of Social Studies at Bard College, Annandale-on-Hudson, NY. He received a Guggenheim Fellowship in 1987. His principal works include *White Racism* (1970, 1984), *The Age of Desire* (1982), *History and Spirit* (1991, 1998), *Red Hunting in the Promised Land* (1994, 1997), and, just completed, *The Enemy of Nature*. Kovel is a frequent contributor to *Monthly Review* and *Capitalism, Nature, Socialism*, of which he also an editor. Currently Kovel is coediting, with James O'Connor, a collection of essays on ecological socialism. A long-time activist, Kovel ran for the United States Senate on the Green Party ticket in 1998.

Andrew Light is Assistant Professor of Philosophy and Environmental Studies at the State University of New York at Binghamton. He has published over two dozen articles on environmental philosophy, political philosophy, and philosophy of film, and is the editor or coeditor of six books, in print or forthcoming, including, with Eric Katz, *Environmental Pragmatism*. He is the founding coeditor, with Jonathan Smith, of the journal *Philosophy and Geography* and Managing Editor of the *Radical Philosophy Review*. He is currently at work on a book on the relationship between environmental philosophy and environmental policy, and on a collection of his own essays on philosophy of film: *Reel Politics: Film, Philosophy, and Social Criticism*. He has previously held positions at Texas A&M University, the University of Alberta, the University of Montana, and Tel Aviv University, and lectured widely in North America, Europe, Australia, and the Middle East.

David Macauley presently teaches in the General Studies Program at New York University and the Philosophy Department at St. John's University. He is the editor of *Minding Nature: The Philosophers of Ecol-*

ogy (New York: Guilford Press, 1996) and the author of a number of articles on ecological philosophy, political theory, and animal issues. A former coeditor of *Lomakatsi,* he completed his doctorate in Philosophy at the State University of New York at Stony Brook in May 1998, after studying in Tübingen, Germany. His dissertation, "Bewildering Order: Toward an Ecology of the Elements in Ancient Greek Philosophy and Beyond," examines the place and displacement of earth, air, fire, and water in classical thought, arguing for the relevance of the elements to environmental theory and practice.

Alan P. Rudy is Assistant Professor of Sociology at Michigan State University in East Lansing, MI. His primary interest is in the political economy of agriculture, the environment, and science and technology. In additional to theoretical work on the historical relation of society, capital, and nature, of which the chapter in this volume is a part, his research focuses on regional studies as a means of exploring the dialectics and contingencies of nature, labor, and community. His dissertation, completed at the University of California, Santa Cruz, in 1995, on the Imperial Valley of southeastern California is being prepared for publication.

David Watson has been an activist for social change since the middle 1960s, and is the author of three books on ecology and politics: *How Deep Is Ecology?* (Ojai, CA: Times Change Press, 1988, under the pseudonym George Bradford), *Beyond Bookchin: Preface for a Further Social Ecology* (Detroit: Black & Red/New York: Autonomedia, 1996), and *Against the Megamachine: Essays on Empire and Its Enemies* (New York: Autonomedia, 1998). A poet and essayist, he has been a member of the editorial collective of the anarchist journal *Fifth Estate* since 1977. His work has appeared regularly in *Fifth Estate* as well as in *New Internationalist, The Alternative Press Review,* and other publications. He is presently working on a series of poems and an anthology, *The Fifth Estate Reader.* He lives in Detroit and earns his living as a high school teacher.